METHOD OF VARIATION OF PARAMETERS FOR DYNAMIC SYSTEMS

SERIES IN MATHEMATICAL ANALYSIS AND APPLICATIONS

The Series in Mathematical Analysis and Applications (SIMAA) is edited by Ravi P. Agarwal, National University of Singapore and Donal O'Regan, University College of Galway, Ireland.

The series is aimed at reporting on new developments in mathematical analysis and applications of a high standard and of current interest. Each volume in the series is devoted to a topic in analysis that has been applied, or is potentially applicable, to the solutions of scientific, engineering and social problems.

Volume 1
Method of Variation of Parameters for Dynamic Systems
V. Lakshmikantham and S.G. Deo

In preparation:
Volume 2   Integral and Integrodifferential Equations: Theory, Methods and Applications
Ravi P. Agarwal and Donal O'Regan

METHOD OF VARIATION OF PARAMETERS FOR DYNAMIC SYSTEMS

V. Lakshmikantham

and

S.G. Deo

Applied Mathematics Program
Florida Institute of Technology, USA

CRC Press
Taylor & Francis Group
Boca Raton London New York

CRC Press is an imprint of the
Taylor & Francis Group, an **informa** business

First published 1998 by Gordon and Breach Science Publishers

Published 2019 by CRC Press
Taylor & Francis Group
6000 Broken Sound Parkway NW, Suite 300
Boca Raton, FL 33487-2742

© 1998 by Taylor & Francis Group, LLC
CRC Press is an imprint of Taylor & Francis Group, an Informa business

First issued in paperback 2019

No claim to original U.S. Government works

ISBN 13: 978-0-367-45577-4 (pbk)
ISBN 13: 978-90-5699-160-9 (hbk)

Visit the Taylor & Francis Web site at
http://www.taylorandfrancis.com

and the CRC Press Web site at
http://www.crcpress.com

British Library Cataloguing in Publication Data

A catalogue record for this book is available from the British Library.

ISSN 1028-8600

Contents

Preface

The method of variation of parameters (MVP) has been a well known and useful tool in the investigation of the properties of solutions of differential equations. Since one can apply the MVP to nonlinear differential equations whose unperturbed parts are linear, this method has gained importance in what is known as linearization techniques. For a long time, it was believed that the MVP is applicable only to nonlinear equations with linear parts. This belief was shaken by Alekseev in 1961 [1] who showed that the MVP can be developed from differential equations which possess no linear parts, that is, one can have nonlinear unperturbed parts, with some extra smoothness, to obtain nonlinear variation of parameter formula (NVPF). Later, it was also shown that it is possible to develop NVPF for scalar differential equations with variable separable type unperturbed part without any extra smoothness conditions. Moreover, different forms of NVPF exist for general nonlinear equations which offer more flexibility. These developments have given rise to a flood of activity in extending, generalizing and refining the MVP to a variety of dynamic systems (continuous as well as discrete).

It is also well-known that Lyapunov's second method is an important and fruitful technique that has gained increasing significance and has given a decisive impetus for modern development of stability analysis of differential equations. It is now well recognized that the concept of Lyapunov-like functions and the theory of differential inequalities can be utilized to investigate qualitative and quantitative properties of a variety of differential equations. Lyapunov-like functions serve as vehicles to transform a given complicated differential system into a relatively simpler comparison system and hence it is enough to study the properties of comparison systems since the properties of the given system can be deduced from it.

Because of the fact that the MVP and comparison method via Lyapunov-like functions are both extremely useful and effective techniques in the investigation of nonlinear problems, it is natural to combine these two approaches in order to utilize the benefits of these two important methods. Such a unification has recently been achieved and awaits further development.

In view of the enormous amount of the literature offering various generalizations, extensions and refinements of the MVP and Lyapunov's method, it is surprising that no attempt has yet been made to bring all the available literature into one volume, which could enable one to compare and contrast the theory and applications, and then make further progress. It is with this spirit, we see the importance of this monograph. Its aim is to present a systematic and unified development of the theory of the MVP, its unification with Lyapunov's method and typical applications of these methods.

We utilize Chapter 1, to introduce the main ideas that are involved in the development of the linear and nonlinear variation of parameters formulae for differential equations and then provide as applications, several typical results concerning qualitative properties. Chapter 2 is devoted to integrodifferential equations, where it can be seen that the development of NVPF leads to new ideas showing the difficulties encountered in the direct extensions.

Delay differential equations, including neutral differential equations, are discussed in Chapter 3 relative to the generalization of MVP. Chapter 4 studies difference equations. In Chapter 5 we discuss differential equations in a Banach space developing linear and nonlinear VPF using semigroup approach. Impulsive differential equations from the contents of Chapter while Chapter 7 deals with stochastic differential equations. Finally, Chapter 8 introduc some miscellaneous topics such as complex differential equations, hyperbolic differenti equations, differential equations with piecewise constant delay, and dynamic systems c measure chains, where the known extensions of MVP are given.

Some important features of the monograph are as follows:

1. it presents a unified approach to the method of linear and nonlinear variation of parameters together with its generalizations, extensions and refinements;
2. it develops the blending of the two important techniques, namely MVP and Lyapuno like functions and comparison principle;
3. it exhibits the role of the MVP in a variety of dynamic systems in one volume;
4. it provides several typical applications of MVP such as exponential dichotomy, boun ary value problems, monotone iterative technique, perturbation theory, controllabili existence of almost periodic solutions, stability analysis and asymptotic behavior; an
5. it incorporates the theory of matrix differential equations relative to MVP where tl usefulness of Kronecker product of matrices is stressed.

We are immensely thankful to Mrs Donn Miller-Kermani for her excellent and painsta ing typing of this manuscript.

1. Ordinary Differential Equations

1.0 INTRODUCTION

This chapter introduces the main ideas that are involved in the development of the method of variation of parameters relative to differential systems whose unperturbed parts are linear as well as nonlinear. The discussion of the method of nonlinear variation of parameters leads to different forms of the variation of parameters formulae (VPF) which offer flexibility in applications. We then provide as applications of the method of variation of parameters several typical results concerning qualitative properties of differential systems. The remaining chapters are devoted to extensions, generalizations and refinements of these ideas to other dynamic systems.

Section 1.1 deals with the classical VPF for differential systems with linear unperturbed parts. We also include, in this section, linear matrix differential systems and obtain corresponding results. Section 1.2 is devoted to nonlinear VPF. We describe first a simple scalar situation, where the unperturbed part is of variable separable nature. Without any additional assumptions other than continuity, we prove VPF for nonlinear differential equations and then extend the results to nonlinear cascade systems. We then develop systematically, nonlinear VPF for differential systems whose unperturbed part is nonlinear with a smooth behavior. We offer several possible VPFs relative to such systems. Finally, we extend these considerations to nonlinear matrix differential equations where it becomes important to utilize Kronecker product of matrices to derive needed VPF.

Section 1.3 incorporates a generalization of nonlinear VPF in terms of Lyapunov-like functions thus unifying the method of variation of parameters with Lyapunov-like functions.

From Sections 1.4 to 1.11, we present various typical qualitative properties of solutions of differential systems whose proofs depend on utilizing suitable VPF. Section 1.4 discusses a boundary value problem (BVP) for a perturbed linear system, whereas Section 1.5 investigates periodic BVP using constructive monotone iterative technique. In Section 1.6, exponential dichotomy of linear systems with nonlinear perturbations, is explored.

Section 1.7 deals with sensitivity analysis of matrix differential equations with linear unperturbed part which is important when the model process contains errors of measurements. Existence of almost periodic solutions of differential systems forms the content of Section 1.8. A variety of problems related to stability and asymptotic behavior are investigated including applications to matrix differential equations, in Section 1.9.

Controllability of linear and nonlinear differential systems is dealt with in Section 1.10.

Finally, in Section 1.11, utilizing the ideas of generalized VPF discussed earlier, we formulate a variational comparison result, which includes the well-known comparison result as a special case. Then employing this result; practical stability properties of perturbed differential systems are proved in terms of the method of variation of Lyapunov functions.

1.1 LINEAR VARIATION OF PARAMETERS

Let us begin by proving some elementary facts about linear differential systems,

$$x' = A(t)x, \qquad (1.1.1)$$

where $A(t)$ is a continuous $n \times n$ matrix on R_+. Let $U(t, t_0)$ be the $n \times n$ matrix whose columns are the n-vector solutions $x(t)$, $x(t)$ being so chosen to satisfy the initial condition $U(t_0, t_0) = I$ (unit matrix). Since each column of $U(t, t_0)$ is the solution of (1.1.1), it is clear that U satisfies the matrix differential equation

$$U' = A(t)U, \ U(t_0, t_0) = I. \qquad (1.1.2)$$

Theorem 1.1.1: *Let $A(t)$ be a continuous $n \times n$ matrix on R_+. Then the fundamental matrix solution $U(t, t_0)$ of (1.1.2) is nonsingular on R_+. More precisely,*

$$\det U(t, t_0) = exp[\int_{t_0}^{t} tr A(s)ds], \quad t \in R_+,$$

where $tr A(t) = \sum_{i=1}^{n} a_{ii}(t)$.

Proof: The proof utilizes the following two facts:

(*i*) $\frac{d}{dt}(\det U(t)) = $ sum of the determinants formed by replacing the elements of one row of $\det U(t, t_0)$ by their derivatives.

(*ii*) The columns of $U(t, t_0)$ are the solutions of (1.1.1).

Simplifying the determinants obtained in (*i*) by the use of (*ii*), one gets

$$\frac{d}{dt}\det U(t, t_0) = tr A(t)\det U(t, t_0).$$

The result follows, since $U(t_0, t_0) = I$.

Theorem 1.1.2: *Let $y(t)$ be a solution of*

$$y' = A(t)y + F(t, y), \qquad (1.1.3)$$

such that $y(t_0) = y_0$, where $F \in C[R_+ \times R^n, R^n]$. If $U(t, t_0)$ is the fundamental matrix solution of (1.1.2), then $y(t)$ satisfies the integral equation

$$y(t) = U(t, t_0)y_0 + \int_{t_0}^{t} U(t, t_0)U^{-1}(s, t_0)F(s, y(s))ds. \qquad (1.1.4)$$

Proof: Defining $y(t) = U(t, t_0)z(t)$ and substituting in (1.1.3), we get

$$U'(t, t_0)z(t) + U(t, t_0)z'(t) = A(t)U(t, t_0)z(t) + F(t, y(t)).$$

This, because of $(1.1.2)$ yields

$$z'(t) = U^{-1}(t, t_0)F(t, y(t))$$

from which we obtain

$$z(t) = y_0 + \int_{t_0}^{t} U^{-1}(s, t_0)F(s, y(s))ds.$$

Now multiplying this equation by $U(t, t_0)$ gives $(1.1.4)$. The proof is complete.

The relation $(1.1.4)$ is the variation of parameters formula (VPF) for the equation $(1.1.2)$.

Corollary 1.1.1: *Let $A(t)$ be a continuous $n \times n$ matrix on R_+ such that every solution $x(t)$ of $(1.1.1)$ is bounded for $t \geq t_0$. Let $U(t, t_0)$ be the fundamental matrix solution of $(1.1.1)$. Then $U^{-1}(t, t_0)$ is bounded if and only if*

$$Re[\int_{t_0}^{t} tr A(s)ds]$$

is bounded below.

If $A(t) = A$ an $n \times n$ constant matrix, then $U(t, t_0) = e^{A(t-t_0)}$ and therefore, the VPF $(1.1.4)$ reduces to

$$y(t) = e^{A(t-t_0)}y_0 + \int_{t_0}^{t} e^{A(t-s)}F(s, y(s))ds, \ \ t \geq t_0.$$

Next, we shall consider the linear matrix differential equation

$$X' = A(t)X + XB(t) \tag{1.1.5}$$

and the corresponding perturbed matrix differential equation

$$Y' = A(t)Y + YB(t) + F(t, Y), \ Y(t_0) = Y_0, \tag{1.1.6}$$

where $A(t)$ and $B(t)$ are continuous $n \times n$ matrices, $F \in C[R_+ \times R^{n \times n}, R^{n \times n}]$ and $X, Y \in R^{n \times n}$. Then corresponding to the VPF $(1.1.4)$, we have the following result.

Theorem 1.1.3: *Let $Y(t)$ be a matrix valued solution of $(1.1.6)$ for $t \geq t_0$ such that $Y(t_0) = Y_0$, $Y_0 \in R^{n \times n}$. Moreover, let $v(t, t_0)$, $w(t, t_0)$ be the matrix solutions of the linear matrix differential equations*

$$v' = A(t)v, \ v(t_0) = I,$$

$$w' = wB(t) \ w(t_0) = I,$$

respectively. Then $Y(t)$ verifies the integral equation

$$Y(t) =$$

$$v(t,t_0)Y_0 w(t,t_0) + \int_{t_0}^{t} v(t,t_0)v^{-1}(s,t_0)F(s,Y(s))w^{-1}(s,t_0)w(t,t_0)ds,$$

$$t \geq t_0. \tag{1.1.7}$$

Proof: It is easy to see that

$$X(t) = v(t,t_0)Y_0 w(t,t_0)$$

is the solution of (1.1.5) with $X(t_0) = Y_0$. Now employing the method of variation of parameters, so that we replace Y_0 by $Y_0(t)$ and $Y(t) = v(t,t_0)Y_0(t)w(t,t_0)$ we arrive at

$$v'(t,t_0)Y_0(t)w(t,t_0) + v(t,t_0)Y_0'(t)w(t,t_0) + v(t,t_0)Y_0(t)w'(t,t_0)$$

$$= A(t)Y(t) + Y(t)B(t) + F(t,y(t)),$$

which yields,

$$v(t,t_0)Y_0'(t)w(t,t_0) = F(t,y(t)).$$

It now follows easily that

$$Y_0'(t) = v^{-1}(t,t_0)F(t,y(t))w^{-1}(t,t_0)$$

from which we get

$$Y_0(t) = Y_0 + \int_{t_0}^{t} v^{-1}(s,t_0)F(s,y(s))w^{-1}(s,t_0)ds.$$

Substituting this expression into the definition of $Y(t)$ gives the desired VPF (1.1.7).

As before, if $A(t) = A$, $B(t) = B$ are constant matrices, the VPF (1.1.7) reduces to

$$Y(t) = e^{A(t-t_0)}Y_0 e^{B(t-t_0)} + \int_{t_0}^{t} e^{A(t-s)}F(s,y(s))e^{B(t-s)}ds, \quad t \geq t_0. \tag{1.1.8}$$

1.2 NONLINEAR VARIATION OF PARAMETERS

The method of variation of parameters was considered for a long time as a technique that can be applied only to perturbed systems with linear unperturbed parts. To extend this method, when the unperturbed part is nonlinear, we need to know the continuity and differentiability of the solutions of unperturbed system with respect to initial values, since such a property is inherent in linear systems. This requirement implies imposing additional restrictions on the nonlinear functions involved. However, when we discuss certain nonlinear unperturbed systems in which variables are separable, we can derive VPF without additional assumptions which is somewhat surprising. We shall begin proving such a simple result first and then consider the general nonlinear unperturbed systems.

Before we proceed to discuss systems with variables separable, let us first develop VPF for scalar perturbed differential equation in which unperturbed part is of variable separable type, and then utilize it to derive VPF for certain systems.

Consider

$$x' = \lambda(t)g(x), \ x(t_0) = y_0, \tag{1.2.1}$$

and

$$y' = \lambda(t)g(y) + F(t,y), y(t_0) = y_0, \tag{1.2.2}$$

where $\lambda \in C[R_+, R]$, $g \in C[(0, \infty), (0, \infty)]$ and $F \in C[R_+ \times R, R]$. We have the following elementary result.

Theorem 1.2.1: *Any solution $x(t)$ of* (1.2.1) *has the form*

$$x(t) = G^{-1}[G(y_0) + \int_{t_0}^{t} \lambda(s)ds], \ t_0 \le t \le T, \tag{1.2.3}$$

where

$$T = sup[t \ge t_0, G(y_0) + \int_{t_0}^{t} \lambda(s)ds \in dom \, G^{-1}],$$

$G(x) = G(y_0) + \int_{y_0}^{x} \frac{ds}{g(s)}$, *and G^{-1} is the inverse function of G.*

Proof: Let $x(t)$ be a solution of (1.2.1). Then

$$\frac{dx(t)}{g(x(t))} = \lambda(t)dt.$$

Consequently, in view of the definition of G,

$$G(x(t)) - G(y_0) = \int_{y_0}^{x(t)} \frac{d\sigma}{g(\sigma)} = \int_{t_0}^{t} \frac{dx(s)}{g(x(s))} = \int_{t_0}^{t} \lambda(t)dt,$$

which yields (1.2.3) as desired.

We now apply the method of variation of parameters to obtain an expression for the solution $y(t)$ of (1.2.2). For this purpose, we set

$$G(y(t)) = \int_{t_0}^{t} \lambda(s)ds + G(\psi(t)), \tag{1.2.4}$$

which on differentiation yields,

$$G_y(y(t))y'(t) = \lambda(t) + G_y(\psi(t))\psi'(t).$$

Using the definition of G, we find that

$$\psi'(t) = \frac{g(\psi(t))F(t,y)}{g(y)}$$

$$= \frac{g(\psi(t))F(t,G^{-1}[G(\psi(t))+\int\limits_{t_0}^{t}\lambda(s)ds])}{g(G^{-1}[G(\psi(t))+\int\limits_{t_0}^{t}\lambda(s)ds])}$$

$$= \omega(t,\psi(t)). \tag{1.2.5}$$

Here $\psi(t)$ is a solution of $\psi' = \omega(t,\psi)$, $\psi(t_0) = y_0$ on some interval I_0 contained in $t_0 \leq t \leq T$. Substituting for $\psi(t)$ in (1.2.4), we get the equation satisfied by a solution $y(t) = y(t,t_0,y_0)$ of (1.2.2), namely,

$$y(t) = G^{-1}\left\{\int\limits_{t_0}^{t}\lambda(s)ds + G\left(y_0 + \int\limits_{t_0}^{t}g[G^{-1}(G(y(s))) - \int\limits_{t_0}^{t}\lambda(\tau)d\tau]F(s,y(s))ds\right)\right\},$$

$$t \in I_0. \tag{1.2.6}$$

We have thus proved the following result.

Theorem 1.2.2: *Let λ, g, F be as given above. Then a solution $y(t,t_0,y_0)$ of (1.2.2) can be exhibited in the form (1.2.6) on I_0 and it can be obtained by the method of variation of parameters.*

Corollary 1.2.1: *If, in addition to the assumptions of Theorem 1.2.2, we assume that $\lambda \geq 0$, $F(t,y) = \sigma(t) \geq 0$ and g is nondecreasing in y, then the following estimate is true*

$$y(t) \leq G^{-1}[\int\limits_{t_0}^{t}\lambda(s)ds + G(y_0 + \int\limits_{t_0}^{t}\sigma(s)ds)], \quad t \in I_0.$$

Example 1.2.1: Consider the Ricatti equation $y' = y^2 + t$, $y(0) = 1$. Then from Corollary 1.2.1, we get

$$G(x) = -\frac{1}{x}, \quad y_0 + \int\limits_{t_0}^{t}\sigma(s)ds = \frac{t^2+2}{2}.$$

Hence it follows that

$$y(t) \leq \frac{t^2+2}{2-2t-t^3} \text{ for } 0 \leq t \leq \frac{3}{4}.$$

We are now ready to extend the foregoing discussion to the special system

$$x' = A(t)g(x), \quad x(t_0) = y_0, \tag{1.2.7}$$

$$y' = [A(t) + B(t)]g(y) + F(t), \quad y(t_0) = y_0, \tag{1.2.8}$$

where x, y, y_0 are n-vectors, $t, t_0 \in R_+$, $F \in C[R_+, R^n]$, $g(x) = (g_1(x_1), \ldots, g_n(x_n))$,

$$A(t) = \begin{bmatrix} \lambda_{11}(t) & 0 & \cdots & 0 \\ 0 & \lambda_{22}(t) & \cdots & 0 \\ 0 & 0 & \cdots & \lambda_{nn}(t) \end{bmatrix}$$

and

$$B(t) = \begin{bmatrix} 0 & 0 & \cdots & 0 & 0 \\ \lambda_{21}(t) & 0 & \cdots & 0 & 0 \\ \cdots & \cdots & \cdots & 0 & 0 \\ \lambda_{n1}(t) & \lambda_{n2}(t) & \cdots & \lambda_{n,n-1}(t) & 0 \end{bmatrix}$$

λ_{ij} being continuous functions on R_+.

Following the proof of Theorem 1.2.1, we can get the vector equation

$$G(x(t)) = G(y_0) + \int_{t_0}^{t} A(s)ds \qquad (1.2.9)$$

where $G(u) = G(u_0) + \int_{u_0}^{u} \frac{ds}{g(s)}$.

We need to determine parameter y_0 as a function of t, say, an n-vector function $\psi(t)$, such that

$$G(y(t)) = G(\psi(t)) + \int_{t_0}^{t} A(s)ds, \quad \psi(t_0) = y_0,$$

where $y(t) = y(t, t_0, y_0)$ is a solution of the perturbed system (1.2.8).

Since (1.2.8) is a special system usually called a cascade system, we can find the solution $y(t)$ of (1.2.8) in successive steps using the cyclic argument. For this purpose, we obtain first an expression for the solution $y_1(t)$ from the equation

$$y_1' = \lambda_{11}(t)g_1(y_1) + F_1(t), \quad y_1(t_0) = y_{01},$$

as in Theorem 1.2.2, substitute it in the equation for $y_2(t)$ so that we have

$$y_2' = \lambda_{22}(t)g_2(y_2) + \lambda_{21}(t)g_1(y_1(t)) + F_2(t), \quad y_2(0) = y_{02},$$

which can then be solved as before treating $\lambda_{21}(t)g_1(y_1(t) + F_2(t)$ as the known perturbation and so on. As a result of this procedure, we arrive at the desired conclusion giving a general expression for the solution $y(t)$ of (1.2.8) in the following form

$$y(t) = G^{-1}\left[\int_{t_0}^{t} B(s)ds + G(y_0 + \int_{t_0}^{t} g[G^{-1}(G(y(s)) - \int_{t_0}^{t} B(\sigma)d\sigma]F(s)ds\right] \quad (1.2.10)$$

for $t \in J$ where J is the smallest set depending on the domains of G^{-1} as in Theorem 1.2.2. In the above discussion, vector relations (1.2.9) and (1.2.10) correspond to the scalar relations (1.2.3) and (1.2.6) respectively.

The foregoing considerations prove the following result.

Theorem 1.2.3: *Let the matrices A, B and vectors g, F be as given above. Then a solution $y(t, t_0, y_0)$ of (1.2.8) can be represented in the form (1.2.10) on a nonempty interval J which is obtained by the method of variation of parameters.*

Let $P(t)$ be any $n \times n$ continuous matrix defined on R_+ and we are given the general nonlinear system of variable separable type

$$y' = P(t)g(y) + F(t)$$

where $g(y) = \{g_1(y_1), \ldots, g_n(y_n)\}$ as before. We now write $P(t) = A(t) + B(t)$, where the matrix $A(t)$ constitutes only the diagonal elements of $P(t)$ and $B(t)$ constitutes non-diagonal elements of $P(t)$. Then the given system takes the form

$$y' = A(t)g(y) + [B(t)g(y) + F(t)], \quad y(t_0) = y_0$$

where the unperturbed system is

$$x' = A(t)g(x), \quad x(t_0) = y_0.$$

Treating $[B(t)g(y) + F(t)]$ as the nonlinear perturbation, it is easy to apply the method of VP described in Theorem 1.2.2 to obtain the representation of solution $y(t, t_0, y_0)$ of $y' = P(t)g(y) + F(t)$, in the vectorial form following (1.2.6). Such a representation may be useful sometimes to obtain qualitative properties of the given system.

Let us next consider the general nonlinear differential systems

$$x' = f(t, x), \quad x(t_0) = x_0, \tag{1.2.11}$$

$$y' = f(t, y) + F(t, y), \quad y(t_0) = x_0, \tag{1.2.12}$$

where $f, F \in C[R_+ \times R^n, R^n]$. We shall develop a variety of VPFs for (1.2.12). Let us first suppose that the solutions $x(t, t_0, x_0)$ of (1.2.11) exist, unique and that $\frac{\partial x}{\partial x_0}(t, t_0, x_0)$ exists and continuous for all $t \geq t_0$. Then the following result provides a VPF for (1.2.12).

Theorem 1.2.4: *Suppose that the system (1.2.11) admits unique solutions $x(t, t_0, x_0)$. Suppose, also, that*

$$\Phi(t, t_0, x_0) \equiv \frac{\partial x}{\partial x_0}(t, t_0, x_0)$$

exists and is continuous for all $t \geq t_0$ and that $\Phi^{-1}(t, t_0, x_0)$ exists for all $t \geq t_0$. If $\phi(t)$ is a solution of (1.2.15) below, then any solution $y(t, t_0, x_0)$ of (1.2.12) satisfies the relation

$$y(t, t_0, x_0) = x(t, t_0, x_0 + \int_{t_0}^{t} \Phi^{-1}(s, t_0, \phi(s))F(s, y(s, t_0, x_0))ds), \quad (1.2.13)$$

as far as $\phi(t)$ exists to the right of t_0.

Proof: Let $x(t, t_0, x_0)$ be any solution of (1.2.11) existing for $t \geq t_0$. The method of variation of parameters requires that we determine a function $\phi(t)$ so that

$$y(t, t_0, x_0) = x(t, t_0, \phi(t)), \ \phi(t_0) = x_0, \quad (1.2.14)$$

is a solution of (1.2.12). Differentiating with respect to t gives us

$$y'(t, t_0, x_0) = x'(t, t_0, \phi(t)) + \tfrac{\partial x}{\partial x_0}(t, t_0, \phi(t))\phi'(t).$$

Thus

$$f(t, y(t, t_0, x_0)) + F(t, y(t, t_0, x_0)) = f(t, x(t, t_0, \phi(t))) + \Phi(t, t_0, \phi(t))\phi'(t),$$

which, because of (1.2.14) and the fact that $\Phi^{-1}(t, t_0, x_0)$ exists, yields

$$\phi'(t) = \Phi^{-1}(t, t_0, \phi(t))F(t, x(t, t_0, \phi(t))), \ \phi(t_0) = x_0. \quad (1.2.15)$$

Solutions of (1.2.15) then determine $\phi(t)$.

If $\phi(t)$ is a solution of (1.2.15) then $y(t, t_0, x_0)$ given by (1.2.14) is a solution of (1.2.12). From (1.2.15), $\phi(t)$ must satisfy the integral equation

$$\phi(t) = x_0 + \int_{t_0}^{t} \Phi^{-1}(s, t_0, \phi(s))F(s, x(s, t_0, \phi(s)))ds.$$

This fact establishes the relation (1.2.13). Notice that $y(t, t_0, x_0)$ exists for those values of $t \geq t_0$ for which $\phi(t)$ exists. The proof is complete.

Corollary 1.2.2: *If $f(t, x) = A(t)x$, then $x(t, t_0, x_0) = \Phi(t, t_0)x_0$, where $\Phi(t, t_0)$ is the fundamental matrix solution of $x' = A(t)x$ with $\Phi(t_0, t_0) = I$. Then the relation (1.2.13) yields a well-known form (1.1.4), namely*

$$y(t, t_0, x_0) = x(t, t_0, x_0) + \int_{t_0}^{t} \Phi(t, t_0)\Phi^{-1}(s, t_0)F(s, y(s, t_0, x_0))ds. \quad (1.2.16)$$

The question whether the relation (1.2.13) can be stated in the form (1.2.16) is answered by the next result.

Theorem 1.2.5: *Under the assumptions of Theorem 1.2.4, the following relation is also valid:*

$$y(t, t_0, x_0) =$$

$$x(t, t_0, x_0) + \int_{t_0}^{t} \Phi(t, t_0, \phi(s))\Phi^{-1}(s, t_0, \phi(s))F(s, y(s, t_0, x_0))ds, \quad (1.2.17)$$

where $\phi(t)$ is any solution of $(1.2.15)$.

Proof: For $t_0 \leq s \leq t$, one gets

$$\frac{d}{ds}x(t, t_0, \phi(s)) = \frac{\partial x}{\partial x_0}(t, t_0, \phi(s))\phi'(s) = \Phi(t, t_0, \phi(s))\phi'(s).$$

On integration from t_0 to t, we obtain

$$x(t, t_0, \phi(t)) = x(t, t_0, x_0) + \int_{t_0}^{t} \Phi(t, t_0, \phi(s))\phi'(s)ds.$$

If $\phi(t)$ is any solution of $(1.2.15)$, then the conclusion $(1.2.17)$ follows in view of the relations $(1.2.14)$ and $(1.2.15)$.

We note that Theorem 1.2.2 can also be deduced from the above discussion.

Remark 1.2.1: Observe that the solution $x(t)$ of $(1.2.1)$ satisfies

$$G(x(t, t_0, x_0)) = G(x_0) + \int_{t_0}^{t} \lambda(s)ds$$

which yields, differentiating with respect to x_0,

$$\frac{\partial G}{\partial x}(x(t, t_0, x_0))\frac{\partial x}{\partial x_0}(t, t_0, x_0) = \frac{\partial G}{\partial x_0}(x_0).$$

This implies, in view of the definition of G,

$$\Phi(t, t_0, x_0) = \frac{\partial x}{\partial x_0}(t, t_0, x_0) = \frac{g(x(t, t_0, x_0))}{g(x_0)}.$$

Hence the equation $(1.2.15)$, in this particular situation, is

$$\phi'(t) = \frac{g(\phi(t))F(t, G^{-1}(G(\phi(t)) + \int_{t_0}^{t}\lambda(s)ds))}{g[G^{-1}\{G(\phi(t)) + \int_{t_0}^{t}\lambda(s)ds\}]}$$

which finally yields $(1.2.6)$.

Remark 1.2.2: The solution $x(t, t_0, x_0)$ of $(1.2.11)$ is related to $\Phi(t, t_0, x_0)$ defined in Theorem 1.2.4 by

$$x(t, t_0, x_0) = [\int_{0}^{1} \Phi(t, t_0, sx_0)ds] \cdot x_0$$

as can be seen by integrating

$$\frac{dx}{ds}(t, t_0, sx_0) = \frac{\partial x}{\partial x_0}(t, t_0, sx_0) \cdot x_0 = \Phi(t, t_0, sx_0) \cdot x_0$$

between $s = 0$ and $s = 1$.

In order to establish what is known as Alekseev's formula for nonlinear equations (1.2.12), we assume that the function f in (1.2.11) is such that f_x exists and is continuous on $R_+ \times R^n$. Before we establish the desired formula we need the following result, which we merely state. In fact, in the next chapter on integrodifferential equations we shall prove the corresponding result of which this will be a special case.

Theorem 1.2.6: *Assume that $f \in C[R_+ \times R^n, R^n]$ and possesses continuous partial derivatives $\partial f / \partial x$ on $R_+ \times R^n$. Let the solution $x_0(t) = x(t, t_0, x_0)$ of (1.2.11) exist for $t \geq t_0$, and let*

$$H(t, t_0, x_0) = \frac{\partial f(t, x(t, t_0, x_0))}{\partial x}.$$

Then

(i) $\Phi(t, t_0, x_0) = \frac{\partial x(t, t_0, x_0)}{\partial x_0}$
exists and is the solution of

$$y' = H(t, t_0, x_0)y \tag{1.2.18}$$

such that $\Phi(t_0, t_0, x_0) = I$.

(ii) $\frac{\partial x(t, t_0, x_0)}{\partial t_0}$
exists, is the solution of (1.2.18), and satisfies the relation

$$\frac{\partial x(t, t_0, x_0)}{\partial t_0} = -\Phi(t, t_0, x_0) \ f(t_0, x_0), \ t \geq t_0. \tag{1.2.19}$$

We shall now consider the nonlinear differential system (1.2.11). The following theorem gives an analog of VPF for the solutions $y(t, t_0, x_0)$ of (1.2.12).

Theorem 1.2.7: *Let $f, F \in C[R_+ \times R^n, R^n]$, and let $\partial f / \partial x$ exist and be continuous on $R_+ \times R^n$. If $x(t, t_0, x_0)$ is the solution of (1.2.11) and existing for $t \geq t_0$, any solution $y(t, t_0, x_0)$ of (1.2.12), with $y(t_0) = x_0$, satisfies the integral equation*

$$y(t, t_0, x_0) = x(t, t_0, x_0) + \int_{t_0}^{t} \Phi(t, s, y(s, t_0, x_0))F(s, y(s, t_0, x_0))ds \tag{1.2.20}$$

for $t \geq t_0$, where $\Phi(t, t_0, x_0) = \partial x(t, t_0, x_0)/\partial x_0$.

Proof: Write $y(t) = y(t, t_0, x_0)$. Then, in view of Theorem 1.2.6, we get

$$\frac{dx(t, s, y(s))}{ds} = \frac{\partial x(t, s, y(s))}{\partial s} + \frac{\partial x(t, s, y(s))}{\partial y} \cdot y'(s)$$

$$= \Phi(t, s, y(s))[y'(s) - f(s, y(s))]. \tag{1.2.21}$$

Noting that $x(t, t, y(t, t_0, x_0)) = y(t, t_0, x_0)$ and $y'(s) - f(s, y(s)) = F(s, y(s))$, by integrating (1.2.21) from t_0 to t, the desired result (1.2.20) follows.

It is natural to ask whether the different forms of VPF given by (1.2.17) and (1.2.20) are equivalent under certain conditions. The answer is affirmative and the next result deals with this question.

Theorem 1.2.8: *Assume that $f_x(t, x)$ exists and is continuous for $(t, x) \in R_+ \times R^n$. Then formulas* (1.2.17) *and* (1.2.20) *are equivalent.*

Proof: We first observe that

$$x(t, s, y(s)) = x(t, t_0, \phi(s)),$$

where $\phi(t)$ is a solution of (1.2.15). Then differentiating with respect to s, we obtain

$$- \Phi(t, s, y(s))f(s, y(s)) + \Phi(t, s, y(s))y'(s) = \Phi(t, t_0, \phi(s))\phi'(s).$$

This fact yields if we substitute for y' and ϕ',

$$\Phi(t, s, y(s))F(s, y(s)) = \Phi(t, t_0, \phi(s))\Phi^{-1}(s, t_0, \phi(s))F(s, y(s)),$$

and as a result, we get

$$\Phi(t, s, y(s)) \equiv \Phi(t, t_0, \phi(s))\Phi^{-1}(s, t_0, \phi(s)).$$

The proof is complete.

The next results obtains an expression which estimates the difference between the solutions $x(t, t_0, x_0)$ and $x(t, t_0, y_0)$ of (1.2.11) with different initial values which will be useful for later discussions.

Theorem 1.2.9: *Let $f \in C[R_+ \times R^n, R^n]$, and $\partial f/\partial x$ exist and be continuous on $R_+ \times R^n$. Assume that $x(t, t_0, x_0)$ and $x(t, t_0, y_0)$ are the solutions of* (1.2.11) *through (t_0, x_0) and (t_0, y_0), respectively, existing for $t \geq t_0$, such that x_0, y_0 belong to a convex subset of R^n. Then, for $t \geq t_0$,*

$$x(t, t_0, y_0) - x(t, t_0, x_0)$$

$$= \left[\int_0^1 \Phi(t, t_0, x_0 + s(y_0 - x_0))ds \right] \cdot (y_0 - x_0). \tag{1.2.22}$$

Proof: Here x_0, y_0 belong to a convex subset of R^n, $x(t, t_0, x_0 + s(y_0 - x_0))$ is defined for $0 \leq s \leq 1$. It follows that

$$\frac{dx(t, t_0, x_0 + s(y_0 - x_0))}{ds} = \Phi(t, t_0, x_0 + s(y_0 - x_0)) \cdot (y_0 - x_0),$$

and hence the integration from 0 to 1 yields (1.2.22). The theorem is proved.

We present below simple but illustrative examples.

Example 1.2.2: Consider

$$x' = e^{-t}x^2, \ x(t_0) = x_0 \quad x_0 \geq 0,$$

and

$$y' = e^{-t}y^2 - \frac{y^2}{2}, \ y(t_0) = x_0.$$

Here

$$x(t, t_0, x_0) = \frac{x_0}{1+x_0(e^{-t}-e^{-t_0})}, \quad t \geq t_0, \tag{1.2.23}$$

and

$$\Phi(t, t_0, x_0) = \frac{1}{[1+x_0(e^{-t}-e^{-t_0})]^2} \ , t \geq t_0.$$

Hence $y(t) = y(t, t_0, x_0)$ satisfies the integral equation

$$y(t, t_0, x_0) = \frac{x_0}{1+x_0(e^{-t}-e^{-t_0})} - \int_{t_0}^{t} \frac{y^2(s)ds}{2[1+y(s)(e^{-t}-e^{-s})]^2}.$$

Observe that only some solutions of (1.2.23) are bounded. For example, for $t_0 = 0$ and $x_0 = 1$, $x(t, 0, 1) = e^t$. However, all solutions $y(t, t_0, x_0)$ tend to zero as $t \to \infty$, since, $y(t, 0, 1) = \frac{2}{2+t+2e^{-t}}$.

Example 1.2.3: Consider the second order equation

$$[r(t)y']' + f(t, y) = p(t), \quad y(t_0) = x_0, \ y'(t_0) = x_1 \tag{1.2.24}$$

where $p, r \in C[R_+, (0, \infty)]$, $f \in C[R_+ \times R, R]$ and f_x exists and is continuous on $R_+ \times R$. The different forms of VPF discussed above are applicable once we realize that the given unperturbed part of (1.2.24) is equivalent to the system

$$x' = \frac{u(t)}{r(t)}, \qquad (x(t_0), u(t_0)) = (x_0, x_1),$$
$$u' = -f(t, x);$$

and the perturbed system is

$$y' = \frac{v(t)}{r(t)}, \qquad (y(t_0), v(t_0)) = (x_0, x_1)$$
$$v' = -f(t, y(t)) + p(t).$$

Let $x(t) = x(t, t_0, x_0, x_1)$ be the solution of the unperturbed system. Suppose, for convenience, that $r(t_0) = 1$. Then $y(t) = y(t, t_0, x_0, x_1)$ is the solution of the perturbed system. We get, in view of Theorem 1.2.7, and the assumptions on f,

$$\frac{dx}{ds}(t, s, y(s), r(s)y'(s))$$

$$= \frac{\partial x}{\partial t_0}(t, s, y(s), r(s)y'(s)) + \frac{\partial x}{\partial x_0}(t, s, y(s), r(s)y'(s))y'(s)$$

$$+ \frac{\partial x}{\partial x_1}(t, s, y(s), r(s)y'(s))[r(s)y'(s)]'$$

$$= \frac{\partial x}{\partial x_1}(t, s, y(s), r(s)y'(s))p(s).$$

One thereafter obtains, integrating from t_0 to t

$$y(t) = x(t) + \int_{t_0}^{t} \frac{\partial x}{\partial x_1}(t, s, y(s), r(s)y'(s))p(s)ds, \ t \geq t_0$$

which is the desired VPF for this case.

Finally, we shall establish VPF for nonlinear matrix differential equations. For this purpose, following directly the known methods of vector differential systems is not applicable. It is therefore convenient to convert the given matrix differential system into corresponding vector differential system, apply the known results for such systems and then try to revert to the framework of matrix differential system. This procedure essentially depends on utilizing the notion of Kronecker product of matrices and its properties together with certain results which provide Kronecker decomposition of matrices. We shall follow such a procedure here.

Consider the matrix differential equation

$$X' = F(t, X), \ X(t_0) = X_0, t_0 \in R_+, \tag{1.2.25}$$

and its associated perturbed system

$$Y' = F(t, Y) + R(t, Y), \ y(t_0) = X_0, \tag{1.2.26}$$

where $F, R \in C[R_+ \times R^{n \times n}, R^{n \times n}]$.

We introduce the $vec(\cdot)$ operator which maps a $m \times n$ matrix $A = (a_{ij})$ onto the vector composed of the columns of A:

$$vec(A) = (a_{11}, \ldots, a_{m1}, a_{12}, \ldots, a_{m2}, \ldots, a_{1n}, \ldots, a_{mn})^T.$$

Then the equivalent vector differential systems can be written as

$$x' = f(t, x), \ x(t_0) = x_0, \tag{1.2.27}$$

$$y' = f(t, y) + r(t, y), \ y(t_0) = x_0, \tag{1.2.28}$$

where $x = vec(X^T)$, $y = vec(Y^T)$, $f = vec(F^T)$, $r = vec(R^T)$, and $f, r \in C[R_+ \times R^{n^2}, R^{n^2}]$.

Let us begin by defining the Kronecker product of matrices.

Definition 1.2.1: If $A \in R^{p \times q}$ and $B \in R^{m \times n}$, then the Kronecker product of A and B, $A \otimes B \in R^{pm \times qn}$, is defined by the matrix

$$A \otimes B = \begin{pmatrix} a_{11}B & a_{12}B & \cdots & \cdots & a_{1q}B \\ a_{21}B & a_{22}B & \cdots & \cdots & a_{2q}B \\ \vdots & \vdots & & & \vdots \\ a_{p1}B & a_{p2}B & \cdots & \cdots & a_{pq}B \end{pmatrix}.$$

If $A, B, C, X \in R^{n \times n}$, then

$$AXB = C \Leftrightarrow (B^T \otimes A)vec(X) = vec(C).$$

Lemma 1.2.1: *Given $G \in R^{n^2 \times n^2}$ with $G_{ij} \in R^{n \times n}$, where*

$$G = \begin{pmatrix} G_{11} & G_{12} & \cdots & \cdots & G_{1n} \\ G_{21} & G_{22} & \cdots & \cdots & G_{2n} \\ \vdots & \vdots & & & \vdots \\ G_{n1} & G_{n2} & \cdots & \cdots & G_{nn} \end{pmatrix},$$

there exist $A, B \in R^{n \times n}$ such that the Kronecker decomposition of G is given by

$$G(A \otimes I) + (I \otimes B^T)$$

if and only if

$$G_{ii} = B^T + A_{ii}I \text{ for } i = 1, \dots, n, \tag{1.2.29}$$

$$G_{ij} = a_{ij}I \text{ for } i, j = 1, \dots, n \text{ with } i \neq j. \tag{1.2.30}$$

Note that one diagonal element of B may be chosen arbitrarily.

The following result offers sufficient conditions for Kronecker decomposition of matrices.

Theorem 1.2.10: *If $F(x)$ has continuous partial derivatives on $R^{n \times n}$, then the $n^2 \times n^2$ matrix $G(x)$ defined by*

$$G(X) = \begin{pmatrix} G_{11}(X) & G_{12}(X) & \cdots & \cdots & G_{1n}(X) \\ G_{21}(X) & G_{22}(X) & \cdots & \cdots & G_{2n}(X) \\ \vdots & \vdots & & & \vdots \\ G_{n1}(X) & G_{n2}(X) & \cdots & \cdots & G_{nn}(X) \end{pmatrix}$$

where

$$vec(G_{ij}) = \frac{\partial F_{j_0}^T}{\partial X_{i_0}^T},$$

that is,

$$G_{ij} = \begin{pmatrix} \dfrac{\partial F_{j1}}{\partial X_{i1}} & \cdots & \cdots & \dfrac{\partial F_{jm}}{\partial X_{i1}} \\ \vdots & & & \vdots \\ \vdots & & & \vdots \\ \dfrac{\partial F_{j1}}{\partial X_{in}} & \cdots & \cdots & \dfrac{\partial F_{jm}}{\partial X_{in}} \end{pmatrix},$$

may be decomposed as $G(X) = (A(X) \otimes I) + (I \otimes B^T(X))$ where $A(X)$, $B(X)$, and I are $n \times n$ matrices if and only if

$$\frac{\partial F_{ij}}{\partial X_{ij}} - \frac{\partial F_{kj}}{\partial X_{kj}} = \frac{\partial F_{il}}{\partial X_{il}} - \frac{\partial F_{kl}}{\partial X_{kl}} \quad \begin{array}{l} \textit{for } i,k=1,\dots,n \textit{ with } i \neq k \\ \textit{and } i,l=1,\dots,n \textit{ with } j \neq l; \end{array} \qquad (1.2.31)$$

$$\frac{\partial F_{ij}}{\partial X_{ik}} = \frac{\partial F_{lj}}{\partial X_{lk}} \quad \begin{array}{l} \textit{for } i,l=1,\dots,n \textit{ with } i \neq j \\ \textit{and } j,k=1,\dots,n \textit{ with } j \neq k; \end{array} \qquad (1.2.32)$$

$$\frac{\partial F_{ij}}{\partial X_{kj}} = \frac{\partial F_{il}}{\partial X_{kl}} \quad \begin{array}{l} \textit{for } i,k=1,\dots,n \textit{ with } i \neq k \\ \textit{and } j,l=1,\dots,n \textit{ with } j \neq l; \end{array} \qquad (1.2.33)$$

$$\frac{\partial F_{ij}}{\partial X_{kl}} = 0 \quad \begin{array}{l} \textit{for } i,k=1,\dots,n \textit{ with } i \neq k \\ \textit{and } j,l=1,\dots,n \textit{ with } j \neq l. \end{array} \qquad (1.2.34)$$

Proof: By (1.2.29) we know that the jth element of the main diagonal of the ith diagonal block is of the form $a_{ii} + b_{jj}$, that is:

$$\frac{\partial F_{ij}}{\partial X_{ij}} = a_{ii} + b_{jj} \text{ for } i, j = 1, \dots, n.$$

The difference between the corresponding elements on the main diagonal of any two diagonal blocks is invariant with respect to the position of the elements on the main diagonal. Hence, for the ith and k^{th} diagonal blocks, we obtain

$$(a_{ii} + b_{jj}) - (a_{kk} + b_{jj}) = a_{ii} - a_{kk} \text{ for } j = 1, \dots, n.$$

this relationship can be satisfied if and only if

$$\frac{\partial F_{ij}}{\partial X_{ij}} - \frac{\partial F_{kj}}{\partial X_{kj}} = \frac{\partial F_{il}}{\partial X_{il}} - \frac{\partial F_{kl}}{\partial X_{kl}} \quad \begin{array}{l} \textit{for } i,k=1,\dots,n; \\ j,l=1,\dots,n. \end{array} \qquad (1.2.35)$$

Similarly, for the jth and l^{th} elements on the main diagonal of any diagonal block we have

$$(a_{ii} + b_{jj}) - (a_{ii} + b_{ll}) = b_{jj} - b_{ll} \text{ for } i = 1, \dots, n,$$

which can be satisfied if and only if

$$\frac{\partial F_{ij}}{\partial X_{ij}} - \frac{\partial F_{il}}{\partial X_{il}} = \frac{\partial F_{kj}}{\partial X_{kj}} - \frac{\partial F_{kl}}{\partial X_{kl}} \quad \begin{array}{l} \textit{for } i,k=1,\dots,n; \\ j,l=1,\dots,n. \end{array} \qquad (1.2.36)$$

Since equations (1.2.35) and (1.2.36) are equivalent, only a single condition is imposed. Equation (1.2.31) is therefore proved.

From (1.2.29) again, the element in the kth row and j^{th} column ($j \neq k$) of G_{ii} is of the form

$$b_{jk} = \frac{\partial F_{ij}}{\partial X_{ik}} \text{ for } i = 1, \ldots, n.$$

The element in this position will be the same in all diagonal blocks if and only if

$$\frac{\partial F_{ij}}{\partial X_{ik}} = \frac{\partial F_{lj}}{\partial X_{lk}} \qquad \begin{array}{l} \text{for } i,l=1,\ldots,n; \\ j,k=1,\ldots,n \text{ (with } j\neq k). \end{array}$$

This proves equation (1.2.32).

Now by equation (1.2.30), each element on the main diagonal of the kth block row and j^{th} block column is

$$a_{ki} = \frac{\partial F_{ij}}{\partial X_{kj}} \text{ for } \begin{array}{l} i,k=1,\ldots,n \text{ (with } i\neq k) \\ j=1,\ldots,n. \end{array}$$

The diagonal elements of any off-diagonal block will be invariant with respect to their position on the main diagonal of the block if and only if

$$\frac{\partial F_{ij}}{\partial X_{kj}} = \frac{\partial F_{il}}{\partial X_{kl}} \qquad \begin{array}{l} \text{for } i,k=1,\ldots,n \text{ (with } i\neq k) \\ j,k=1,\ldots,n. \end{array}$$

This is the equation (1.2.33).

Finally, from (1.2.30), all elements off the main diagonal of G_{ki} are equal to zero, that is:

$$\frac{\partial F_{ij}}{\partial X_{kl}} = 0 \qquad \begin{array}{l} \text{for } i,k=1,\ldots,n \text{ (with } i\neq k); \\ j,l=1,\ldots,n \text{ (with } j\neq l). \end{array}$$

The proof of the theorem is therefore complete.

We are now in a position to prove the VPF for matrix differential equation (1.2.26).

Theorem 1.2.11: *Assume that $f(t, x)$ in equation (1.2.27) has continuous partial derivatives on $R_+ \times R^{n^2}$ and $x(t, t_0, x_0)$ is the unique solution of equation (1.2.27) for $t \geq t_0$, and let*

$$G(t, t_0, x_0) = \frac{\partial f}{\partial x}(t, x(t, t_0, x_0)).$$

If there exist $n \times n$ matrices $A(t, t_0, x_0)$ and $B(t, t_0, x_0)$ such that

$$G = (A \otimes I) + (I \otimes B^T)$$

then

(i) $\Phi(t, t_0, x_0) = \frac{\partial x(t,t_0,x_0)}{\partial x_0}$ *exists and is the fundamental matrix solution of the variational equation*

$$\varphi' = G(t, t_0, x_0)\varphi$$

such that $\Phi(t_0, t_0, x_0) = I$; and therefore

$$\Phi(t, t_0, x_0) = W(t, t_0, x_0) \otimes Z^T(t, t_0, x_0)$$

where W and Z are solutions of

$$W' = A(t, t_0, x_0)W, W(t_0) = I, \qquad (1.2.37)$$

and

$$Z' = ZB(t, t_0, x_0), Z(t_0) = I, \qquad (1.2.38)$$

respectively;

(*ii*) *any solution $Y(t, t_0, X_0)$ of (1.2.26) satisfies the integral equation*

$$Y(t, t_0, X_0) = X(t, t_0, X_0)$$

$$+ \int_{t_0}^{t} W(t, s, Y(s, t_0, X_0))R(s, Y(s, t_0, X_0))Z(t, s, Y(s, t_0, X_0))ds \text{ for } t \geq t_0.$$

Proof:

(*i*)   We have the existence of $\Phi(t, t_0, x_0)$ by Theorem 1.2.6. By hypothesis in this case, we get

$$\varphi' = G(t, t_0, x_0)\varphi = [A(t, t_0, x_0) \otimes I + I \otimes B^T(t, t_0, x_0)]\varphi$$

with the initial value

$$\varphi(t_0, t_0, x_0) = e, \ e = vec(I^T),$$

which has the solution

$$\varphi(t, t_0, x_0) = W(t, t_0, x_0)IZ(t, t_0, x_0) \qquad (1.2.39)$$

where W and Z are the solutions of (1.2.37) and (1.2.38) respectively, and I is the $n^2 \times n^2$ identity matrix. However, the matrix equation (1.2.39) is equivalent to

$$\varphi(t, t_0, x_0) = (W(t, t_0, x_0) \otimes Z^T(t, t_0, x_0))e.$$

Therefore,

$$\Phi(t, t_0, x_0) = W(t, t_0, x_0) \otimes Z^T(t, t_0, x_0). \qquad (1.2.40)$$

(*ii*) Employing the relations (1.2.27), (1.2.28) and substituting for Φ the right-hand side of (1.2.40), we get

$$y(t, t_0, x_0) = x(t, t_0, x_0) +$$

$$\int_{t_0}^{t} [W(t, s, y(s, t_0, x_0)) \otimes Z^T(t, s, y(s, t_0, x_0))]r(s, y(s, t_0, x_0))ds$$

for $t \geq t_0$ where $y(t, t_0, x_0)$ is any solution of (1.2.28).

Now, if we define $X(t, t_0, X_0)$, $Y(t, t_0, X_0)$, $F(t, X)$ and $R(t, Y)$ by $x = vec(X^T)$, $y = vec(Y^T)$, $f = vec(F^T)$, and $r = vec(R^T)$, then the system

(1.2.27) and (1.2.28) are equivalent to the systems (1.2.25) and (1.2.26). Thus we have that

$$Y(t, t_0, X_0) = X(t, t_0, X_0)$$

$$+ \int_{t_0}^{t} W(t, s, Y(s, t_0, X_0)) R(s, Y(s, t_0, X_0)) Z(t, s, Y(s, t_0, X_0)) ds$$

for $t \geq t_0$, where $X(t, t_0, X_0)$ is the unique solution of (1.2.25) for $t \geq t_0$. The proof is complete.

1.3 VPF IN TERMS OF LYAPUNOV-LIKE FUNCTIONALS

We have seen in Sections 1.1 and 1.2 several forms of VPF relative to linear, nonlinear differential systems including matrix differential equations. In this section, we wish to generalize the VPF, in terms of Lyapunov-like functions so as to be more flexible when we employ the corresponding VPF. We do this in a general setting of comparing the solutions of two different differential systems so that the perturbed differential system becomes a special case.

Consider the two different differential systems given by

$$x' = f(t, x), \ x(t_0) = x_0, \tag{1.3.1}$$

and

$$y' = F(t, y), y(t_0) = y_0, \tag{1.3.2}$$

where $f, F \in C[R_+ \times R^n, R^n]$. Relative to these two systems, we have the following general result.

Theorem 1.3.1: *Assume that*
(i)   the solution $x(t, t_0, x_0)$ of (1.3.1) exists, continuously differentiable with respect to (t_0, x_0) for $t \geq t_0$;
(ii) $V \in C^1[R_+ \times R^n, R^N]$ and $y(t) = y(t, t_0, x_0)$ is any solution of (1.3.2).
Then the following generalized VPF is true for $t \leq t_0$,

$$V(t, y(t)) = V(t_0, x(t)) + \int_{t_0}^{t} V_s\Big\{(s, x(t, s, y(s)))$$

$$+ V_x(s, X(t, s, y(s))) \Big[\tfrac{\partial x}{\partial t_0}(t, s, y(s)) + \tfrac{\partial x}{\partial x_0}((t, s, y(s)) F(s, y(s)) \Big] \Big\} ds.$$

Proof: Set $m(s) = V(s, x(t, s, y(s)))$ for $t_0 \leq s \leq t$. Then

$$\frac{dm(s)}{ds} = V_s(s, x(t, s, y(s))$$

$$+ V_x(s, x(t, s, y(s))) \Big[\tfrac{\partial x}{\partial t_0}(t, s, y(s)) + \tfrac{\partial x}{\partial x_0}(t, s, y(s)) F(s, y(s)) \Big],$$

and hence integrating from t_0 to t, we get

$$m(t) - m(t_0) = \int_{t_0}^{t} \Big\{ V_s(s, x(t, s, y(s)))$$

$$V_x(s, x(t, s, y(s))) \Big[\frac{\partial x}{\partial t_0}(t, s, y(s)) + \frac{\partial x}{\partial x_0}(t, s, y(s)) F(s, y(s)) \Big] \Big\} \, ds.$$

Since $m(t) = V(t, y(t))$ and $m(t_0) = V(t_0, x(t))$, the desired formula (1.3.3) is valid completing the proof.

Remark 1.3.1: Several remarks are now in order.
(1) If $F(t, y) = f(t, y) + R(t, y)$, $f_x(t, y)$ exists and is continuous for $t \geq t_0$, and $V(t, x) = x$, then (1.3.3) reduces to Alekseev's VPF (1.2.20).
(2) If $f(t, y) = A(t)y + R(t, y)$ and $V(t, x) = x$, where $A(t)$ is an $n \times n$ continuous s matrix, then (1.3.3) is equivalent to (1.1.4).
(3) If $V(t, x) = x \cdot x = |x|^2$ and $F(t, y) = f(t, y) + R(t, y)$, we get from (1.3.3) the expression

$$|y(t)|^2 = |x(t)|^2 + 2 \int_{t_0}^{t} x(t, s, y(s)) \frac{\partial x}{\partial x_0}(t, s, y(s)) R(s, y(s)) \, ds,$$

which is useful in deriving estimates on solutions $y(t)$ in terms of norms.

We shall discuss this approach in more detail in a later section.

1.4 BOUNDARY VALUE PROBLEMS

Having developed various VPF's for linear and nonlinear differential equations including matrix differential equations, we shall, in the rest of the chapter, proceed to discuss typical applications of the method of variation of parameters. With this motivation, we study in this section, boundary value problem for perturbed linear system given by

$$x' = A(t)x + F(t, x), \quad t \in J = [0, T], \tag{1.4.1}$$

$$Ux = r, \tag{1.4.2}$$

where $A(t)$ is an $n \times n$ continuous matrix on J, $F \in C[J \times R^n, R^n]$, $U: C[0, T] \to R^n$, where $C[0, T]$ is the vector space of all bounded continuous functions $f: [0, T] \to R^n$. Let $|f|_\infty = \sup_J |f(t)|$. We shall use without further mention $|\cdot|$ to denote the norm of a vector as well as the corresponding norm of a matrix.
Let $\Phi(t)$ denote the fundamental matrix for the system

$$x' = A(t)x \tag{1.4.3}$$

such that $\Phi(0) = I$.   The general solution of (1.4.3) is $x(t) = \Phi(t)x_0$, where $x_0 \in R^n$, is arbitrary.   Then $U(\Phi(\cdot)x_0) = \Psi x_0$ for every $x_0 \in R^n$ where Ψ is the matrix whose columns are the values of U on the corresponding columns of $\Phi(t)$. The general solution of (1.4.1) in view of Theorem 1.1.2 is given by

$$x(t) = \Phi(t)x_0 + \int_0^t \Phi(t)\Phi^{-1}(s)F(s, x(s))ds, \ \ t \in [0, T]$$

$$= \Phi(t)x_0 + p(t, x). \tag{1.4.4}$$

This solution $x(t)$ with $x(0) = x_0$ satisfies (1.4.2) if and only if

$$Ux = r = \Psi x_0 + Up(\cdot , x).$$

This equation in x_0 has a unique solution in R^n for some $r \in R^n$ if and only if

$$x_0 = \Psi^{-1}[r - Up(\cdot , x)]. \tag{1.4.5}$$

Thus the problem (1.4.1), (1.4.2) will have a solution on $[0, T]$ if a function $x(t)$, in view of (1.4.4) and (1.4.5), can be found satisfying the integral equation

$$x(t) = \Phi(t)\Psi^{-1}[r - Up(\cdot , x)] + p(t, x). \tag{1.4.6}$$

We are now in a position to prove the following existence theorem for the BVP (1.4.1) and (1.4.2) using Schauder-Tychonov fixed point theorem.

Theorem 1.4.1: *Let*

$$q(t) = \max_{| u | \le \alpha} \{ | \Phi^{-1}(t)F(t, u) | \}, \ \ t \in [0, T], \ \alpha > 0.$$

Define the operator

$$Kf = \Psi^{-1}[r - Up(\cdot , x)]$$

for every $f \in B^\alpha$, where B^α is the closed ball of $C[0, T]$ with the center at zero and radius α. Suppose that

$$L = \max_{t \in [0, T]} | \Phi(t) |, \ M = \sup_{g \in B^\alpha} | Kg |, \ N = \int_0^T q(t)dt.$$

Then the problem (1.4.1), (1.4.2) has at least one solution, if $L(M + N) \le \alpha$.

Proof: We shall show that the operator $V: B^\alpha \to C[0, T]$, defined by

$$(Vu)(t) = \Phi(t)\left[Ku + \int_0^t \Phi^{-1}(s)F(s, u(s))ds \right],$$

has a fixed point in B^α. For this purpose, let $t, t_1 \in [0, T]$ be given. Then we obtain,

$$| (Vu)(t) - (Vu)(t_1) | = | \Phi(t) \left[Ku + \int_0^t \Phi^{-1}(s) F(s, u(s)) ds \right]$$

$$- \Phi(t_1) \left[Ku + \int_0^{t_1} \Phi^{-1}(s) F(s, u(s)) ds \right] | \qquad (1.4.7)$$

$$\leq M | \Phi(t) - \Phi(t_1) | + N | \Phi(t) - \Phi(t_1) | + | \Phi(t_1) | | \int_t^{t_1} q(s) ds |$$

$$\leq (M + N) | \Phi(t) - \Phi(t_1) | + L | \int_t^{t_1} q(s) ds | .$$

Let $\epsilon > 0$ be given. Then there exists a $\delta(\epsilon) > 0$ satisfying

$$| \Phi(t) - \Phi(t_1) | < \frac{\epsilon}{2(M+N)}, \quad | \int_0^{t_1} q(s) ds | < \frac{\epsilon}{2L}, \qquad (1.4.8)$$

for every $t, t_1 \in [0, T]$ with $| t - t_1 | < \delta(\epsilon)$. This fact is clear from the uniform continuity of the function $\Phi(t)$ and the function

$$h(t) \equiv \int_0^t q(s) ds$$

on the interval $[0, T]$. Inequalities (1.4.7) and (1.4.8) imply the equicontinuity of the set $V B^\alpha$. The inclusion relation $V B^\alpha \subset B^\alpha$ follows from $L(M + N) \leq \alpha$. Hence $V B^\alpha$ is relatively compact. Now, we prove that V is continuous on B^α. In fact, let $\{u_m\}_{m=1}^\infty \subset B^\alpha, u \in B^\alpha$ satisfy

$$| u_m - u |_\infty \to 0 \text{ as } m \to \infty.$$

Then it follows that

$$| V u_m - V u |_\infty \leq L [\{ K u_m - K u |$$

$$+ \int_0^T | \Phi^{-1}(s) [F(s, u_m(s)) - F(s, u(s))] | ds] \qquad (1.4.9)$$

$$\leq L (L | \Phi^{-1} | | u | + 1)$$

$$\times \int_0^T | \Psi^{-1}(s) [F(s, u_m(s)) - F(s, u(s))] | ds.$$

The integrand of the second integral in (1.4.9) converges uniformly to zero. Hence, $|Vu_m - Vu|_\infty \to 0$ as $m \to \infty$. By the Schauder-Tychonov Theorem, there exists a fixed point $x \in B^\alpha$ of the operator V. This function $x(t)$, $t \in [0, T]$ is a solution of the problem (1.4.1), (1.4.2). The proof is complete.

Alternately, under a different hypothesis, we establish the following existence result for the BVP (1.4.1), (1.4.2).

Theorem 1.4.2: *Let B^α be as in Theorem 1.4.1 for a fixed $\alpha > 0$. Assume that there exists a constant $k > 0$ such that for every $x_0 \in B_\alpha$ the solution $x(t, 0, x_0)$ of (1.4.1) with $x(0) = x_0$ exists on $[0, T]$, is unique and satisfies*

$$\sup_{\substack{t \in [0,T] \\ x_0 \in B_\alpha}} |x(t, 0, x_0)| \leq k.$$

Let $q(t)$, N and L be as in Theorem 1.4.1, and

$$|\Psi^{-1}|(|r| + |U|LN) \leq \alpha.$$

Then there exists a solution to the problem (1.4.1), (1.4.2).

Proof: We wish to apply Brouwer's fixed point theorem. To this end, consider the operator $Q: B_\alpha \to R^n$ defined by

$$Qu = \Psi^{-1}(r - U p_1(\,\cdot\,, u))$$

where

$$P_1(t, u) = \int_0^t \Phi(t)\Phi^{-1}(s)F(s, x(s, 0, u))ds.$$

Observe that $QB_\alpha \subset B_\alpha$. To prove the continuity of Q, we show the continuity of $x(t, 0, u)$ with respect to u. Let $u_m \in B_\alpha$, $m = 1, 2, \ldots$, $u \in B_\alpha$ be such that $|u_m - u| \to 0$ as $m \to \infty$ and let $x_m(t)$, $x(t)$ be the solutions of

$$x' = A(t)x + F(t, x),\ x(0) = u_m,$$

$$x' = A(t)x + F(t, x),\ x(0) = u,$$

respectively. Then our assumptions imply that $|x_m(t)| \leq k$ and $|x(t)| \leq k$ for all $t \in [0, T]$, $m = 1, 2, \ldots$. The inequality

$$|x_m'(t)| \leq k\sup_{t \in [0,T]}\{|A(t)|\} + \sup_{\substack{t \in [0,T] \\ |u| \leq k}}\{|F(t, u)|\}$$

proves that $\{x_m(t)\}$ is also equicontinuous. By the Arzelá-Ascoli Theorem, there exists a subsequence $\{x_j(t)\}_{j=1}^\infty$ of $\{x_m(t)\}$ such that $x_j(t) \to \bar{x}(t)$ as $j \to \infty$

uniformly on $[0, T]$. Here $\overline{x}(t)$ is some function in $C[0, T]$. Taking limits as $j \to \infty$ in

$$x_j(t) = x_j(0) + \int_0^t A(s)x_j(s)ds + \int_0^t F(s, x_j(s))ds,$$

we obtain

$$\overline{x}(t) = u + \int_0^t A(s)\overline{x}(s)ds + \int_0^t F(s, \overline{x}(s))ds.$$

Hence, by uniqueness, $\overline{x}(t) \equiv x(t)$. In fact, we could have started with any subsequence of $\{x_m(t)\}$ instead of $\{x_m(t)\}$ itself. Here we have actually shown the following: every subsequence of $\{x_m(t)\}$ contains a subsequence converging uniformly on $x(t)$ on $[0, T]$. This implies the uniform convergence of $\{x_m(t)\}$ to $x(t)$ on $[0, T]$. Equivalently, if $u_m \in B_\alpha$, $m = 1, 2, \ldots,$ $u \in B_\alpha$ satisfy $|u_m - u| \to 0$ as $m \to \infty$, then

$$| x(t, 0, u_m) - x(t, 0, u) | \to 0 \text{ uniformly on } [0, T].$$

This proves the continuity of the function $x(t, 0, x_0)$ in $x_0 \in B_\alpha$. Note that it is uniform with respect to $t \in [0, T]$. Let $\{u_m\}$, u be as above. It then follows that

$$| Qu_m - Qu | \leq | \Psi^{-1} | \, | U | \, L \int_0^T | \Phi^{-1}(s)[F(s, x(s, 0, u_m))$$

$$- F(s, x(s, 0, u))]ds. \qquad (1.4.11)$$

The integrand in $(1.4.11)$ tends uniformly to zero as $m \to \infty$. Thus, $| Q_m - Q_u | \to 0$ as $m \to \infty$. This proves the continuity of Q on B_α. Brouwer's Theorem implies the existence of some $x_0 \in B_\alpha$ with the property $Qx_0 = x_0$. This vector x_0 is the initial value of a solution to the problem $(1.4.1)$, $(1.4.3)$. The proof is complete.

1.5 MONOTONE ITERATIVE TECHNIQUE

This section is devoted to the study of periodic boundary value problem (PBVP) for a differential system employing monotone iterative techniques for obtaining extremal solutions of PBVP

$$u' = f(t, u), u(0) = u(2\pi), \qquad (1.5.1)$$

where $f \in C[[0, 2\pi] \times R^n, R^n]$.

We need the following comparison result for our discussion. The inequalities between vectors are understood componentwise, throughout the book.

Lemma 1.5.1: *Let* $m \in C^1[[0, 2\pi], R^n]$ *and* $m'(t) \leq -Mm(t), 0 < t < 2\pi$, *where* $M = [M_{ij}]$ *is an* $n \times n$ *constant matrix such that* $M_{ij} \leq 0$, $i \neq j$, $M_{ii} > 0$ *and*

$\sum_{j=1}^{n} M_{ij} + M_{ii} \geq 0, \; i = 1, \ldots, n.$ Then $m(2\pi) \geq m(0)$ *implies that* $m(t) \leq 0$ *on* $[0, 2\pi]$.

Proof: Suppose that the claim is not true. Then there exists a $t_0 \in [0, 2\pi]$ and an $\epsilon > 0$ such that for at least one i, $(i = 1, \ldots, n)$ $m_i(t)$ attains maximum at t_0 so that we have

$$m_i(t_0) = \epsilon, \, m_i(t) \leq \epsilon, \, t \in [0, 2\pi] \text{ and for } t \in [0, 2\pi]$$

$$m_j(t_0) \leq 0, \; j \neq i.$$

(i)   Let $t_0 \in (0, 2\pi]$. Then $m_i'(t_0) \geq 0$ and

$$0 \leq m_i'(t_0) \leq -(M_{i1}m_1(t_0) + \ldots + M_{in}m_n(t_0))$$

$$\leq - \left[\sum_{\substack{j=1 \\ j \neq i}}^{n} M_{ij} + M_{ii} \right] \epsilon < 0$$

in view of the hypothesis. This is a contradiction.

(ii)   If $t_0 = 0$, then we have $m_i(2\pi) \geq m_i(0) = \epsilon$ and consequently,

$$0 \leq m_i'(2\pi) \leq -[M_{i1}m_1(2\pi) + \ldots + M_{in}m_n(2\pi)]$$

$$\leq - \left[\sum_{\substack{j=1 \\ j \neq i}}^{n} M_{ij} + M_{ii} \right] \epsilon < 0.$$

Again we arrive at a contradiction. The proof is therefore complete.

We now assume the following hypotheses:

(H_1) $\alpha, \beta \in C^1[[0, 2\pi], R^n]$, $\alpha(t) \leq \beta(t)$ on $[0, 2\pi]$ and for $0 < t \leq 2\pi$

$$\alpha'(t) \leq f(t, \alpha(t)), \; \alpha(0) \leq \alpha(2\pi),$$

$$\beta'(t) \geq f(t, \beta(t)), \; \beta(0) \geq \beta(2\pi),$$

(H_2) $f(t, u) - f(t, v) \geq -M(u - v)$ whenever $\alpha(t) \leq v \leq u \leq \beta(t), t \in [0, 2\pi]$, where M is a $n \times n$ constant matrix such that $M_{ij} \leq 0$, $M_{ii} > 0$, $i \neq j$ and $\sum_{j=1}^{n} M_{ij} + M_{ii} > 0$, $i = 1, \ldots, n.$

We now prove the main result of this section.

Theorem 1.5.1: *Let* (H_1) *and* (H_2) *hold. Then there exist monotone sequences* $\{\alpha_n(t)\}$, $\{\beta_n(t)\}$ *with* $\alpha_0 = \alpha$, $\beta_0 = \beta$ *such that* $\lim_{n \to \infty} \alpha_n(t) = \rho(t)$, $\lim_{n \to \infty} \beta_n(t) =$

$r(t)$ *uniformly and monotonically on* $[0, 2\pi]$ *and that* ρ *and* r *are the minimal and maximal solutions of the PBVP* (1.5.1) *respectively.*

Proof: For each $\eta \in C[[0, 2\pi], R^n]$ such that $\alpha(t) \leq \eta(t) \leq \beta(t)$ with $\eta(0) \leq \eta(2\pi)$, consider the PBVP

$$u' = G(t, u), \ u(0) = u(2\pi),   \tag{1.5.2}$$

where $G(t, u) = f(t, \eta(t)) - M(u - \eta(t))$. By using the VPF obtained in Theorem 1.1.2, we get

$$u(t) = e^{-Mt}u(0) + \int_0^t e^{-M(t-s)}[f(s, \eta(s)) + M\eta(s)]ds.   \tag{1.5.3}$$

Since at $t = 2\pi$,

$$u(2\pi) = e^{-2M\pi}[u(0) + \int_0^{2\pi} e^{Ms}[f(s, \eta(s)) + M\eta(s)]ds],$$

it follows that, in view of the fact $[I - e^{-2M\pi}]^{-1}$ exists, by (H_1)

$$u(2\pi) = u(0) = [I - e^{2M\pi}]^{-1}e^{-2M\pi}\int_0^{2\pi} e^{Ms}[f(s, \eta(s)) + M\eta(s)]ds$$

is uniquely determined. Hence the PBVP (1.5.2) possesses unique solution and it is obtained by substituting the value of $u(0)$ in (1.5.3).

Defining a mapping A by $A\eta = u$ where $u = u(t)$ is the unique solution of PBVP (1.5.2) for η. We show that
(a) $\alpha \leq A\alpha, \beta \geq A\beta$;
(b) A is a monotone increasing operator on $[\alpha, \beta]$.

To prove (a), set $A\alpha = \alpha_1$ where α_1 is the unique solution of the PBVP (1.5.2) with $\eta = \alpha$. Set $p = \alpha - \alpha_1$. Then

$$p' = \alpha' - \alpha_1' \leq f(t, \alpha) - f(t, \alpha) + M(\alpha - \alpha_1) = -Mp$$

and $p(2\pi) = \alpha(2\pi) - \alpha_1(2\pi) \geq \alpha(0) - \alpha_1(0) = p(0)$.

By Lemma 1.5.1, it follows that $p \leq 0$ on $[0, 2\pi]$ which proves that $\alpha \leq A\alpha$. Similarly, we can show that $\beta \geq A\beta$.

To prove (b), let $\eta_1, \eta_2 \in C[[0, 2\pi], R^n]$ such that $\alpha(t) \leq \eta_2(t) \leq \eta_1(t) \leq \beta(t)$, $t \in [0, 2\pi]$ with $\eta_1(0) \leq \eta_1(2\pi)$ and $\eta_2(0) \leq \eta_2(2\pi)$. Suppose that $u_1 = A\eta_1$ and $u_2 = A\eta_2$. Setting $p = u_1 - u_2$, we get

$$p' = f(t, \eta_1) - M(u_1 - \eta_1) - f(t, \eta_2) + M(u_2 - \eta_2)$$

$$\leq M(\eta_2 - \eta_1) - M(u_1 - \eta_1) + M(u_2 - \eta_2) = -Mp.$$

Further $p(2\pi) = p(0)$. Hence by Lemma 1.5.1, we conclude that $p(t) \geq 0$ on $[0, 2\pi]$ proving (b).

Define the sequences $\{\alpha_n(t)\}$, $\{\beta_n(t)\}$ with $\alpha = \alpha_0$, $\beta = \beta_0$ such that $\alpha_n = A\alpha_{n-1}$, $\beta_n = A\beta_{n-1}$. We conclude that

$$\alpha_0 \leq \alpha_1 \leq \dots \leq \alpha_n \leq \beta_n \leq \dots \leq \beta_1 \leq \beta_0 \text{ on } [0, 2\pi].$$

Hence by following the standard argument, we get

$$\lim_{t \to \infty} \alpha_n(t) = \rho(t) \text{ and } \lim_{\eta \to \infty} \beta_n(t) = r(t)$$

uniformly and monotonically on $[0, 2\pi]$. Since α_n, β_n are given by

$$\alpha_n' = f(t, \alpha_{n-1}) - M(\alpha_n - \alpha_{n-1}), \ \alpha_n(0) = \alpha_n(2\pi),$$

$$\beta_n' = f(t, \beta_{n-1}) - M(\beta_n - \beta_{n-1}), \ \beta_n(0) = \beta_n(2\pi),$$

which satisfy the corresponding integral equations of the type (1.5.3), it is easy to see that ρ, r are solutions of PBVP (1.5.1).

To show that ρ, r are the minimal and maximal solutions of the PBVP (1.5.1), we need to show that if $u(t)$ is any solution of (1.5.1) such that $\alpha(t) \leq u(t) \leq \beta(t)$ on $[0, 2\pi]$, then $\alpha \leq \rho \leq u \leq r \leq \beta$ on $[0, 2\pi]$. To do this, suppose that for some n $\alpha_n(t) \leq u(t) \leq \beta_n(t)$, on $[0, 2\pi]$. Set $\rho(t) = \alpha_{n+1}(t) - u(t)$. We then get

$$p' = f(t, \alpha_n) - M(\alpha_{n+1} - \alpha_n) - f(t, u)$$

$$\leq M(u - \alpha_n) - M(\alpha_{n-1} - \alpha_n) = -Mp.$$

Further, $p(2\pi) = \alpha_{n+1}(2\pi) - u(2\pi) = \alpha_{n+1}(0) - u(0) = p(0)$, and therefore by Lemma 1.5.1, $\alpha_{n+1}(t) \leq u(t)$ on $[0, 2\pi]$. Similarly, we can show that $u(t) \leq \beta_{n+1}(t)$ on $[0, 2\pi]$ and hence $\alpha_{n+1} \leq u(t) \leq \beta_{n+1}(t)$, $t \in [0, 2\pi]$. Since $\alpha_0 = \alpha \leq u \leq \beta = \beta_0$, this proves by induction that for all n, $\alpha_n(t) \leq u(t) \leq \beta_n(t)$ on $[0, 2\pi]$. Taking the limit as $n \to \infty$, we conclude that $\alpha \leq \rho \leq u \leq r \leq \beta$ on $[0, 2\pi]$. The proof is complete.

1.6 EXPONENTIAL DICHOTOMY

This section is devoted to the splitting of the solution space of the linear differential systems which induces interesting properties for the solutions of nonlinear perturbations of such solutions. Consider the linear system

$$x' = A(t)x \tag{1.6.1}$$

where A is an $n \times n$ continuous matrix and its perturbation

$$y' = A(t)y + F(t, y), \tag{1.6.2}$$

where $F \in C[R \times R^n, R^n]$. We let, as before, $|f|_\infty = \sup_{t \in R} |f(t)|$ where $f \in C[R, R^n]$.

Let P_1 be a projection matrix and $R^n = R_1 \oplus R_2$, where $R_1 = P_1 R^n$ and $R_2 = P_2 R^n$. We show that the matrix P_1 or P_2 may induce a "split" in the space of solutions of (1.6.1). This means that solutions of (1.6.1) with initial condition $x_0 \neq 0$ in R_1 tend to zero as $t \to \infty$, while the solutions with initial condition $x_0 \neq 0$ in R_2 have norms that tend to infinity as $t \to \infty$. A corresponding situation holds, when the roles of R_1 and R_2 are reversed, in R_-. This split in solution space is caused by the property of "exponential splitting" defined below. We identify the projections P_1, P_2 with the corresponding bounded linear operators. We need the following definitions.

Definition 1.6.1: The *angular distance* $\alpha(R_1, R_2)$ between the subspaces $R_1 \neq \{0\}$, $R_2 \neq \{0\}$ is defined as

$$\alpha(R_1, R_2) = inf\{ \mid u_1 + u_2 \mid ; u_k \in R_k, \mid u_k \mid = 1, k = 1, 2\}. \qquad (1.6.3)$$

The following lemma establishes the basic relationship between the norms of the projections P_1, P_2 and the function $\alpha(R_1, R_2)$.

Lemma 1.6.1: *Let $R^n = R_1 \oplus R_2$ and let P_1, P_2 be the projections of R_1, R_2, respectively. Then if $\mid P_k \mid > 0$, $k = 1, 2$, we have*

$$\frac{1}{|P_k|} \leq \alpha(R_1, R_2) \leq \frac{2}{|P_k|}, \quad k = 1, 2.$$

Proof: Let $\mu > \alpha(R_1, R_2)$, where μ is a positive constant, and let $u_1 \in R_1$, $u_2 \in R_2$ be such that $\mid u_k \mid = 1$, $k = 1, 2$, and $\mid u_1 + u_2 \mid < \mu$. Then, if $u = u_1 + u_2$, we have $P_k u = u_k$, $k = 1, 2$, and

$$1 = \mid u_k \mid = \mid P_k u \mid \leq \mid P_k \mid \mid u \mid < \mu \mid P_k \mid.$$

The left-side inequality follows. The right-side inequality can be proved once we realize that

$$\alpha(R_1, R_2) \leq \left| \frac{P_1 u}{|P_1 u|} + \frac{P_2 u}{|P_2 u|} \right|,$$

where $u \in R^n$ is such that $P_k u \neq 0$, $k = 1, 2$.

Let $\Phi(t)$ denote the fundamental matrix solution of (1.6.1) such than $\Phi(0) = I$. Then we can define the exponential split and dichotomy as follows.

Definition 1.6.2: Let A be an $n \times n$ continuous matrix. We say that system (1.6.1) possesses an *exponential splitting* if there exist two positive numbers H, m_0 and a projection matrix P with the following properties:
(*i*) Every solution $x(t) \equiv \Phi(t) x_0$ of (1.6.1) with $x_0 \in R_1$ satisfies

$$\mid x(t) \mid \leq H \, exp\{ - m_0(t - s)\} \mid x(s) \mid, \quad t \geq s. \qquad (1.6.4)$$

(*ii*) Every solution $x(t) \equiv \Phi(t) x_0$ of (1.6.1) with $x_0 \in R_2$ satisfies

$$\mid x(t) \mid \leq H \, exp\{ - m_0(s - t)\} \mid x(s) \mid, \quad s \geq t. \qquad (1.6.5)$$

(iii) $| P_k | \neq 0, k = 1, 2$, and if we denote by $P_1(t)$, $P_2(t)$ the projections $\Phi(t)P_1\Phi^{-1}(t)$, $\Phi(t)P_2\Phi^{-1}(t)$, respectively, and let $R_1(t) \equiv P_1(t)R^n$, $R_2(t) \equiv P_2(t)R^n$, then there exists a constant $\beta > 0$ such that

$$\alpha(R_1(T), R_2(t)) \geq \beta, t \in R.$$

Next, we introduce the concept of an exponential dichotomy. The existence of an exponential dichotomy for the system (1.6.1) (with $P_1 \neq 0, I$) is equivalent to the existence of an exponential splitting. This is shown in Theorem 1.6.1 below.

Definition 1.6.3: Let A be an $n \times n$ continuous matrix. We say that the system (1.6.1) possesses an *exponential dichotomy* if there exist two positive constants H_1, m_0 and a projection matrix P with the following properties:

$$| \Phi(t)P_1\Phi^{-1}(x) | \leq H_1 \, exp\{ - m_0(t - s)\}, t \geq s, \tag{1.6.6}$$

$$| \Phi(t)P_2\Phi^{-1}(s) | \leq H_1 \, exp\{ - m_0(s - t)\}, s \geq t.$$

Theorem 1.6.1: *The system* (1.6.1) *possesses an exponential dichotomy with* $P_1 \neq 0, I$ *if and only if* (1.6.1) *possesses an exponential splitting.*

Proof: Assume that the system (1.6.1) possesses an exponential dichotomy with $P_1 \neq 0, I$ and constants H_1, m_0 as in (1.6.6). Let $x(t)$ be a solution of (1.6.1) with $x(0) = x_0 = P_1x_0 \in R_1$. Then we have $x(s) = \Phi(s)x_0$. Hence we get

$$| x(t) | = | \Phi(t)P_1x_0 | = | \Phi(t)P_1\Phi^{-1}(s)x(s) |$$

$$\leq H_1 \, exp\{ - m_0(t - s)\} | x(s) |, t \geq s.$$

Similarly, the inequality (1.6.5) is proved.

To show that $\alpha(R_1(t), R_2(t))$ has a positive lower bound, it is sufficient to note that by virtue of Lemma 1.6.1 that $| P_k(t) | \leq H_1, t \in R$.

Conversely, assume that the system (1.6.1) possesses an exponential splitting with H, m_0 as in (1.6.4). Let $x(t)$ be a solution of (1.6.1) with $x(0) = P_1\Phi^{-1}(s)u$, for a fixed $(s, u) \in R \times R^n$. Then from (1.6.4) we obtain

$$| \Phi(t)P_1\Phi^{-1}(s)u | = | x(t) |$$

$$\leq H \, exp\{ - m_0(t - s)\} | x(s) |$$

$$= H \, exp\{ = - m_0(t - s)\} | \Phi(s)P_1\Phi^{-1}(s)u |$$

$$\leq M H \, exp\{ - m_0(t - s)\} | u |, t \geq s,$$

where the constant M is an upper bound for $| P_1(t) |$.

Here we have used the fact that the existence of a positive lower bound for $\alpha(R_1(t), R_2(t))$ is equivalent to the boundedness of the projections $P_1(t)$, $P_2(t)$ on R. Thus, the first inequality in (1.6.6) is true.

The second inequality in (1.6.6) is proved in a similar way. The proof is complete.

Now, assume that (1.6.1) possesses an exponential splitting. It is easy to see that every solution $x(t)$ of (1.6.1) with $x(0) = P_1 x(0)$ satisfies

$$\lim_{t \to \infty} |x(t)| = 0.$$

Let $x(t)$ be a solution of (1.6.1) such that $x(0) \in R_2$, $x(0) \neq 0$. Then from (1.6.5), we obtain

$$|x(s)| \geq (1/H)exp\{m_0(s - t)\} |x(t)|, \; s \geq t,$$

which yields

$$\lim_{s \to \infty} |x(s)| = +\infty.$$

If $x(t)$ is now any solution of (1.6.1) with $P_2 x(0) \neq 0$, then $x(0) = x_1(0) + x_2(0)$ with $x_1(0) \in R_1$ and $x_2(0) \in R_2$. It follows that

$$x(t) = \Phi(t)x(0) = \Phi(t)x_1(0) + \Phi(t)x_2(0) \equiv x_1(t) + x_2(t),$$

with $x_1(t)$, $x_2(t)$ solutions of (1.6.1). From the above considerations and

$$|x(t)| \geq |x_2(t)| - |x_1(t)|,$$

it follows that $|x(t)| \to +\infty$ as $t \to \infty$. Clearly, R_1 is precisely the space of all initial conditions of solutions of (1.6.1) which are bounded on R_+. The situation is reversed on the interval R_-.

The foregoing discussion proves the fact that, in the presence of an exponential splitting, the system (1.6.1) can have only one bounded solution on R_-, the zero solution.

The following theorem ensures the existence of a bounded solution on R of the system (1.6.2). This solution has some interesting stability properties. We need the following definition.

Definition 1.6.4: The zero solution of the system (1.6.2) is *negatively unstable* if there exists a number $r > 0$ with the following property: every other solution $x(t)$ of (1.6.2) defined on an interval $(-\infty, a]$, for any number a, satisfies

$$\sup_{t \leq a} |x(t)| > r.$$

Theorem 1.6.2: *Consider the system* (1.6.2) *under the following conditions:*
(i)   *A be an $n \times n$ continuous matrix on R and such that the system* (1.6.1) *possesses an exponential dichotomy.*
(ii)   *$F: R \times R^n \to R^n$ is continuous and, for some constant $r > 0$, there exists a function $\beta: R \to R_+$, continuous and such that*

$$|F(t, u) - F(t, v)| \leq \beta(t)|u - v|$$

for every $(t, u, v) \in R \times \overline{B_r(0)} \times \overline{B_r(0)}$, *where* $B_r(0) = \{x \in R^n, \mid x \mid < r\}$.

(iii)

$$\rho = H_1 \sup_{t \in R} \left\{ \int_{-\infty}^{0} exp\{m_0 s\} \beta(s + t) ds + \int_{0}^{\infty} exp\{-m_0 s\} \beta(s + t) ds \right\} < 1.$$

(iv)

$$\left| \int_{-\infty}^{t} \Phi(t) P_1 \Phi^{-1}(s) F(s, 0) ds - \int_{t}^{\infty} \Phi(t) P_2 \Phi^{-1}(s) F(s, 0) ds \right|$$

$$< \frac{r(1-\rho)}{2}, \, t \in R.$$

Then
(i)   *there exists a unique solution* $x(t)$, $t \in R$, *of the system* (1.6.2) *satisfying*
      $\mid x \mid_\infty \leq r$.
(ii) *If* $P = I$, $F(t, 0) \equiv 0$, *and* (iii) *holds without necessarily the second integral,*
      then the zero solution of (1.6.2) *is negatively unstable.*
(iii) *Let* $P = I$ *and let* (iii) *hold without necessarily the second integral. Suppose*
      that

$$\lambda = \lim_{t \to \infty} \sup \left\{ (1/t) \int_{0}^{t} \beta(s) ds \right\} < m_0 / H_1 \qquad (1.6.7)$$

Then there exists a constant $\delta > 0$ *with the property: if* $x(t)$ *is the solution of Conclusion* (i) *and* $y(t)$, $t \in [0, T)$, $T \in (0, \infty)$, *is another solution of* (1.6.2) *with*

$$\mid x(0) - y(0) \mid \leq \delta, \qquad (1.6.8)$$

then $y(t)$ *exists on* R_+,

$$\mid y(t) \mid \leq r, \, t \in R_+, \qquad (1.6.9)$$

and

$$\lim_{t \to \infty} \mid x(t) - y(t) \mid = 0. \qquad (1.6.10)$$

Moreover, if $F(t, 0) \equiv 0$, *then the zero solution of* (1.6.2) *is asymptotically stable.*

Proof: (i) We consider the operator U defined as follows:

$$(Uf)(t) = \int_{-\infty}^{t} \Phi(t) P_1 \Phi^{-1}(s) F(s, f(s)) ds - \int_{t}^{\infty} \Phi(t) P_2 \Phi^{-1}(s) F(s, f(s)) ds \quad (1.6.11)$$

This operator satisfies $U B^r \subset B^r$, where

$$B^r = \{f \in C_n(R); \mid f \mid_\infty \leq r\}.$$

We have

$$| Uf |_\infty \le r(1+\rho)/2 < r, \qquad\qquad (1.6.12)$$

and

$$| U f_1 - U f_2 |_\infty \le \rho | f_1 - f_2 |_\infty, \; f_1, f_2 \in B^r.$$

By the Banach Contraction Principle, U has a unique fixed point x in the ball B^r. It is easy to see that the function $x(t)$ is a solution to the system (1.6.2) on R.

Let $y(t)$, $t \in R$, be another solution of the system (1.6.2) with $| y |_\infty \le r$, and let $g(t) \equiv (Uy)(t)$. Then the function $g(t)$ satisfies the equation

$$x' = A(t)x + F(t, y(t)). \qquad\qquad (1.6.13)$$

However, (1.6.6) implies that (1.6.13) can have only one bounded solution on R. Hence, $g(t) = y(t)$, $t \in R$, and y is a fixed point of U in B^r. This says that $y(t) \equiv x(t)$. Thus, $x(t)$ is unique.

(ii) Now, let $P = I$, $F(t, 0) \equiv 0$, and let (iii) hold without necessarily the second integral. Then zero is the only solution of (1.6.2) in the ball B^r. Assume that $y(t)$, $t \in (-\infty, a]$, is some solution of (1.6.2) such that

$$\sup_{t \le a} | y(t) | \le r.$$

It is easy to observe, that the operator U^a, defined by

$$(U^a f)(t) \equiv \int_{-\infty}^{t} \Phi(t)\Phi^{-1}(s)F(s, f(s))ds,$$

has a unique fixed point x in the ball

$$B_a^r = \{f \in C(-\infty, a]; \; | f |_\infty \le r\}.$$

We must have $x(t) = 0$, $t \in (-\infty, a]$. The function $y(t)$ satisfies the equation

$$y(t) = \Phi(t)\left[\Phi^{-1}(t_0)y(t_0) + \int_{t_0}^{t} \Phi^{-1}(s)F(s, y(s))ds\right] \qquad (1.6.14)$$

for any $t_0, t \in (-\infty, a]$ with $t_0 \le t$. We fix t in (1.6.14) and take the limit of the right-hand side as $t_0 \to -\infty$. This limit exists as a finite vector because

$$| \Phi(t)\Phi^{-1}(t_0)y(t_0) | \le rH \exp\{ -m_0(t - t_0)\}, \; t \ge t_0.$$

We find

$$y(t) = \Phi(t) \int_{-\infty}^{t} \Phi^{-1}(s) F(s, y(s)) ds.$$

Consequently, y is a fixed point for the operator U^a in B_a^r. This proves that $y(t) = 0$, $t \in (-\infty, a]$.

It follows that every solution $y(t)$ of (1.6.2), defined on an interval $(-\infty, a]$, must be such that $|y(t_m)| > r$, for a sequence $\{t_m\}_{m=1}^{\infty}$ with $t_m \to -\infty$ as $m \to \infty$. Therefore, the zero solution of (1.6.2) is negatively unstable.

To prove (iii), let the assumptions given in (iii) be satisfied (without necessarily $F(t, 0) \equiv 0$) and let the positive number $\delta < r(1 - \rho)/2$ satisfy the condition

$$|x(0) - y(0)| < \delta,$$

where $x(t)$ is the solution in Conclusion (i) and $y(t)$ is another solution of (1.6.2) defined on the interval $[0, T)$, $T \in (0, \infty)$. Then there exists a sufficiently small neighborhood $[0, T_1) \subset [0, T)$ such that $|y(t)| < r$ for all $t \in [0, T_1)$. Here we have used the fact that $|x|_{\infty} < r(1 + \rho)/2$ from (1.6.12). For such values of t, use the VPF to obtain

$$|x(t) - y(t)| \leq |\Phi(t)\Phi^{-1}(0)| \; |x(0) - y(0)|$$

$$+ \int_0^t |\Phi(t)\Phi^{-1}(s)| \, |\beta(s)| \, |x(s) - y(s)| \, ds.$$

Using the dichotomy (1.6.6), we further obtain

$$exp\{m_0 t\} \, |x(t) - y(t)|$$

$$\leq H_1 \left[|x(0) - y(0)| + \int_0^t exp\{m_0 s\} \beta(s) \, |x(s) - y(s)| \, ds \right]. \quad (1.6.15)$$

An application of Gronwall's Inequality to (1.6.15) gives

$$|x(t) - y(t)| \leq H_1 |x(0) - y(0)| \, exp\left\{ -m_0 t + H_1 \int_0^t \beta(s) ds \right\} \quad (1.6.16)$$

for all $t \in [0, T_1)$. From the definition of λ in Conclusion (iii), we obtain that if $\epsilon > 0$ is such that $\lambda + 2\epsilon < m_0/H_1$, then there exists an interval $[t_0, \infty)$, $t_0 > 0$, such that

$$(1/t) \int_0^t \beta(s) ds < \lambda + \epsilon < (m_0/H_1) - \epsilon, t \geq t_0.$$

This fact yields

$$H_1 \int\limits_0^t \beta(s)ds - m_0 t < -\epsilon H_1 t, \, t \geq t_0.$$

Now, choose

$$\delta < min\{r(1-\rho)/(2H_1 M), r(1-\rho)/2\},$$

where

$$M = \sup_{t \geq 0}\left\{exp\left\{-m_0 t + H_1\int\limits_0^t \beta(s)ds\right\}\right\}.$$

For such δ, in view of (1.6.16), we get

$$|x(t) - y(t)| < r(1-\rho)/2, \, t \in [0, T). \tag{1.6.17}$$

In case (1.6.17) is false, then setting

$$T_2 = inf\{t \in [0, T); \, |x(t) - y(t)| = r(1-\rho)/2\},$$

we obtain from (1.6.16)

$$|x(T_2) - y(T_2)| < r(1-\rho)/2,$$

which is a contradiction. It follows that $y(t)$ is continuable to $+\infty$ and that $|y|_\infty < r$. If we further assume that $F(t, 0) = 0$, $t \in R_+$, then we must take $x(0) = 0$, $t \in R_+$. In this case (1.6.16) implies the asymptotic stability of the zero solution of (1.6.2). This completes the proof.

1.7 SENSITIVITY ANALYSIS

In this section we discuss sensitivity analysis of linear matrix differential equation with nonlinear perturbation. Such equations are called, in the literature, Lyapunov systems which arise in a number of areas such as control engineering, dynamical systems and optimal filters. Consider the Lyapunov system

$$X'(t) = A(t)X(t) + X(t)B(t) + F(t, X(t)), \tag{1.7.1}$$

where $A(t), B(t), X(t)$ are square matrixes of order n and $F: [a, b] \times R^{n \times n} \to R^{n \times n}$ is continuous and Lipschitzian in x. The general solution is established in terms of the fundamental matrix solutions of $\Phi' = A(t)\Phi$ and $\Psi' = \Psi B(t)$.

The main reason for the investigation of sensitivity analysis of (1.7.1) lies in the fact that the mathematical models of processes are contaminated with measurement errors. It is therefore necessary to study the set of all solutions of admissible perturbations of the parameters.

Suppose that the matrices A and B in (1.7.1) undergo perturbation $A \to A^+ = A + \Delta A, B \to B^+ = B + \Delta B$. Then the perturbed system of (1.7.1) is

$$(X^+)' = A^+(t)X^+(t) + (X^+(t)B^+(t) + F(t, X^+), \qquad (1.7.2)$$

where we denote $X^+ = X + \Delta X$, X being the solution of (1.7.1). Hence, it follows that

$$\Delta X' = [A(t) + \Delta A(t)]\Delta X + \Delta X[B(t) + \Delta B(t)] + g(t, \Delta X), \qquad (1.7.3)$$

where

$$g(t, \Delta X) = \Delta A X + X \Delta B + F(t, X + \Delta X) - F(t, X). \qquad (1.7.4)$$

Let Y be a fundamental matrix solution of $p' = (A + \Delta A)P$ and Z be a fundamental matrix solution of $p' = (B^T + \Delta B^T)P$, then by the VPF obtained in Theorem 1.1.3, any solution of (1.7.3) is given by

$$\Delta X = Y(t)CZ^T(t) + Y(t)\left[\int_a^t Y^{-1}(s)g(s, \Delta X(s))(Z^T)^{-1}(s)ds\right]Z^T(t) \qquad (1.7.5)$$

where C is an $n \times n$ constant matrix. Suppose that

$$|g(t, \Delta X)| \leq |X|\delta_A + \delta_B|X| + k_1|\Delta X|$$

where $\delta_A = |\Delta X|$, $\delta_B = |\Delta B|$ and k_1 is the Lipschitz constant for F. Further, set $\Phi(t, s) = Y(t)Y^{-1}(s)$, $\Psi(t, s) = Z(t)Z^{-1}(s)$ and $\Psi^T(t, s) = (Z^T)^{-1}(s)Z^T(t)$, and consider the operator equation related to (1.7.5)

$$\Delta X(t) \equiv T(\Delta X)(t) \equiv$$
$$\Phi(t, t_1)X_1\Psi^T(t, t_1) + \int_a^t \Phi(t, s)g(s, \Delta X(s))\Psi^T(t, s)dt. \qquad (1.7.6)$$

Suppose that

$$\lambda \doteq \sup_t \{ |\Phi(t, t_1)| \ |\Psi(t, t_1)| \},$$

and

$$\mu = \sup_t \int_a^t \{ |\Phi(t, s)| \ |\Psi(t, s)| ds.$$

We then get

$$|T(\Delta X)(t)| \leq \lambda|X_1| + \mu[\delta_A + \delta_B]|X| + k_1|\Delta X|$$

$$\leq k_0 + k_1|\Delta X|,$$

where

$$k_0 = \lambda|X_1| + \mu[\delta_A + \delta_B]|X|.$$

The above inequality yields

$$| T(\Delta X)(t) - T(\Delta Y)(t) | \le k_1 | \Delta X - \Delta Y |$$

proving that T is a contraction map when $k_1 < 1$. The application of Banach fixed point theorem provides a unique fixed point which is the unique solution of the perturbed nonlinear Lyapunov differential equation. We therefore have the following result.

Theorem 1.7.1: *Let F in (1.7.1) satisfy Lipschitz condition and the matrices A and B are subjected to perturbations ΔA and ΔB. Then solution $X(t)$ of (1.7.1) gets perturbed to $X(t) + \Delta X(t)$, where $\Delta X(t)$ satisfies the equation (1.7.6) and is uniquely determined.*

As a particular case of the equation (1.7.1), consider the matrix Ricatti equation

$$X'(t) = A(t) + B(t)X(t) + X(t)C(t) + X(t)D(t)X(t) \qquad (1.7.7)$$

where X, A, B, C, D are $n \times n$ matrices continuous on $[a, b]$. Assume that A and D are positive definite symmetric matrices. The equation (1.7.7) is one of the basic equations of nonlinear analysis and in the synthesis of optimal continuous filters. Let $X(a) = I$. Following the same notation for perturbations as in (1.7.2) and similar procedure, we finally get as in (1.7.6)

$$(T\Delta X)(t) = \Phi(t, t_0)\Delta X_0 \psi^T(t_0, t) + \int_a^t \Phi(t, s)g(t, \Delta X(s))\Psi^T(s, t)dt \qquad (1.7.8)$$

where $\Delta X(a_0) = \Delta X_0$ and

$$g(t, \Delta X(t)) = X\Delta DX + \Delta BX + X\Delta C + \Delta A + (\Delta B + \Delta XD)\Delta X$$

$$+ \Delta X(\Delta C + \Delta DX) + X\Delta X^2 + \Delta X\Delta D\Delta X, \qquad (1.7.9)$$

in view of (1.7.7).

Now assume that

$$| g(t, \Delta X) | \le | X |^2 \delta_D + | X | \delta_B + | X | \delta_C + \delta_A$$

$$+ (\delta_B + | D | \delta_X)\delta_X + \delta_X(\delta_C + \delta_D | X |) + X\delta_{X^2} + \delta_D\delta_{X^2}.$$

Denote

$$\mu_1 = \max_t | \Phi(t, a) |, \mu_2 = \max_t | \Psi(a, t) |$$

$$\nu_1 = \sup_t \left\{ \int_a^t | \Phi(s, t) | ds \right\}, \nu_2 = \sup_t \left\{ \int_a^t | \Psi(s, t) | ds \right\}$$

$$\mu = \mu_1\mu_2 \text{ and } \nu = \nu_1\nu_2. \qquad (1.7.10)$$

Now because of (1.7.8) to (1.7.10), it follows that

$$| T(\Delta X)(t) | \leq h(| X |) \equiv a_0 + a_1 | \Delta X | + a_2 | \Delta X |^2,$$

where

$$a_0 = a_0(\delta) = \nu \delta_A + \mu \delta_{X_0} + \mu | X |^2 \delta_D + | X | (\delta_B + \delta_C),$$

$$a_1 = a_1(\delta) = \nu(\delta_B + \delta_C + \delta_D | X |),$$

$$a_2 = a_2(\delta) = \nu(| X | + | D | + \delta_D), \qquad (1.7.11)$$

and further

$$| T(\Delta X(t)) - T(\Delta Y(t)) | \leq | a_1 + 2a_2 r | \ | \Delta X - \Delta Y |$$

where $r = \max\limits_{t}\{ | X(t) | , | Y(t) | \}$.

Now, whenever $h'(r) = a_1 + 2a_2 r < 1$, T is a contraction map. Hence there exists a unique solution $X^+ = X + \Delta X$ of the perturbed nonlinear Ricatti differential equation. Hence, we have the following result.

Theorem 1.7.2: *Let $S\rho$ be the set of continuous matrix-valued functions $\Delta X: [a, b] \to R^{n \times n}$ such that $| \Delta X | < \rho$, $\rho > 0$, $h(\rho) \leq \rho$, $h'(\rho) < 1$ and further suppose that $a_0, a_1,$ and a_2 are the numbers defined as in (1.7.11). Then there exists a solution $X^+ = X + \Delta X$ of the perturbed Ricatti equation (1.7.7) whenever $a_1 + 2a_2 r < 1$.*

Example 1.7.1: Consider the linear quadratic optimization problem

$$x'(t) = A(t)x(t) + B(t)u(t), \ x(a) = x_0, \qquad (1.7.12)$$

$$y(t) = F(t)x(t),$$

$$J(u) = x^T(b)Gx(b) + \int\limits_{a}^{b} [y^T(t)y(t) + u^T(t)R(t)u(t)]dt \qquad (1.7.13)$$

where $x \in R^n$, $u \in R^n$, $y \in R^b$ are the state, control and output vectors respectively. The matrices $A, B, F,$ and R are continuous matrix functions with appropriate dimensions with $R(t)$ a positive definite symmetric matrix and the matrix $G^T = G$ is nonnegative definite. The solution of the problem (1.7.12), (1.7.13) has the representation

$$u(t) = -R^{-1}(t)B^T(t)P(t)x(t),$$

where $P(t)$ is the unique nonnegative solution of the matrix Ricatti equation

$$P'(t) = P(t)S(t)P(t) - P(t)A(t) - A^T(t)P(t) - Q(t)$$

and $S = BR^{-1}B^T$, $Q = F^T F$. The minimal value (1.7.13) corresponding to the control (1.7.12) is

$$J\,min = x_0^T P(t_0)x_0.$$

Thus the linear quadratic optimization problem finally turns out to be a particular case of the general matrix Ricatti differential equation (1.7.7). Theorem 1.7.2 is now applicable.

1.8 EXISTENCE OF ALMOST PERIODIC SOLUTIONS

This section deals with the existence of almost periodic solutions of differential equations. We present a typical application in which VPF is profitably employed.

An almost periodic function is the one which can be uniformly approximated on the real line by trigonometric polynomials.

Let

$$T(x) = \sum_{k=1}^{n} c_k e^{i\lambda_k x}, \tag{1.8.1}$$

where c_k are complex numbers and λ_k real numbers, and $T(x)$ is a complex trigonometric polynomial. Setting $c_k = a_k + ib_k$, we see that

$$Re\{T(x)\} = \sum_{k=1}^{n} [a_k \cos \lambda_k x - b_k \sin \lambda_k x],$$

and

$$Im\{T(x)\} = \sum_{k=1}^{n} [b_k \cos \lambda_k x + a_k \sin \lambda_k x].$$

Thus real and imaginary parts of $T(x)$ are real trigonometric polynomials.

Definition 1.8.1: A complex valued function $f(x)$ defined for $-\infty < x < +\infty$ is called almost periodic, if, for any $\epsilon > 0$, there exists a trigonometric polynomial $T_\epsilon(x)$ such that

$$|f(x) - T_\epsilon(x)| < \epsilon, \quad -\infty < x < \infty.$$

Clearly, any trigonometric polynomial is an almost periodic function. Further, any periodic function is almost periodic. However, the converse is not true.

Let $f(x)$ be a complex-valued function defined for $-\infty < x < +\infty$. For any $\epsilon > 0$, there exists a number $\ell(\epsilon) > 0$ with the property that any interval of length $\ell(\epsilon)$ of the real line contains at least one point with x-coordinate ξ, such that

$$|f(x + \xi) - f(x)| < \epsilon, \quad -\infty < x < \infty.$$

A function f possessing this property is almost periodic. The number ξ is called ϵ-translation number of $f(x)$. In (1.8.1), when $c_k \neq 0$, then λ_k are called Fourier exponents of the function $f(x)$. Further, it is also known that the derivative of an almost periodic function is almost periodic if and only if it is bounded on the real line.

We shall consider systems of the form

$$\frac{dy_i}{dt} = \sum_{j=1}^{n} a_{ij} y_j + f_i(t), \quad i = 1, 2, \ldots, n, \tag{1.8.2}$$

where a_{ij} are complex numbers and $f_i(t)$ are almost periodic (complex-valued) functions.

Arguments similar to those that follow can be made in the case of systems of the same form where a_{ij} and $f_i(t)$ are real.

A solution of system (1.8.2) will be called almost periodic, if all its components $y_1, y_2, \ldots, y_n$ are almost periodic functions of (1.8.2).

Of course, to find almost periodic solutions of systems such as (1.8.2) we must look among their bounded solutions. Since the systems of the form (1.8.2) have in general unbounded solutions, one cannot expect that all the solutions be almost periodic. We shall show however that the bounded solutions, if such solutions exist, are almost periodic.

We need the following result from matrix theory which we assume for our discussion.

Lemma 1.8.1: *Given an arbitrary square matrix* $A = (a_{ij})$, *there exists a matrix* $\alpha = (\alpha_{ij})$ *of the same order, such that*
(i) $\det \alpha \neq 0$;
(ii) $\alpha^{-1} A \alpha$ *is triangular, i.e.*

$$\alpha^{-1} A \alpha = \begin{pmatrix} \lambda_1 & b_{12} & \cdots & b_{1n} \\ 0 & \lambda_2 & \cdots & b_{2n} \\ \cdots & \cdots & \cdots & \cdots \\ 0 & 0 & \cdots & \lambda_n \end{pmatrix},$$

where $\lambda_1, \ldots, \lambda_n$ *are the eigenvalues of* A.

We can now prove the following result.

Theorem 1.8.1: *If the functions* $f_i(t), i = 1, 2, \ldots, n$, *are almost periodic, then any bounded solution (on the whole real line) of the system* (1.8.2) *is almost periodic.*

Proof: Let us first observe that one can assume without loss of generality that A is a triangular matrix. Indeed, if A were not triangular, then we would introduce some unknown functions by the change of variables

$$y_i = \sum_{j=1}^{n} \alpha_{ij} z_j, \tag{1.8.3}$$

the matrix $\alpha = (\alpha_{ij})$, being chosen so that $\alpha^{-1} A \alpha$ is triangular. Differentiating both sides of (1.8.3), substituting in (1.8.2), and then solving with respect to dz_i/dt, we obtain a system of the form

$$\frac{dz_1}{dt} = \lambda_1 z_1 + b_{12} z_2 + \ldots + b_{1n} z_n + f_1^*(t),$$

$$\frac{dz_2}{dt} = \qquad \lambda_2 z_2 + \ldots + b_{2n} z_n + f_2^*(t), \qquad (1.8.4)$$

$$\ldots \ldots \ldots \ldots \ldots \ldots \ldots \ldots$$

$$\frac{dz_n}{dt} = \qquad \lambda_n z_n + f_n^*(t),$$

where the functions $f_i^*(t)$ are uniquely determined by the relations

$$\sum_{j=1}^{n} \alpha_{ij} f_j^*(t) = f_i(t), \quad i = 1, 2, \ldots, n. \qquad (1.8.5)$$

Furthermore, if $y_1, \ldots, y_n$ is a bounded solution of system (1.8.2), then the corresponding solution $z_1, \ldots, z_n$ of system (1.8.4) is also bounded and conversely. This follows immediately from (1.8.3) and from the fact that $\alpha = (\alpha_{ij})$ has an inverse. Finally (1.8.5) shows that $f_i^*(t)$ are almost periodic, if and only if $f_i(t)$ are almost periodic.

Let us consider the equation

$$\frac{dz}{dt} = \lambda z + f(t), \qquad (1.8.6)$$

where λ is a complex number and $f(t)$ is an almost periodic function. Let us show that any bounded solution of such an equation is almost periodic. Consider $\lambda = \mu + i\nu$. We distinguish the following cases: (i) $\mu > 0$; (ii) $\mu < 0$; (iii) $\mu = 0$.

The general solution of equation (1.8.6) is

$$z(t) = e^{\lambda t} \left[C + \int_0^t e^{-\lambda \tau} f(\tau) d\tau \right], \qquad (1.8.7)$$

where C is an arbitrary constant.

In case (i), $|e^{\lambda t}| = e^{\mu t} \to +\infty$ as $t \to +\infty$. For $z(t)$ to be bounded on the real line, one must have $C + \int_0^t e^{-\lambda \tau} f(\tau) d\tau \to 0$ as $t \to \infty$. This means that we must take in (1.8.7)

$$C = -\int_0^\infty e^{-\lambda t} f(t) dt. \qquad (1.8.8)$$

We note that the improper integral in (1.8.8) is convergent, since

$$|e^{-\lambda t} f(t)| \leq \sup |f(t)| e^{-\mu t}, \quad t \geq 0.$$

Thus, the unique bounded solution of equation (1.8.6) could be

$$z_0(t) = - \int_t^\infty e^{\lambda(t-\tau)} f(\tau) d\tau. \tag{1.8.9}$$

We have however

$$|z_0(t)| \leq M e^{\mu t} \int_t^\infty e^{-\mu\tau} d\tau = \frac{M}{\mu}, \tag{1.8.10}$$

where $M = \sup |f(t)|$, $-\infty < t < +\infty$. From $(1.8.10)$ it follows that $z_0(t)$ is bounded. We then have

$$|z_0(t+\xi) - z(t_0) \leq \frac{1}{\mu} \sup_x |f(t+\xi) - f(t)|, \quad -\infty < t < +\infty.$$

This inequality proves that any $\epsilon\mu$-translation number of $f(t)$ is an ϵ-translation number of $z_0(t)$.

In case (ii), one proceeds analogously. We see that

$$z_1(t) = \int_{-\infty}^t e^{\lambda(t-\tau)} f(\tau) d\tau \tag{1.8.11}$$

is the unique bounded solution of equation $(1.8.6)$. We have

$$|z_1(t)| \leq \frac{M}{|\mu|} \tag{1.8.12}$$

and as in the previous case, any $\epsilon |\mu|$-translation number of $f(t)$ is an ϵ-translation number of $z_1(t)$.

Finally, let us consider case (iii). Since

$$z(t) = e^{i\nu t}\left[C + \int_0^t e^{-i\nu\tau} f(\tau) d\tau\right]. \tag{1.8.13}$$

it follows from $(1.8.13)$ that $z(t)$ is bounded if and only if

$$\int_0^t e^{-i\nu\tau} f(\tau) d\tau$$

is bounded on the real line. Since the function under the integral sign is almost periodic, we see that this indefinite integral is almost periodic, if it is bounded. Thus, as a consequence of $(1.8.13)$, any bounded solution (and therefore, any solution) is almost periodic. The assertion of the theorem is proven for equations of the form $(1.8.6)$.

Let us return now to system $(1.8.4)$ and assume that $z_1, \ldots, z_n$ is a bounded solution of this system. From the last equation of a system of the form $(1.8.6)$ it follows that $z_n(t)$ is almost periodic. We substitute now $z_n(t)$ obtained from the last equation in the equation preceding the last, which becomes therefore of the form $(1.8.6)$. From the above considerations it follows that $z_{n-1}(t)$ is almost periodic and so on. Thus, applying n times successively the assertion proven above for equations of the form (1.8.6), it will follow that the bounded solution considered is almost periodic. The proof of Theorem 1.8.1 is now complete.

We must note however, that Theorem 1.8.1 asserts the almost periodicity of any bounded solution under the hypothesis that such a solution exists. It indicates no condition which will ensure the existence of bounded solutions. We shall state a theorem which will specify when a system of the form (1.8.2) possesses bounded solutions.

Theorem 1.8.2: *If the functions $f_i(t)$, $i = 1, 2, \dots, n$, are almost periodic and the matrix $A = (a_{ij})$ has no eigenvalues with real part zero, then the system (1.8.2) admits a unique almost periodic solution.*

If $y_1(t), \dots, y_n(t)$ is the almost periodic solution and
$$M = \max_i\{\sup | f_i(t) | \}, \quad -\infty < t < +\infty,$$

then

$$| y_i(t) | \le KM, \quad -\infty < t < +\infty, i = 1, 2, \dots, n,$$

where K is a positive constant depending only on the matrix A.

Remarks 1.8.1: Note that the constant K might depend on the transformation (1.8.3), i.e., on the matrix $\alpha = (\alpha_{ij})$ for which $\alpha^{-1}A\alpha$ has a triangular form. Once we fix the matrix α, K depends only on A. It is important that it does not depend on $f_i(t)$ any more.

Theorem 1.8.2 is not valid if the condition $Re\lambda_i \neq 0$ is not satisfied. More precisely, let us show by an example that Theorem 1.8.2 is not true, if there exists at least one eigenvalue λ_k for which $Re\lambda_k = 0$. Without loss of generality consider that $\lambda_k = \lambda_n = i\nu$, where ν is real. The last equation of system (1.8.4) can then be written

$$\frac{dz_n}{dt} = i\nu z_n + f_n^*(t). \tag{1.8.14}$$

Since $f_n^*(t)$ is an arbitrary almost periodic function, we take

$$f_n^*(t) = e^{i\nu t}.$$

One sees the general integral of equation (1.8.14) is

$$z_n(t) = (t + C)e^{i\nu t}.$$

Consequently the equation in z_n has no bounded solution on the whole real line. In the theory of oscillations, one frequently encounters equations such as

$$\frac{d^2y}{dt^2} + 2A\frac{dy}{dt} + By = f(t), \tag{1.8.15}$$

where $f(t)$ is an almost periodic function. Most frequently $f(t)$ is a sine function or a finite sum of such functions. If the ratios of the periods are rational numbers, then $f(t)$ is periodic; in the contrary case, $f(t)$ is almost periodic.

The linear system which is equivalent to equation (1.8.15) is

$$\frac{dy}{dt} = z,$$

$$\frac{dz}{dt} = -By - 2Az + f(t).$$

The characteristic roots are $-A \pm \sqrt{A^2 - B}$. If we assume A and B real, $A \neq 0$, then the real parts of the characteristic roots are distinct from zero. Indeed, the real part is $-A$ for both roots if $A^2 - B \leq 0$; if $A^2 - B > 0$, then the roots are real, distinct and none of them is zero. Thus, by Theorem 1.8.2, equation (1.8.15) has a unique almost periodic solution. In the case $A = 0$, $B \neq 0$, any bounded solution of equation (1.8.15) will be almost periodic as follows from Theorem 1.8.1. This last case includes equations of the form

$$\frac{d^2y}{dt^2} + \omega^2 y = f(t).$$

The above considerations are valid if $f(t)$ is real valued and then the solution is also real.

1.9 STABILITY AND ASYMPTOTIC BEHAVIOR

This section deals with a variety of results related to stability, Lipschitz stability, asymptotic equilibrium and asymptotic equivalence, utilizing various VPFs. We begin with some typical results relative to stability. Consider the differential equations

$$x' = f(t, x), \quad x(t_0) = x_0, \tag{1.9.1}$$

and its perturbed equation

$$y' = f(t, y) + F(t, y), \quad y(t_0) = x_0 \tag{1.9.2}$$

where $t_0 \in R_+$, $x_0 \in R^n$, $f, F \in C[R_+ \times R^n, R^n]$. We shall first obtain sufficient conditions for uniform stability of the zero solution of (1.9.2).

Theorem 1.9.1: *Assume the conditions in Theorem* 1.2.4. *Let the equations* (1.9.1), (1.9.2) *admit trivial solutions. Suppose that*
(i) $|\Phi^{-1}(t, t_0, x_0)F(t, x(t, t_0, x_0))| \leq g(t, |x_0|)$ *where* $g \in C[R_+ \times R_+, R_+]$, $g(t, 0) \equiv 0$, *and the trivial solution of*

$$u' = g(t, u), \quad u(t_0) = u_0 \geq 0 \tag{1.9.3}$$

is uniformly stable,
(ii) the trivial solution of (1.9.1) *is uniformly stable.*
Then the trivial solution of (1.9.2) *is uniformly stable.*

Proof: By Theorem 1.2.4, any solution $y(t, t_0, x_0)$ of (1.9.2) satisfies

$$y(t, t_0, x_0) = x(t, t_0, \phi(t)), \tag{1.9.4}$$

where $\phi(t)$ is a solution of (1.2.15). The assumption (i) then leads to, setting $m(t) = |\phi(t)|$, the inequality

$$D^+ m(t) \leq g(t, m(t)),$$

which, by comparison theorem, gives the estimate

$$|\phi(t)| = m(t) \leq r(t, t_0, |x_0|), \quad t \geq t_0, \tag{1.9.5}$$

where $r(t, t_0, u_0)$ is the maximal solution of (1.9.3).

Because of (1.9.4), we obtain

$$|y(t, t_0, x_0)| = |x(t, t_0, \phi(t))|. \tag{1.9.6}$$

By assumption (ii), given $\epsilon > 0$, $t_0 \in R_+$ there exists a $\delta_1(e) > 0$ satisfying

$$|x(t, t_0, x_0)| < \epsilon, \quad t \geq t_0, \text{ if } |x_0| < \delta_1(\epsilon).$$

By the fact that $u \equiv 0$ of (1.9.3) is uniformly stable, we conclude that, given $\delta_1(\epsilon) > 0$, $t_0 \in R^+$, there exists a $\delta(\epsilon) > 0$ such that $r(t, t_0, u_0) < \delta_1(\epsilon)$, $t \geq t_0$, if $0 \leq u_0 < \delta(\epsilon)$. Thus the relations (1.9.5) and (1.9.6) yield that $|x_0| < \delta(\epsilon)$ implies $|y(t, t_0, x_0)| < \epsilon, t \geq t_0$. The proof is complete.

Theorem 1.9.2: *In addition to the assumptions of Theorem* 1.2.4, *suppose that*
(i) $|\Phi(t, t_0, x_0| \leq K$ *and* $|\Phi^{-1}(t, t_0, x_0)| \leq K$, $t \geq t_0$, *provided*
  $|x_0| \leq \gamma$,
(ii) $|F(t, y)| \leq g(t, |y|)$, *where* $g \in C[R_+ \times R_+, R_+]$, $g(t, 0) \equiv 0$, $g(t, u)$ *is nondecreasing in* u, *and the trivial solution of*

$$u' = K^2 g(t, u), \quad u(t_0) = u_0 \geq 0,$$

  is stable.
Then the trivial solution (1.9.2) *is stable.*

Proof: The solution $x(t, t_0, x_0)$ of (1.9.1), as seen in Remark 1.2.2, is related to $\Phi(t, t_0, x_0)$ by

$$x(t, t_0, x_0) = \left(\int_0^1 \Phi(t, t_0, sx_0) ds \right) \cdot x_0.$$

Therefore, assumption (i) yields

$$|x(t, t_0, x_0)| \leq K|x_0|, \text{ for } |x_0| \leq \gamma.$$

Now suppose there exists a $t_1 > t_0$ such that $|\phi(t)| < \gamma$ for $t_0 \leq t < t_1$ and $|\phi(t_1)| = \gamma$. Then for $t_0 \leq t \leq t_1$, it follows from (1.2.15) that

$$| \phi(t) | \leq | x_0 | + K \int_{t_0}^{t} g(s, | x(s, t_0, \phi(s)) |)ds$$

$$< | x_0 | + K \int_{t_0}^{t} g(s, K | \phi(s) |)ds.$$

Hence,

$$K | \phi(t) | \leq K | x_0 | + K^2 \int_{t_0}^{t} g(s, K | \phi(s) |)ds.$$

This fact leads us to

$$K | \phi(t) | \leq r(t, t_0, u_0), \quad t_0 \leq t \leq t_1,$$

where $r(t, t_0, u_0)$ is the maximal solution of

$$u' = K^2 g(t, u), \quad u(t_0) = u_0 \geq 0,$$

such that $K | x_0 | \leq u_0$.

By assumption (ii), given $\gamma K > 0$ there exists a $\delta_1 > 0$ such that $r(t, t_0, u_0) < \gamma K$, $t \geq t_0$, whenever $u_0 < \delta_1$. Thus if $| x_0 | < \delta_1 / K$, then $| \phi(t) | < \gamma$ for all $t \geq t_0$. This contradicts $| \phi(t_1) | = \gamma$. Thus $| \phi(t) | < \gamma$, $t \geq t_0$. Now (1.2.17) means that

$$| y(t, t_0, x_0) | \leq K | x_0 | + K^2 \int_{t_0}^{t} g(s, | y(s) |)ds.$$

Thus

$$| y(t, t_0, x_0) | \leq r(t, t_0, u_0),$$

where $r(t, t_0, u_0)$, is the maximal solution of

$$u' = K^2 g(t, u), \quad u(t_0) = K | x_0 | .$$

Given $\epsilon > 0$, there exists a $\delta(\epsilon, t_0)$ such that $| x_0 | < \delta$ implies $r(t, t_0, K | x_0 |) < \epsilon$, $t \geq t_0$. Therefore, the null solution of (1.9.2) is stable. The proof is complete.

We now proceed to study the asymptotic equivalence property related to two differential systems. Let

$$x' = f_1(t, x), \quad x(t_0) = x_0, \tag{1.9.7}$$

$$y' = f_2(t, y), \quad y(t_0) = y_0. \tag{1.9.8}$$

where f_1 and $f_2 \in C[R_+ \times R^n, R^n]$. We need the following definition.

Definition 1.9.1: The differential systems (1.9.7) and (1.9.8) are said to be asymptotically equivalent if, for every solution $y(t)$ of (1.9.8) $[x(t)$ of (1.9.7)] there exists a solution $x(t)$ of (1.9.7) $[y(t)$ of (1.9.8)] such that $[x(t) - y(t)] \to 0$ as

$t \to \infty$. We are now in a position to prove the following result on asymptotic equivalence.

Theorem 1.9.3: *Suppose that*
(i) $f_1, f_2 \in C[R_+ \times R^n, R^n]$, $\frac{\partial f_1}{\partial x}$ *exists and is continuous on* $R_+ \times R^n$,
(ii) $\Phi_1(t, t_0, x_0)$ *is the fundamental matrix solution of the variational system*

$$z_1' = \frac{\partial f_1(t, x(t, t_0, x_0))}{\partial x} z_1, \quad \Phi_1(t_0, t_0, x_0) = I,$$

(iii) for a given solution $y(t)$ *of* (1.9.8) *existing on* $[t_0, \infty)$,

$$\int_t^\infty \Phi_1(t, s, y(s))[f_2(s, y(s)) - f_1(s, y(s))]ds \to 0 \tag{1.9.9}$$

as $t \to \infty$.
Then, there exists a solution $x(t)$ *of* (1.9.7) *on* $[t_0, \infty)$ *satisfying the relation*

$$\lim_{t \to \infty} [x(t) - y(t)] = 0. \tag{1.9.10}$$

Proof: Let $y(t) = y(t, t_0, y_0)$ be a given solution of (1.9.8) existing on $[t_0, \infty)$. Define a function $x(t)$ as follows:

$$x(t) = y(t) + \int_t^\infty \Phi_1(t, s, y(s))[f_2(s, y(s)) - f_1(s, y(s))]ds. \tag{1.9.11}$$

Since the integral converges by assumption (*iii*), it follows that $x(t)$ is well defined, and, consequently (1.9.10) is satisfied. It therefore suffices to prove that $x(t)$ is a solution of (1.9.7). For this purpose, we observe, as in Theorem 1.2.7 that

$$\frac{dx(t, s, y(s))}{ds} = \Phi_1(t, s, y(s))[f_2(s, y(s)) - f_1(s, y(s))]. \tag{1.9.12}$$

Here use is made of the relation (1.2.18) and the fact that

$$\frac{\partial x(t, s, y(s))}{\partial x_0} = \Phi_1(t, s, y(s)).$$

The relations (1.9.11) and (1.9.12) lead us to

$$x(t) = y(t) + \int_t^\infty \frac{dx(t, s, y(s))}{ds} ds$$

$$= y(t) + \lim_{T \to \infty} \int_t^T \frac{dx(t, s, y(s))}{ds} ds$$

$$= y(t) + \lim_{T \to \infty} x(t, T, y(T)) - y(t)$$

$$= \lim_{T \to \infty} x(t, T, y(T)).$$

The continuity of $f_1(t, x)$ now asserts that

$$\lim_{T \to \infty} f_1(t, x(t, T, y(T))) = f_1(t, \lim_{T \to \infty} x(t, T, y(T)))$$

$$= f_1(t, x(t)). \tag{1.9.13}$$

Moreover, we have

$$\frac{df_1(t, x(t, s, y(s)))}{ds} = \frac{\partial f_1(t, x(t, s, y(s)))}{\partial x} \cdot \frac{dx(t, s, y(s))}{ds}. \tag{1.9.14}$$

Let us differentiate (1.9.11) and use (1.9.12) to get

$$x'(t) = f_2(t, y(t)) - \Phi_1(t, t, y(t))[f_2(t, y(t)) - f_1(t, y(t))]$$

$$+ \int_t^\infty \frac{\partial \Phi_1(t, s, y(s))}{\partial t}[f_2(s, y(s)) - f_1(s, y(s))]ds$$

$$= f_1(t, y(t)) + \int_t^\infty \frac{\partial f_1(t, x(t, s, y(s)))}{\partial x} \cdot \frac{dx(t, s, y(s))}{ds} ds.$$

In view of (1.9.14), we obtain

$$x'(t) = f_1(t, y(t)) + \int_t^\infty \frac{df_1(t, x(t, s, y(s)))}{ds} ds$$

$$= f_1(t, y(t)) + \lim_{T \to \infty} \int_t^T \frac{df_1(t, x(t, s, y(s)))}{ds} ds$$

$$= f_1(t, y(t)) + \lim_{T \to \infty} f_1(t, x(t, T, y(T))) - f_1(t, y(t))$$

$$= \lim_{T \to \infty} f_1(t, x(t, T, y(T))).$$

The relation (1.9.13) implies that $x(t)$ is a solution of (1.9.7) with

$$x_0 = y_0 + \int_{t_0}^\infty \Phi_1(t_0, s, y(s))[f_2(s, y(s)) - f_1(s, y(s))]ds.$$

On the other hand, if $x(t)$ is a solution of (1.9.7), we can show exactly in a similar way that there exists a solution $y(t)$ of (1.9.8) existing on $[t_0, \infty)$ such that (1.9.10) holds. The conditions (ii) and (iii) above need to be stated for $\Phi_2(t, t_0, x_0)$ for this purpose. The proof is complete.

The asymptotic equivalence property may also be derived from the VPF proved in Theorem 1.2.4. For this purpose, consider the systems (1.9.1) and (1.9.2) where $f, F \in C[R_+ \times D, R^n]$, $D \subset R^n$ is a convex set. We need the following definitions.

Definition 1.9.2: We shall say that the differential system (1.9.1) has *asymptotic equilibrium* if every solution of the system (1.9.2) tends to a finite limit vector ξ as $t \to \infty$ and to every constant vector ξ there is a solution $x(t)$ of (1.9.1) on $t_0 \le t < \infty$ such that $\lim_{t \to \infty} x(t) = \xi$.

The following theorem gives sufficient conditions for the system (1.9.1) to have asymptotic equilibrium. The proof is omitted.

Theorem 1.9.4: *Let $f \in C[R_+ \times D, R^n]$ and*

$$| f(t,x) | \le g(t, | x |), \ (t,x) \in J \times D, \qquad (1.9.15)$$

where $g \in C[R_+ \times D, R_+]$ and monotone nondecreasing in u for each $t \in R_+$. Assume that all solutions $u(t)$ of

$$u' = g(t,u), \ u(t_0) = u_0 > 0 \qquad (1.9.16)$$

are bounded on $[t_0, \infty)$. Then the system (1.9.1) has asymptotic equilibrium.

We consider next a result on asymptotic equivalence.

Theorem 1.9.5: *Assume the hypothesis of Theorem 1.2.4. Let $\Phi(t, t_0, x_0)$ be bounded for $t \ge t_0$ and $x_0 \in D \subset R^n$, D being an open set. Further, let*

$$| \Phi^{-1}(t, t_0, x_0 F(t, x(t, t_0, x_0))) | \le g(t, x_0), \qquad (1.9.17)$$

where g is as in (1.9.16). Then, given $y(t, t_0, y_0)$ a solution of (1.9.2), there exists a solution $x(t, t_0, x_0)$ of (1.9.1) satisfying

$$\lim_{t \to \infty} y(t, t_0, y_0) = x(t, t_0, x_0) \qquad (1.9.18)$$

Proof: Any solution $y(t, t_0, y_0)$ of (1.9.2) has the form

$$y(t, t_0, y_0) = x(t, t_0, v(t))$$

where $v(t)$ is a solution of (1.2.15) with $v(t_0) = y_0$. The estimate (1.9.18) guarantees $v(t)$ has asymptotic equilibrium by Theorem 1.9.4. Thus, let $v(t)$ denote the solution of (1.9.18) with $v(t_0) = y_0$ and take

$$x_0 = \lim_{t \to +\infty} v(t).$$

Now Theorem 1.2.9 together with the boundedness of $\Phi(t, t_0, x_0)$ gives

$$| y(t, t_0, y_0) - x(t, t_0, x_0) | = | x(t, t_0, v(t)) - x(t, t_0, x_0) |$$

$$\le \sup_{t \ge t_0} | \Phi(t, t_0, x_0) | \ | v(t) - x_0 |$$

$$= K | v(t) - x_0 |,$$

from which it follows that $y(t, t_0, x_0) \to x(t, t_0, x_0)$ as $t \to +\infty$.

It is remarked that this only shows in one direction the condition for asymptotic equivalence. A complete asymptotic equivalence result can be obtained if the boundedness of $\Psi(t, t_0, y_0) = (\partial/\partial y_0)y(t, t_0, y_0)$ is assumed.

The asymptotic equivalence would then follow if (1.9.1) was considered as a perturbation of (1.9.2).

We now establish the relationship between Theorems 1.9.3 and 1.9.5, since they reach to the same conclusion under different conditions.

In order to prove the equivalence of two conclusions, identify f_1 of (1.9.7) with f of (1.9.1) and f_2 of (1.9.8) with $f + F$ of (1.9.2). Then the relation (1.9.11) becomes

$$x(t, t_0, x_0) = y(t, t_0, y_0) + \int_t^\infty \Phi(t, s, y(s))F(s, y(s))ds$$

$$= x(t, t_0, v(t)) + \int_t^\infty \Phi(t, s, y(s))F(s, y(s))ds$$

where $v(t_0) = y_0$. Solving this equation for $v(t)$, we get

$$v(t) = x(t_0, t, y(t)).$$

Let

$$x_0 = \lim_{t \to \infty} v(t) = \lim_{t \to \infty} x(t_0, t, y(t)).$$

Since $\Phi(t, t_0, x_0)$ is bounded, we get

$$\mid x(t, t_0, v(T)) - x(t, t_0, x_0) \mid \leq K \mid v(T) - x_0 \mid.$$

Note that $x(t, T, y(T)) = x(t, t_0, v(T))$ so that $\lim_{T \to \infty} x(t, T, y(T)) = \lim_{T \to \infty} x(t, t_0, v(T)) = x(t, t_0, x_0)$. Integrate the relation

$$\frac{d}{ds} x(t, t_0, v(s)) = \Phi(t, t_0, v(s))v(s)$$

between t to T to yield

$$x(t, T, y(T)) = y(t, t_0, y_0) + \int_t^T \Phi(t, t_0, v(s))\Phi^{-1}(s, t_0, v(s))F(s, y(s))ds.$$

Taking the limit as $t \to \infty$ yields,

$$x(t, t_0, x_0) = y(t, t_0, y_0) + \int_t^\infty \Phi(t, t_0, v(s))\Phi^{-1}(s, t_0, v(s))F(s, y(s))ds. \quad (1.9.19)$$

From (1.9.10), we have $x(t, t_0, x_0) - y(t, t_0, y_0) \to 0$ as $t \to \infty$. Hence

$$\lim_{t \to \infty} \int_t^\infty \Phi\left(t, t_0, v(s\right)\Phi^{-1}(s, t_0, v(s))F(s, y(s))ds = 0.$$

From Theorem 1.2.8, it therefore follows that

$$\lim_{t \to \infty} \int_t^{\infty} \Phi(t, s, y(s)) F(s, y(s)) ds = 0. \tag{1.9.20}$$

One can now easily see the equivalence of the conditions (1.9.19) and (1.9.20) with those of (1.9.11) and (1.9.9) respectively.

Finally we study the Lipschitz stability properties of the zero solutions of the matrix differential systems (1.2.25) and (1.2.26) under the hypothesis of Kronecker product decompositions of the matrix of partial derivatives of $g(t, x)$ as stated in Theorem 1.2.10. We need the following definitions.

Definition 1.9.3: The zero solution of the equation (1.2.25) is said to be
(L_1) uniformly Lipschitz stable if there exist $M > 0$ and $\delta > 0$ satisfying

$$|X(t, t_0, X_0)| \leq M |X_0| \text{ for } t \geq t_0, \text{ whenever } |X_0| \leq \delta;$$

(L_2) asymptotically Lipschitz stable if there exist $M > 0$, $\alpha > 0$ and $\delta > 0$ satisfying

$$|X(t, t_0, X_0)| \leq M |X_0| e^{-\alpha(t-t_0)} \text{ for } t \geq t_0, \text{ whenever } |X_0| \leq \delta;$$

(L_3) uniformly Lipschitz stable in variation if there exist $M_1 > 0$, $M_2 > 0$ and $\delta > 0$ satisfying

$$|W(t, t_0, X_0)| \leq M_1 \text{ and } |Z(t, t_0, X_0)| \leq M_2 \text{ for } t \geq t_0 \geq 0$$

provided $|X_0| \leq \delta$ where the matrices $W(t, t_0 X_0)$ and $Z(t, t_0, X_0)$ are given in Theorem 1.2.11.

Now, we have the following new results.

Theorem 1.9.6: *If the zero solution of* (1.2.25) *is uniformly Lipschitz stable in variation, and if the hypotheses of Theorem* 1.2.11 *are satisfied, then the zero solution of* (1.2.25) *is uniformly Lipschitz stable.*

Proof: Following Remark 1.2.2 and noting that

$$\Phi(t, t_0, X_0) = W(t, t_0, X_0) \otimes Z^T(t, t_0, X_0),$$

it follows that

$$X(t, t_0, X_0) = \left[\int_0^1 W(t, t_0, sX_0) \otimes Z^T(t, t_0, sX_0) ds \right] X_0.$$

Hence,

$$|X(t, t_0, X_0)| \leq |X_0| \int_0^1 |W(t, t_0, sX_0) \otimes Z^T(t, t_0, sX_0| ds$$

$$= |X_0| \int_0^1 |W(t, t_0, sX_0)| \, |Z^T(t, t_0, sX_0)| \, ds$$

$$\leq K |X_0|, \text{ for all } t \geq t_0 \geq 0,$$

provided $|X_0| < \delta$ where $K = M_1 \cdot M_2$.

The following theorem gives sufficient conditions for the uniform Lipschitz stability of the perturbed matrix differential equations.

Theorem 1.9.7: *Assume that the hypotheses of Theorem 1.2.11 are satisfied, and*
(i)   the zero solution of (1.2.25) is uniformly Lipschitz stable in variation;

(ii) $|R(t, Y(t))| \leq \gamma(t) |Y(t)|$ provided that $\gamma(t) > 0$ and $\int_{t_0}^{\infty} \gamma(t)dt < \infty$ for

all $t_0 \geq 0$.
Then the zero solutions of (1.2.26) is uniformly Lipschitz stable.

Proof: By Theorem 1.2.11, the solutions of (1.2.25) and (1.2.26) with the same initial values are related by

$$Y(t, t_0, X_0) = X(t, t_0, X_0)$$

$$+ \int_{t_0}^t W(t, s, Y(s, t_0, X_0)) R(s, Y(s, t_0, x_0)) Z(t, s, Y(s, t_0, X_0)) ds.$$

The assumptions (i) and (ii), yield

$$|Y(t, t_0, X_0)| \leq |X(t, t_0, X_0)|$$

$$+ \int_{t_0}^t |W(t, s, Y(s, t_0, X_0))| \, |R(s, Y(s, t_0, X_0))| \, |Z(t, s, Y(s, t_0, X_0))| \, ds$$

$$\leq L |X_0| + M_1 M_2 \int_{t_0}^t \gamma(s) |Y(s, t_0, X_0| \, ds.$$

Now, by Gronwall's inequality one gets

$$|Y(t, t_0, X_0| \leq L |X_0| e^{M_1 M_2 \int_{t_0}^t \gamma(s)ds}$$

$$\leq L |X_0| e^{M_1 M_2 \int_{t_0}^{\infty} \gamma(s)ds}$$

$$\leq LN(t_0 |X_0| = \widetilde{L} |X_0|,$$

provided $|X_0| \le \delta$ for $t \ge t_0$, where $N(t_0) = e^{M_1 M_2 K(t_0)}$, $K(t_0) = \int\limits_{t_0}^{\infty} \gamma(s)ds$ and $\widetilde{L} = LN(t_0)$.

The proof is complete.

Theorem 1.9.8: *In addition to the hypotheses of Theorem 1.2.11, assume that*
(i)   the zero solution of (1.2.25) is asymptotically Lipschitz stable;
(ii) there exists $M > 0$ such that for $t \ge t_0 \ge 0$

$$|W(t, t_0, X_0)| \ |Z(t, t_0, X_0)| \le Me^{-\alpha(t-t_0)}$$

whenever $|X_0| < \delta$;

(iii) $|F(t, Y(t))| \le \gamma(t)|Y(t)|$ where $\gamma(t) \ge 0$ and $\int\limits_{t_0}^{\infty} \gamma(t)dt < \infty$ for all
        $t_0 \ge 0$.
Then the zero solution of (1.2.26) is asymptotically Lipschitz stable.

Proof: By Theorem 1.2.11 and the assumptions (i) and (ii), we have

$$|Y(t, t_0, X_0| \le |X(t, t_0, X_0)|$$

$$+ \int\limits_{t_0}^{t} |W(t, s, Y(s, t_0, X_0))| \ |F(s, Y(s, t_0, X_0))| \ |Z(t, s, Y(s, t_0, X_0))ds$$

$$\le L|X_0|e^{-\alpha(t-t_0)} + M \int\limits_{t_0}^{t} e^{-\alpha(t-s)}\gamma(s)|Y(s, t_0, X_0)| \ ds.$$

Setting $V(t) = e^{\alpha(t-t_0)}Y(t, t_0, X_0)$, we obtain

$$|V(t)| \le L|X_0| + M \int\limits_{t_0}^{t} \gamma(s)|V(s)| Ds$$

and using Gronwall's inequality, one gets

$$|V(t)| \le L|X_0|e^{M\int\limits_{t_0}^{\infty} \gamma(s)ds}$$

$$\le L|X_0|e^{MK(t_0)} = \widetilde{L}|X_0|, \ t \ge t_0,$$

provided $|X_0| < \delta$, where $\widetilde{L} = Le^{MK(t_0)}$. This proves that the zero solution of the differential system $V' = G(t, V(t)), V(t_0) = X_0$ is uniformly Lipschitz stable which implies asymptotic Lipschitz stability of the zero solution of (1.2.26). The theorem is proved.

1.10 CONTROLLABILITY

We shall present in this section sufficient conditions for the controllability of general nonlinear systems. For this purpose, we need to develop first controllability conditions for linear control systems and then utilize such conditions to extend the results to nonlinear systems. Consider

$$x'(t) = A(t)x(t) + B(t)u(t), \ t \in J, \tag{1.10.1}$$

where A is $n \times n$ and B is $n \times m$ continuous matrices on J, where $J = [t_0, t_1]$. Let $u(t)$ be continuous on J.

Definition 1.10.1: The system $(1.10.1)$ is controllable from (t_0, x_0) to (t_1, x_1) if, for some $u(t)$ on J, the solution of $(1.10.1)$ with $x(t_0) = x_0$ is such that $x(t_1) = x_1$, where t_1 and x_1 are preassigned terminal time and state respectively. If $(1.10.1)$ is controllable for all x_0 at $t = t_0$ and for all x_1 at $t = t_1$, it is called completely controllable (c.c.) on J.

Theorem 1.10.1: *The system $(1.10.1)$ is c.c. if and only if the $n \times n$ symmetric controllability matrix*

$$v(t_0, t_1) = \int_{t_0}^{t_1} \Phi(t_0, \sigma)B(\sigma)B^T(\sigma)\Phi^T(t_0, \sigma)d\sigma, \tag{1.10.2}$$

where $\Phi(t, t_0)$ is the fundamental solution of $x' = A(t)x$ such that $\Phi(t_0, t_0) = I$, is nonsingular, that is, for some positive constants c we have $\det v(t_0, t_1) \geq c$.
The control function

$$u(t) = -B^T(t)\Phi^T(t_0, t)v^{-1}(t_0, t_1)[x_0 - \Phi(t_0, t_1)x_1], \tag{1.10.3}$$

defined for $t_0 < t < t_1$ transfers $x(t_0) = x_0$ to $x(t_1) = x_1$.

Proof: Assume that $v(t_0, t_1)$ given in $(1.10.2)$ is nonsingular. Then $u(t)$ given in $(1.10.3)$ is well-defined. We know that the solution of $(1.10.1)$ with $x(t_0) = x_0$ is given by

$$x(t) = \Phi(t, t_0)[x_0 + \int_{t_0}^{t} \Phi(t_0, \sigma)B(\sigma)u(\sigma)d\sigma].$$

Substituting the value of $u(t)$ defined on $(1.10.3)$ in the expression of the solution $x(t)$, it is not difficult to see that we get $x(t_1) = x_1$, which is the desired value. Hence the system $(1.10.1)$ is c.c.

Conversely, suppose that $(1.10.1)$ is c.c. Let α be an arbitrary constant column vector. Since v is symmetric, we can construct the quadratic form

$$\alpha^T v \alpha = \int_{t_0}^{t} \theta^T(\sigma, t_0)\theta(\sigma, t_0)d\sigma$$

$$= \int\limits_{t_0}^{t_1} |\, \theta(\sigma, t_0) \,|^{\,2} d\sigma \geq 0,$$

where $\theta(\sigma, t_0) = B^T(\sigma)\Phi^T(t_0, \sigma)\alpha$. Hence $v(t_0, t_1)$ is positive semi-definite.

Suppose that there exists some $\widetilde{\alpha} \neq 0$ such that $\widetilde{\alpha}^{\,T} v(t_0, t_1)\widetilde{\alpha} = 0$. Then, in view of (1.10.3), $\int\limits_{t_0}^{t_1} |\, \widetilde{\theta}\,(\sigma, t_0) \,|^{\,2} d\sigma = 0$, which means that $\widetilde{\theta}\,(\sigma, t_0) \equiv 0$ on $[t_0, t_1]$. Since (1.10.1) is c.c., there exists a control $v(t)$ (say) such that $x(t_1) = 0$ if $x(t_0) = \widetilde{\alpha}$. Observe that

$$x_1 = \Phi(t_1, t_0[\widetilde{\alpha} + \int\limits_{t_0}^{t_1} \Phi(t_0, \sigma)B(\sigma)v(\sigma)d\sigma] = 0.$$

Hence, it follows that

$$\widetilde{\alpha} = -\int\limits_{t_0}^{t_1} \Phi(t_0, \sigma)B(\sigma)v(\sigma)d\sigma.$$

Consequently,

$$|\,\widetilde{\alpha}\,|^{\,2} = \widetilde{\alpha}^{\,T}\widetilde{\alpha} = -\int\limits_{t_0}^{t_1} v^T(\sigma)B^T(\sigma)\Phi^T(t_0, \sigma)\widetilde{\alpha} \; d\sigma,$$

$$= -\int\limits_{t_0}^{t_1} v^T(\sigma)\widetilde{\theta}\,(\sigma, t_0)\sigma = 0.$$

This fact contradicts the assumption that $\widetilde{\alpha} \neq 0$. Hence $v(t_0, t_1)$ is positive definite and consequently nonsingular. The proof is complete.

The conclusion of the previous theorem is exploited in the study of the following nonlinear control process given by

$$x' = A(x, u, t)x + B(x, u, t)u + f(x, u, t), \tag{1.10.4}$$

where x, f are n-vectors, u is a m-vector, and $A(x, u, t)$, $B(x, u, t)$ are $n \times n$, $n \times m$ matrices respectively. Assume that A, B and f are continuous with respect to all the three arguments. The aim here is to provide a theory to ensure that the system state can be transferred from $x(t_0) = x_0$ to x_1 in the allotted time $t_1 - t_0$. For this purpose, assume that

$$|\, A(x, u, t) \,| \leq Ma, \; |\, B(x, u, t) \,| \leq Mb, \tag{1.10.5}$$

$$|\, f(x, u, t) \,| \leq g(|\, x \,|, |\, u \,|, t), \tag{1.10.6}$$

where Ma, Mb are some positive constants and $g(\alpha, \beta, t)$ is a continuous function with respect to its arguments and is nondecreasing in $\alpha > 0$, $\beta > 0$.

Let $x_0, x_1 \in R^n$ be given. Presently, we do not know if a control u exists to achieve our goal since (1.10.4) is nonlinear. Hence, in place of (1.10.4), let us consider the related system

$$x' = A(y, v, t)x + B(y, v, t)u + f(y, v, t) \qquad (1.10.7)$$

where y, v are some given vector functions. Then (1.10.7) is a linear system. Recalling the conclusions of Theorem 1.10.1, the necessary and sufficient condition for (1.10.7) to be c.c., is that

$$det\, G(t_0, t_1, y, v) \geq c, \qquad (1.10.8)$$

for some positive constant c, where

$$G(t_0, t_1, y, v) = \int_{t_0}^{t_1} H(t_0, \sigma, y, v)H^T(t_0, \sigma, y, v)d\sigma, \qquad (1.10.9)$$

$$H(t_0, t, y, v) = \Phi(t_0, t, y, v)B(y, v, t),$$

$\Phi(t, t_0, y, v)$ being the fundamental solution of the system

$$z' = A(y, v, t)z, \quad z(t, t, y, v) = I.$$

The solution of the system (1.10.7), when $x(t_0) = x_0$, is given by

$$x(t) = \Phi(t, t_0, y, v)x_0 + \int_{t_0}^{t} \Phi(t, \sigma, y, v)B(y, v, \sigma)u(\sigma)d\sigma$$

$$+ \int_{t_0}^{t} \Phi(t, \sigma, y, v)f(y, v, \sigma)d\sigma. \qquad (1.10.10)$$

When condition (1.10.8) holds, then one of the controls which steer the state (1.10.10) to a given x_1 at time t_1 is provided by

$$u(t) = u(t, t_0, x_0, t_1, x_1, y, v)$$

$$= H^T(t_0, t, y, v)G^{-1}(t_0, t_1, y, v)$$

$$\times [\Phi^{-1}(t_1, t_0, y, v)x_1 - x_0 - \int_{t_0}^{t_1} \Phi(t_0, \sigma, y, v)f(y, v, \sigma)d\sigma]. \qquad (1.10.11)$$

The expression (1.10.11) is in fact corresponding version of (1.10.4). In case the vectors y, v in (1.10.7) agree with x, u which result from (1.10.10) and (1.10.11), then these vectors are also solutions of the nonlinear problem (1.10.4). These will ensure the controllability of the system (1.10.4).

The above discussion leads to the fact that the controllability problem (1.10.4) is an existence problem to be determined by the fixed point technique. Observe that the solution of (1.10.7) is given by

$$x(t) = \Phi(t, t_0, y, v)\{x_0 + \int_{t_0}^{t} \Phi(t_0, \sigma, y, v)B(y, v, \sigma)v(\sigma)d\sigma$$

$$+ \int_{t_0}^{t} \Phi(t_0, \sigma, y, v)f(y, v, \sigma)d\sigma\}. \tag{1.10.12}$$

The nonlinear equations (1.10.12), (1.10.11) can be viewed as transferring the vectors $(y, v) \in C[J, R^{n+m}]$ to the vectors $(x, u) \in C[J, R^{n+m}]$. Thus we have an operator F such that

$$(x, u) = F(y, v)$$

$F = (F_1, F_2)$, F_1 being given by (1.10.12) and F_2 given by (1.10.11). The operator F is clearly continuous on $C[J, R^{n+m}]$. Now under suitable conditions, we can show that there exists a closed, bounded convex subset S of $C[J, R^{n+m}]$, which is transferred into itself by F. Hence we conclude that there will exist at least one fixed point for F by the application of the Schauder's fixed point theorem. We have proved the following result.

Theorem 1.10.2: *If there exists a closed bounded convex subset S of $C[J \times R^{n+m}]$ such that the operator F is invariant for S, then the nonlinear system (1.10.4) which satisfies condition (1.10.8) is completely controllable on J.*

Define the set

$$S = \{(y, v) \in C[J, R^{n+m}], \ |y(t)| \le \alpha(t), \ |v(t)| \le \beta(t)\}, \tag{1.10.13}$$

where we need to determine the existence of $\alpha(t)$ and $\beta(t)$ so that the set S possesses the desired property. From (1.10.12) and the assumptions (1.10.5), (1.10.6), we get

$$|x(t)| \le |\Phi(t, t_0, y, v)| \{ |x_0| + \int_{t_0}^{t} |H(t_0, \sigma, y, v)| \ |v(\sigma)| \ d\sigma$$

$$+ \int_{t_0}^{t} |\Phi(t_0, \sigma, y, v)| \ |f(y(\sigma), v(\sigma), \sigma)d\sigma\},$$

$$\le exp(Ma(t_1 - t_0))\{ |x_0 + exp(Ma(t_1 - t_0)Mb\int_{t_0}^{t} v(\sigma)d\sigma$$

$$+ exp(Ma(t_1 - t_0)) \int_{t_0}^{t} g(\,|\,y(\sigma)\,|\,, N(\sigma)\,|\,, \sigma) d\sigma\},$$

$$\leq a_0 + a_1 \int_{t_0}^{t} \beta(\sigma) d\sigma + a_2 \int_{t_0}^{t} g(\alpha(\sigma), \beta(\sigma), \sigma) d\sigma, \qquad (1.10.14)$$

where

$$|\,\Phi(t, t_0, y, v)\,| \leq exp(\int_{t_0}^{t} |\,A(y, v, \sigma)\,|\,d\sigma) \leq exp(Ma(t_1 - t_0)),$$

$$|\,H(t_0, \sigma, y, v)\,| \leq |\,\Phi(t_0, \sigma, y, v)\,|\,\,|\,B(y, v, \sigma)\,| \leq exp(Ma(t_1 - t_0))Mb,$$

and

$$a_0 = exp(Ma(t_1 - t_0))\,|\,x_0\,|\,,$$

$$a_1 = exp(2Ma(t_1 - t_0))Mb,$$

$$a_2 = exp(2Ma(t_1 - t_0)).$$

Similarly, from (1.10.11), we obtain

$$|\,u(t)\,| \leq b_0 + b_1 \int_{t_0}^{t_1} g(\alpha(\sigma), \beta(\sigma), \sigma) d\sigma, \qquad (1.10.15)$$

where

$$b_0 = |\,H^T(t_0, t, y, v)G^{-1}(t_0, t_1, y, v)\{\Phi^{-1}(t_1, t_0, y, v)x_1 - x_0\}\,|$$

$$b_1 = |\,H^T(t_0, t, y, v)G^{-1}(t_0, t_1, y, v)\,|\,\,|\,\Phi(t_0, \sigma, y, v)\,|\,.$$

Observe that a_0, a_1, a_2, b_0, b_1 are well-defined and in order that the set S in (1.10.13) to possess desire properties as given in Theorem 1.10.1, it is sufficient that the right-hand side of the equations (1.10.14) and (1.10.15) are smaller than $\alpha(t)$ and $\beta(t)$ respectively. We therefore have proved the following result.

Theorem 1.10.2: *For the system* (1.10.4), *satisfying* (1.10.5), (1.10.6) *and the condition* (1.10.8), *to be c.c. on* $[t_0, t_1]$, *it is sufficient that the relations*

$$a_0 + a_1 \int_{t_0}^{t} \beta(\sigma) d\sigma + a_2 \int_{t_0}^{t} g(\alpha(\sigma), \beta(\sigma), \sigma) d\sigma \leq \alpha(t), \qquad (1.10.16)$$

$$b_0 + b_1 \int_{t_0}^{t} g(\alpha(\sigma), \beta(\sigma), \sigma) d\sigma \leq \beta(t). \qquad (1.10.17)$$

have at least one nonnegative solution$(\alpha(t), \beta(t))$ *for any* $a_0, b_0 > 0$ *and for some* a_1, a_2, b_1 *defined by the system of equations.*

Since equation (1.10.4) *is sufficiently general, the conclusion of Theorem* 1.10.2 *gets much simplified in the special cases namely* (i) *f does not depend on u and* (ii) *f does not depend on x. We omit the details.*

The above results are obtained by using the comparison principle and include solutions for a large class of control problems. Let, for example, f in (1.10.4) be uniformly bounded. Then, from (1.10.6), we have

$$g(\,|\,x\,|\,,\,|\,u\,|\,,t) = c_1, \text{ a constant.}$$

Then, from (1.10.17)

$$\beta(t) \geq b_0 + b_1 + c_1(t_1 - t_0)$$

and from (1.10.16) together with this $\beta(t)$

$$\alpha(t) \geq a_0 + a_1 + \int_{t_0}^{t_1} \beta(\tau)d\tau + a_2 c_1(t - t_0).$$

We therefore get the following result.

Corollary 1.10.1: *Consider the nonlinear control system* (1.10.4) *satisfying the conditions* (1.10.5) *and* (1.10.8). *If the nonlinear function* $f(x, u, t)$ *is uniformly bounded, then the system* (1.10.4) *is c.c.*

1.11 VARIATIONAL COMPARISON RESULT

As indicated in Section 1.3, we shall utilize, in this section, the method of generalized VPF to develop a comparison theorem that connects the solutions of perturbed and unperturbed differential systems in a manner useful in the theory of perturbations. The comparison result blends, in a sense, the two approaches namely, the method of Lyapunov functions and the method of variation of parameters, and consequently provides a flexible mechanism to preserve the nature of perturbations. The results that are given show that the usual comparison theorem in terms of a vector Lyapunov function is included as a special case and that perturbation theory could be studied in a more fruitful way.

Consider the two differential systems

$$x' = f(t, x),\ x(t_0) = x_0, \tag{1.11.1}$$

$$y' = F(t, y),\ y(t_0) = x_0, \tag{1.11.2}$$

where $f, F \in C[R_+ \times R^n, R^n]$. Relative to the system (1.11.1), we assume the following hypothesis:

(H) the solutions $x(t, t_0, x_0)$ of (1.11.1) exist for all $t \geq t_0$, unique and continuous with respect to the initial data and $|\,x(t, t_0, x_0)\,|$ is locally Lipschitzian in x_0. For any $V \in C[R_+ \times R^n, R_+^N]$ and any fixed $t \in [t_0, \infty)$, we define

$$D_-V(s, x(t, s, y))$$

$$= \lim_{h \to 0^-} \inf \frac{1}{h} \left[V(s + h, x(t, s + h, y + hF(s, y))) - V(s, x(t, s, y)) \right]$$

for $t_0 < s \leq t$ and $y \in R^n$.

The following comparison result which relates the solutions of (1.11.2) to solutions (1.11.1) is a useful result for subsequent discussion.

Theorem 1.11.1: *Assume that the hypothesis (H) holds. Suppose that*
(i) *$V \in C[R_+ \times R^n, R_+^N]$, $V(s, y)$ is locally Lipschitzian in y and for $t_0 < s \leq t$, $y \in R^n$,*

$$D_-V(s, x(t, s, y)) \leq g(t, V(s, x(t, s, y))); \qquad (1.11.3)$$

(ii) *$g \in C[R_+ \times R_+^N, R^N]$, $g(t, u)$ is quasimonotone nondecreasing in u and the maximal solution $r(t, t_0, u_0)$ of*

$$u' = g(t, u), \quad u(t_0) = u_0 \geq 0, \qquad (1.11.4)$$

exists for $t \geq t_0$.
Then, if $y(t) = y(t, t_0, x_0)$ is any solution of (1.11.2), we have

$$V(t, y(t, t_0, x_0)) \leq r(t, t_0, u_0), \ t \geq t_0 \qquad (1.11.5)$$

provided $V(t_0, x(t, t_0, x_0)) \leq u_0$.

Proof: Let $y(t) = y(t, t_0, x_0)$ be any solution of (1.11.2). Set

$$m(s) = V(s, x(t, s, y(s))), \ t_0 \leq s \leq t,$$

so that $m(t_0) = V(t_0, x(t, t_0, x_0))$. Then, using the hypothesis (H) and (i), it is easy to obtain

$$D_-m(s) \leq g(s, m(s)), \ t_0 \leq s \leq t,$$

which yields the estimate

$$m(s) \leq r(t, t_0, u_0), \ t_0 \leq s \leq t, \qquad (1.11.6)$$

provided $m(t_0) \leq u_0$. Since $m(t) = V(t, x(t, t, y(t))) = V(t, y(t, t_0, x_0))$, the desired result (1.11.5) follows from (1.11.6) by setting $s = t$.
Taking $u_0 = V(t_0, x(t, t_0, x_0))$, the inequality becomes

$$V(t, y(t, t_0, x_0)) \leq r(t, t_0, V(t_0, x(t, t_0, x_0))), \ t \geq t_0$$

which establishes the connection between the solutions of the systems (1.11.1), (1.11.2) and (1.11.4).

Remark 1.11.1: A number of remarks are now in order.

(a) In Theorem 1.11.1, choose $f(t, x) \equiv 0$ so that $x(t, t_0, x_0) \equiv x_0$. Then, it follows that

$$D_- V(s, y) = \lim_{h \to 0^-} \inf \frac{1}{h} [V(s + h, y + hF(s, y)) - V(s, y)],$$

which is the usual definition of generalized derivative of the Lyapunov function relative to the system (1.11.2).

(b) Let $f(t, x) = A(t)x$, where $A(t)$ is an $n \times n$ continuous matrix. Then, we know that solution $x(t, t_0, x_0)$ of (1.11.1) is of the form $\Phi(t, t_0)x_0$, where $\Phi(t, t_0)$ is the fundamental matrix solution of $x' = A(t)x$, $\Phi(t_0, t_0) = I$. The hypothesis (H) holds. Let $g(t, u) \equiv 0$. Then (1.11.5) reduces to

$$V(t, y(t, t_0, x_0)) \leq V(t_0, \Phi(t, t_0)x_0), \; t \geq t_0.$$

On the other hand, if $g(t, u) = Bu$, $B = (b_{ij})$ is an $n \times n$ matrix such that $b_{ij} > 0$, $i \neq j$, we get a sharper estimate

$$V(t, y(t, t_0, x_0)) \leq V(t_0, \Phi(t, t_0)x_0)exp(B(t - t_0)), \; t \geq t_0. \qquad (1.11.7)$$

Clearly, the relation (1.11.7) helps in improving the behavior of solutions of (1.11.2) relative to the solutions of (1.11.1). This is an asset in perturbation theory when $F(t, x) = f(t, x) + R(t, x)$, where $R(t, x)$ is the perturbation term.

(iii) Suppose that $f(t, x)$ is nonlinear, continuous, $f_x(t, x)$ exists and continuous on $R_+ \times R^n$. Then the results of Theorem 1.2.6 hold true. In case, $V(s, x)$ is also assumed to be differentiable, then

$$\frac{d}{ds} V(s, x(t, s, y)) = V_s(s, x(t, s, y)) + V_x(s, x(t, s, y)) \cdot \Phi(t, s, x) \cdot R(t, x)$$

where $R(t, x) = F(t, x) - f(t, x)$.

(iv) When the solutions of (1.11.1) are known, a possible Lyapunov function for (1.11.1) is

$$W(s, y) = V(s, x(t, s, y)).$$

As an application of the comparison Theorem 1.11.1, we shall prove a typical result on practical stability. We need the following definition.

Definition 1.11.1: The system (1.11.1) is said to be

(i) practically stable if given (λ, A) with $0 < \lambda < A$, we have $| x_0 | < \lambda$ implies $| x(t) | < A, t \geq t_0$ for some $t_0 \in R_+$,

(ii) practically quasistable if given $(\lambda, B, T) > 0$ and some $t_0 \in R_+$, we have $| x_0 | < \lambda$ implies $| x(t) | < B, t \geq t_0 + T$,

(iii) strongly practically stable if (i) and (ii) hold simultaneously.

The other definitions of practical stability can be formulated similarly.

Theorem 1.11.2: *Assume that* (H) *holds and* (i) *of Theorem* 1.11.1 *is verified. Suppose that* $g \in C[R \times R_+^N, R^N]$, $g(t, u)$ *is quasi-monotone nondecreasing in* u *and for* $(t, x) \in R_+ \times S(A)$, $S(A) = \{x \in R^n, \ | x | \ < A\}$,

$$b(\ | x | \ ) \leq V_0(t, x) \leq a(\ | x | \ ),$$

where $V_0(t, x) = \sum_{i=1}^{n} V_i(t, x)$, $a, b \in C[[0, A], R_+]$ *and increasing with* $a(0) = b(0) = 0$, *and that* $a(\lambda) < b(A)$. *If the unperturbed system* (1.11.1) *is* (λ, λ) *practically stable, the practical stability properties of* (1.11.1) *imply the corresponding practical stability properties of the perturbed system* (1.11.2).

Proof: Assume that (1.11.4) is strongly practically stable. Then, given $(\lambda, A, B, T) > 0$ such that $\lambda < A, B < A$,

$$\sum_{i=1}^{N} u_i(t, t_0, u_0) < b(A), \ t \geq t_0 \ \text{if} \sum_{i=1}^{N} u_{0i} < a(\lambda) \tag{1.11.8}$$

and

$$\sum_{i=1}^{N} u_{0i} < a(\lambda) \ \text{implies} \sum_{i=1}^{N} u_i(t, t_0, u_0) < b(B), t \geq t_0 + T. \tag{1.11.9}$$

Since (1.11.1) is (λ, λ) practically stable, we have

$$| x(t, t_0, x_0) | \ < \lambda, t \geq t_0, \text{if} \ | x_0 | \ < \lambda.$$

We claim that $| x_0 | \ < \lambda$ also implies that $| y(t, t_0, x_0) | \ < A, \ t \geq t_0$, where $y(t, t_0, x_0)$ is any solution of (1.11.2).

Suppose that this is false. Then, there exists a solution $y(t, t_0, x_0)$ of (1.11.2) with $| x_0 | \ < \lambda$ and a $t_1 > t_0$ satisfying

$$| y(t_1, t_0, x_0) | \ = A \text{ and } | y(t, t_0, x_0) | \ \leq A \text{ for } t_0 \leq t \leq t_1.$$

By Theorem 1.11.1, we obtain

$$V(t, y(t, t_0, x_0)) \leq r(t; t_0, V(t_0, x(t, t_0, x_0)), t_0 \leq t \leq t_1.$$

As a result, we get

$$b(A) \leq V_0(t_1, y(t_1, t_0, x_0)) \leq \sum_{i=1}^{N} r_i(t_1, t_0, a(\ | x(t_1, t_0, x_0) | \ ))$$

$$\leq \sum_{i=1}^{N} r_i(t_1, t_0, a(\lambda)) < b(A).$$

This contradiction proves that

$$|x_0| < \lambda \text{ implies } |y(t)| < A, t \geq t_0.$$

To prove strong practical stability, we have

$$b(|y(t, t_0, x_0)|) \leq V_0(t, y(t, t_0, x_0))$$

$$\leq \sum_{i=1}^{N} r_i(t, t_0, V(t_0, x(t, t_0, x_0)))$$

for all $t \geq t_0$, if $|x_0| < \lambda$. It then follows that

$$b(|y(t, t_0, x_0)|) \leq \sum_{i=1}^{N} r_i(t, t_0, a(\lambda)), t \geq t_0.$$

Now (1.11.9) yields strong practical stability of the system (1.11.2). The proof is completed.

Set $F(t, y) = f(t, y) + R(t, y)$ in (1.11.2). We see that (1.11.1) is only practically stable while the system (1.11.2) is strongly practically stable. This improvement in the behavior is due to the smooth perturbation term $R(t, y)$.

Let us present a simple but illustrative example.

Example 1.11.1:

$$x' = -x, \ x(t_0) = x_0$$

$$y' = e^{-t}y^2 \ y(t_0) = y_0$$

have solutions given by

$$x(t, t_0, x_0) = x_0 e^{-(t-t_0)}, \ t \geq t_0$$

$$y(t, t_0, y_0) = \frac{y_0}{1+y_0(e^{-t}-e^{-t_0})}, \ t \geq t_0.$$

The fundamental solutions of the corresponding variational equations are

$$\psi(t, t_0, x_0) = e^{-(t-t_0)},$$

$$\phi(t, t_0, y_0) = \frac{1}{[1+y_0(e^{-t}-e^{-t_0})]^2}.$$

Choose $V_1(t, x) = x^2$ and $V_2(t, y) = y^2$. Hence

$$V_1'(t, x) = 2x(t, s, x)\psi(t, s, x)L(s, y, x)$$

$$V_2'(t, y) = 2y(t, s, y)\phi(t, s, y)R(s, y, x),$$

where L and R are perturbations. The perturbed differential system is given by

$$x' = -x + L(t, y, x), \; x(t_0) = x_0,$$

$$y' = e^{-t}y^2 + R(t, y, x), \; y(t_0) = y_0.$$

Choose $L(t, y, x) = -\frac{xy^2}{2}$ and $R(t, y, x) = -\frac{y^2}{2}$. Then it is easy to compute

$$g_1(t, V_1, V_2) = -V_1 V_2$$

$$g_2(t, V_1, V_2) = -V_2^{3/2}$$

so that the comparison system reduces to

$$u_1' = -u_1 u_2, \; u_1(t_0) = u_{10},$$

$$u_2' = -u_1^{3/2}, \; u_2(t_0) = u_{20}.$$

Choose $u_{10} = V_1(t_0, x(t, t_0, x_0))$, $u_{20} = V_2(t_0, y(t, t_0, x_0))$. It then follows that

$$u_1(t, t_0, u_0) = u_{10} exp\left[-\frac{2u_{20}(t-t_0)}{2+u_{20}^{1/2}(t-t_0)}\right], \; t \geq t_0$$

$$u_2(t, t_0, u_0) = \frac{4 u_{20}}{\left[2+u_{20}^{1/2}(t-t_0)\right]^2}.$$

Noting that
$$u_{20} = y^2(t, t_0, x_0) = \frac{x_0^2}{[1+x_0(e^{-t}-e^{-t_0})]^2}$$

and
$$u_{10} = x^2(t, t_0, x_0) = x_0^2 e^{-2(t-t_0)},$$

it follows that

$$| \overline{x} (t, t_0, x_0, y_0) |^{\,2}$$

$$\leq | x_0 |^{\,2} e^{-2(t-t_0)} \frac{|y_0|^2(t-t_0)}{[1+x_0(e^{-t}-e^{-t_0}+\left(\frac{t-t_0}{2}\right))][1+y_0(e^{-t}-e^{-t_0})]},$$

and
$$| \overline{y} (t, t_0, x_0, y_0) | \leq \frac{|y_0|^2}{[1+y_0(e^{-t}-e^{-t_0})+\left(\frac{t-t_0}{2}\right)]^2} \quad t \geq t_0.$$

It is clear that $\overline{x} (t, t_0, x_0, y_0)$, $\overline{y} (t, t_0, x_0, y_0) \to 0$ as $t \to \infty$. Observe that solutions $x(t, t_0, x_0)$, $y(t, t_0, y_0)$ of the unperturbed system do not enjoy such fine properties. In fact, choose $t_0 = 0$ and $x_0 = 1$, we get

$$x(t, t_0, x_0) = e^{-t} \text{ and } y(t, t_0, x_0) = e^t, \; t \geq 0.$$

1.12 NOTES

Most of the results of Section 1.1 may be found in the classical books of Coddington and Levinson [1], Coppel [1], Hartman [1], Lakshmikantham and Leela [1]. Theorem 1.1.3 is taken from Bellman [4].

Theorems 1.2.1 and 1.2.2 have been adopted from Lakshmikantham [1] while Theorem 1.2.3 is new. Also see Chandra and Lakshmikantham [1]. The results contained in Theorems 1.2.4 and 1.2.5 are due to Lord and Mitchell [1]. Theorem 1.2.6 is taken from Lakshmikantham and Leela [1]. Theorem 1.2.7 is a VPF due to Alekseev [1] which together with Theorem 1.2.9 is taken from Lakshmikantham and Leela [1]. Theorem 1.2.8 is the work of Lord and Mitchell [1]. For the VPF related to nonlinear matrix differential equation, see Fausett and Köksal [1]. For Kronecker product of two matrices refer to Horn [1], Barnett [1] and Bellman [1].

Theorem 1.3.1 is taken from Ladde, Lakshmikantham and Leela [1]. The generalized comparison principle was first studied by Ladde [1]. The VPF using Lyapunov-like functions is the work of Grabner [1], Grabner and Knapp [1], Wanner and Reitberger [1].   See also Aftabizadeh [1], Rajlakshmi and Sivasundaram [1], and Lakshmikantham, Matrosov, and Sivasundaram [1].
For results on VPFs see the monograph by Lakshmikantham, Leela and Martynyuk [1].

There is an extensive literature in which the VPF's are profitably used to derive qualitative properties of differential equations.

The discussion relative to BVPs given in Theorems 1.4.1 and 1.4.2 is taken from Kartsatos [3].   See also Kartsatos [1], Agarwal and Usmani [1], Murthy and Sivasundarm [1], and Ruan [1] for further details. Monotone iterative technique presented in Section 1.5 is new. This technique is extensively studied in Ladde, Lakshmikantham and Vatsala [1], Theorems 1.6.1 and 1.6.2 have been adopted from Kartsatos [3]. See also Coppel [1] for additional results.

Theorems 1.7.1 and 1.7.2 are given in Murthy and Shaw [1]. For matrix Riccatti equations, refer to the work of Davison and Maki [1], Konstantinore, Petkov, and Christos [1] and Murthy, Prasad and Srinivas [1]. The results on almost periodic functions described in Sections 1.8 are taken from Corduneanu [1]. See also Malkin [2].

The contents appearing in Theorems 1.9.1 and 1.9.2 are due to Lord [1]. The result on asymptotic equivalence given in Theorem 1.9.3 are taken from Lakshmikantham and Leela [1] while those in Theorem 1.9.5 are adopted from Lord [1]. For the application of matrix VPF for deriving Lipschitz stability properties, see Fausett and Köksal [1].

For a variety of results on different types of stability properties, asymptotic equivalence, boundedness, where linear and nonlinear VPFs are extensively used, refer to the work of Brauer [1-4], Onuchic [1], Marlin and Struble [1], Strauss and Yorke [1], Strauss [1], Malkin [1] and Brauer and Strauss [1].

The results on controllability included in Section 1.10 are adopted from Yamamoto [1]. For additional results on control problems refer to Yamamoto and Sugiura [1, 2] and Kostantinov, Cristov and Petkov [1], Kartsatos [1].
The Theorem 1.11.1 is adopted from Lakshmikantham, Matrosov, and Sivasundaram [1]. Also see Ladde [1], Ladde, Lakshmikantham and Leela [1] and Rajlakshmi and Sivasundarm [1] for similar results.

2. Integrodifferential Equations

2.0 INTRODUCTION

This chapter is devoted to the development of the method of variation of parameters for integrodifferential systems (IDS) of Volterra type and related topics. We shall also discuss a method of finding an equivalent linear differential system for a given linear Volterra integrodifferential system and then extend this technique to linear matrix integrodifferential systems.

In Section 2.1, we shall first study the VPF for linear IDS in terms of resolvent function and then offer sufficient conditions to determine an equivalent linear differential system for linear IDS. The idea involved in this transformation is also utilized for linear matrix IDS of Volterra type.

Nonlinear VPF is investigated in Section 2.2. For this purpose, we first consider continuous dependence and differentiability of solutions relative to the initial conditions and employ these results to formulate nonlinear VPF, which as is to be expected, has a different form compared to the nonlinear VPF for nonlinear differential systems.

Section 2.3 deals with the monotone iterative technique that provides monotone sequences which converge to the extremal solutions in a sector determined by the lower and upper solutions of the given problem. We present in Section 2.4, a variety of stability properties with respect to linear and nonlinear IDS utilizing the methods developed in Sections 2.1 and 2.2, proving necessary comparison theorem.

Finally, we include a Section 2.5, controllability results for linear and nonlinear IDS showing by an example that linear IDS is controllable while the corresponding linear differential system is not.

2.1 LINEAR VARIATION OF PARAMETERS

Let us consider the linear integrodifferential equation (IDE)

$$x'(t) = A(t)x(t) + \int_{t_0}^{t} k(t,s)x(s)ds + F(t), x(t_0) = x_0, \qquad (2.1.1)$$

where $A(t)$, $k(t,s)$ are $n \times n$ continuous matrices for $t \in R_+$ and $(t,s) \in R_+^2$ respectively and $F \in C[R_+, R^n]$. We set for $t_0 \leq s \leq t < \infty$,

$$\psi(t,s) = A(t) + \int_{s}^{t} k(t,\sigma)d\sigma, \qquad (2.1.2)$$

and

$$R(t,s) = I + \int\limits_s^t R(t,\sigma)\psi(\sigma,s)ds, \qquad (2.1.3)$$

where I is the identity matrix and $k(t,s) = \psi(t,s) = R(t,s) = 0$ for $s > t \geq t_0$. We shall prove the following result relative to linear IDE (2.1.1).

Theorem 2.1.1: *Assume that $A(t), k(t,s)$ are continuous $n \times n$ matrices for $t \in R_+, (t,s) \in R_+^2$ and $F \in C[R_+, R^n]$. Then the solution of (2.1.1) is given by*

$$x(t) = R(t,t_0)x_0 + \int\limits_{t_0}^t R(t,s)F(s)ds, \quad x(t_0) = x_0, \qquad (2.1.4)$$

where $R(t,s)$ is the unique solution of

$$\frac{\partial R(t,s)}{\partial s} + R(t,s)A(s) + \int\limits_s^t R(t,\sigma)k(\sigma,s)d\sigma = 0, \qquad (2.1.5)$$

$R(t,t) = I$.

Proof: The function $\psi(t,s)$ in (2.1.2) is continuous. Hence it is obvious that the function $R(t,s)$ in (2.1.3) exists. Further, $\frac{\partial R(t,s)}{\partial s}$ also exists and satisfies (2.1.5). Let $x(t)$ denote the solution of (2.1.1) for $t \geq t_0$. Set $p(s) = R(t,s)x(s)$ to get

$$p'(s) = \frac{\partial R(t,s)}{\partial s}x(s) + R(t,s)x'(s)$$

$$= \frac{\partial R(t,s)}{\partial s}x(s) + R(t,s)\left[A(s)x(s) + \int\limits_{t_0}^s k(s,u)x(u)du + F(s)\right].$$

Integrating between t_0 and t, we have

$$p(t) - p(t_0) = \int\limits_{t_0}^t \left[\frac{\partial R(t,s)}{\partial s}x(s) + R(t,s)A(s)x(s) + R(t,s)F(s)\right]ds$$

$$+ \int\limits_{t_0}^t R(t,s)\left[\int\limits_{t_0}^s k(s,u)x(u)du\right]ds.$$

Employing Fubini's theorem, it follows that

$$x(t) - R(t,t_0)x_0 = \int\limits_{t_0}^t \left[\frac{\partial R}{\partial s} + R(t,s)A(s) + \int\limits_s^t R(t,u)k(u,s)du\right]x(s)ds$$

$$+ \int\limits_{t_0}^t R(t,s)F(s)ds.$$

This relation together with (2.1.5) proves (2.1.4).

Conversely, suppose that $y(t)$ is the solution of (2.1.4) with $y(t_0) = x_0$ existing for $t_0 \le t < \infty$. It then follows that

$$\int_{t_0}^{t} R(t,s)y'(s)ds = R(t,t)y(t) - R(t,t_0)x_0 - \int_{t_0}^{t} \frac{\partial R(t,s)}{\partial s} y(s)ds$$

$$= \int_{t_0}^{t} R(t,s)F(s)ds - \int_{t_0}^{t} \frac{\partial R(t,s)}{\partial s} y(s)ds.$$

From (2.1.5) and Fubini's theorem, one obtains

$$\int_{t_0}^{t} R(t,s) \left[y'(s) - A(s)y(s) - \int_{t_0}^{s} k(s,u)y(u)du - F(s) \right] ds = 0.$$

Observe that $R(t,s)$ is a nonzero continuous function for $t_0 \le s \le t < \infty$. We then arrive at

$$y'(t) - A(s)y(s) - \int_{t_0}^{s} k(s,u)y(u)du - F(s) = 0,$$

which implies that $y(t)$ is the solution of (2.1.1). The proof is complete.

Here $R(t,s)$ is the differentiable resolvent function and $f(t)$ is the source function. The formula (2.1.4) expressed in terms of these functions is known as VPF for (2.1.1).

We note that the matrix differential equation (2.1.5) is equivalent to the matrix integral equation of Volterra-type (2.1.3). This equation can be solved explicitly in special situations. The following example illustrates this point.

Example 2.1.1: In the equation (2.1.3), assume that

$$\Psi(t,s) = B^{-1}(t)B(s), (t,s) \in R_+^2, \ \det B(t) \neq 0.$$

Multiply by $B^{-1}(s)$ both sides of (2.1.3) to get

$$R(t,s)B^{-1}(s) = B^{-1}(s) + \int_{s}^{t} R(t,\sigma)B^{-1}(\sigma)d\sigma.$$

Setting

$$Q(t,s) = B^{-1}(s) + Q_1(t,s),$$

where

$$Q(t,s) = R(t,s)B^{-1}(s) \text{ and } Q_1(t,s) = \int_{s}^{t} R(t,\sigma)B^{-1}(\sigma)d\sigma,$$

yields

$$\frac{d}{ds}[Q_1(t,s)] + Q_1(t,s) = B^{-1}(s); \quad Q_1(t,t) = 0,$$

and hence

$$Q_1(t,s) = -\int_s^t \exp\left(I(t-\sigma)\right)B^{-1}(\sigma)d\sigma.$$

Substituting the value of $Q_1(t,s)$, and simplifying, we get finally,

$$R(t,s) = I - \int_s^t \exp\left(I(t-\sigma)\right)\Psi(\sigma,s)d\sigma, \quad R(t,t) = I,$$

which is the explicitly solution of (2.1.3) or equivalently (2.1.5).

We shall next discuss a method of finding an equivalent linear differential system given a linear IDS (2.1.1). The following result addresses this question.

Theorem 2.1.2: *Assume that there exists an $n \times n$ continuous matrix function $L(t,s)$ on R_+^2 such that $L_s(t,s)$ exists, is continuous and satisfies*

$$k(t,s) + L_s(t,s) + L(t,s)A(s) + \int_s^t L(t,u)k(u,s)du = 0,$$

$$L(t,t) = I, \tag{2.1.6}$$

where $A(t)$, $k(t,s)$ are continuous $n \times n$ matrices on R_+ and R_+^2 respectively. Then the initial value problem for the linear IDE

$$u'(t) = A(t)u(t) + \int_{t_0}^t k(t,s)u(s)ds + F(t), u(t_0) = x_0, \tag{2.1.7}$$

where $F \in C[R_+, R^n]$, is equivalent to the initial value problem for linear differential equation

$$v'(t) = B(t)v(t) + L(t,t_0)x_0 + H(t), \quad v(t_0) = x_0, \tag{2.1.8}$$

where $B(t) = A(t) - L(t,t)$ and $H(t) = F(t) + \int_{t_0}^t L(t,s)F(s)ds$.

Proof: Let $u(t)$ be any solution of (2.1.7) existing on $[t_0, \infty)$. Defining $p(s) = L(t,s)u(s)$, we see that

$$p'(s) = L_s(t,s)u(s) + L(t,s)u'(s)$$

$$= L_s(t,s)u(s) + L(t,s)\left[A(s)u(s) + \int_{t_0}^s k(s,\tau)u(\tau)d\tau + F(s)\right].$$

Integrating, we get

$$L(t,t)u(t) - L(t,t_0)x_0 = \int_{t_0}^{t} [L_s(t,s) + L(t,s)A(s)]u(s)ds$$

$$+ \int_{t_0}^{t} L(t,s) \left[\int_{t_0}^{s} k(s,\theta)u(\theta)d\theta \right] ds.$$

Since,

$$\int_{t_0}^{t} L(t,s) \left(\int_{t_0}^{s} k(s,\theta)u(\theta)d\theta \right) ds = \int_{t_0}^{t} \left(\int_{s}^{t} L(t,u)k(u,s)du \right) u(s)ds,$$

by Fubini's theorem, the relations (2.1.6) and (2.1.7) yield

$$L(t,t)u(t) - L(t,t_0)x_0$$

$$= \int_{t_0}^{t} \left[L_s(t,s) + L(t,s)A(s) + \int_{s}^{t} L(t,u)k(u,s)du \right] u(s)ds + H(t) - F(t)$$

$$= - \int_{t_0}^{t} k(t,s)u(s)ds + H(t) - F(t)$$

$$= - u'(t) + A(t)u(t) + H(t).$$

Hence, $u(t)$ satisfies (2.1.8).

Let $v(t)$ be any solution of (2.1.8) which exists on $[t_0, \infty)$. Define

$$z(t) = v'(t) - A(t)v(t) - \int_{t_0}^{t} k(t,s)v(s)ds - F(t).$$

Then proving $z(t) \equiv 0$ is equivalent to showing that $v(t)$ is the solution of (2.1.7). Substitute v' from (2.1.8) and use (2.1.6) and Fubini's theorem to obtain

$$z(t) = - \left[L(t,t)v(t) - L(t,t_0)x_0 - \int_{t_0}^{t} L_s(t,s)v(s)ds \right]$$

$$+ \int_{t_0}^{t} L(t,s) \left[A(s)v(s) + F(s) + \int_{t_0}^{s} k(s,\tau)v(\tau)d\tau \right] ds.$$

Note that $\frac{d}{ds}(L(t,s)v(s)) = L_s(t,s)v(s) + L(t,s)v'(s)$ which after integration gives

$$L(t,t)v(t) - L(t,t_0)x_0 = \int_{t_0}^{t} [L_s(t,s)v(s) + L(t,s)v'(s)]ds.$$

Employing (2.1.7), we then get

$$z(t) = \int_{t_0}^{t} L(t,s) \left[-v'(s) + A(s)v(s) + \int_{t_0}^{s} k(s,u)v(u)du + F(s) \right] ds$$

$$= -\int_{t_0}^{t} L(t,s)z(s)ds,$$

which implies $z(t) \equiv 0$, because of uniqueness of solutions of Volterra linear integral equations. The proof is complete.

Remark 2.1.2: In some situations, it is convenient to reduce (2.1.7) to the form

$$v'(t) = B(t)v(t) + \int_{t_0}^{t} \Phi(t,s)v(s)ds + H(t), v(t_0) = x_0, \qquad (2.1.9)$$

where

$$B(t) = A(t) - L(t,t),$$

$$H(t) = F(t) + L(t,t_0)x_0 + \int_{t_0}^{t} \Phi(t,s)F(s)ds,$$

and $\Phi(t,s)$ is suitably chose.

In Theorem 2.1.2, if the relation (2.1.6) is replaced by

$$\Phi(t,s) = k(t,s) + L_s(t,s)A(s) + \int_{s}^{t} L(t,u)k(u,s)du,$$

where $\Phi(t,s)$ is an $n \times n$ continuous matrix on R_+^2, then it is easy to see from the proof, that (2.1.7) is equivalent to (2.1.9).

Example 2.1.2: The matrix integrodifferential equation (2.1.6) possesses a solution under the given hypothesis. It can be explicitly solved for some special situations.

Observe that the equation (2.1.6) is equivalent to matrix integral equation

$$L(t,s) = I + \int_{s}^{t} [k(t,\sigma) + L(t,\sigma)\Psi(\sigma,s)]d\sigma$$

where $\psi(t,s) = A(t) + \int_{s}^{t} k(t,\sigma)d\sigma.$

Let $\psi(t,s) = B^{-1}(t)B(s)$. Multiply the equation in L by $B^{-1}(s)$ to get

$$L(t,s)B^{-1}(s) = B^{-1}(s) + \int_s^t [k(t,\sigma)B^{-1}(s) + L(t,\sigma)B^{-1}(\sigma)]d\sigma.$$

Set $Q(t,s) = L(t,s)B^{-1}(s)$ and

$$Q_1(t,s) = \int_s^t [k(t,\sigma)B^{-1}(s) + Q(t,\sigma)]d\sigma.$$

Then

(i) $Q_1(t,t) = 0.$

(ii) $\frac{d}{ds}[Q_1(t,s)] = -Q(t,s) - k(t,s)B^{-1}(s) + \int_s^t k(t,\sigma)d\sigma[B^{-1}(s)]',$

(iii) $Q(t,s) = B^{-1}(s) + Q_1(t,s).$

Substituting (iii) in (ii) for $Q(t,s)$, we get

$$\frac{d}{ds}[Q_1(t,s)] + Q_1(t,s) = -B^{-1}(s) - k(t,s)B^{-1}(s) + \int_s^t k(t,\sigma)d\sigma[B^{-1}(s)]',$$

which has a solution

$$Q_1(t,s)$$

$$= \int_s^t \exp(I(t-\sigma))\left\{ B^{-1}(\sigma) + k(t,\sigma)B^{-1}(\sigma) - \left(\int_\sigma^t k(t,\tau)d\tau \right) [B^{-1}(\sigma)] \right\}d\sigma.$$

In view of (iii) and the definition of Q, it follows that

$$L(t,s) = I + \int_s^t \{\exp(I(t-\sigma))[B^{-1}(\sigma) + k(t,\sigma)B^{-1}(\sigma)$$

$$- \left(\int_\tau^t k(t,\tau)d\tau \right) \{B^{-1}(\sigma)\}'\}d\sigma \times B(s);$$

$$= I + \int_s^t \{\exp(I(t-\sigma)\{B^{-1}(\sigma)B(s) + k(t,\sigma)B^{-1}(\sigma)B(s)$$

$$- \left(\int_\tau^t k(t,\tau)d\tau \right) \{B^{-1}(\sigma)\}'B(s)\}d\sigma.$$

$$= I + \int_s^t \{\exp(I(t-\sigma))\left\{ \Psi(\sigma,s) + k(t,\sigma)\Psi(\sigma,s) - \left(\int_\tau^t k(t,\tau)d\tau \right) \Psi_\sigma(\sigma,s) \right\}d\sigma,$$

$$L(t, t) = I.$$

When A is an $n \times n$ constant matrix and $k(t, s) = k(t - s)$, a convolution matrix in (2.1.1), then the resulting equation with $t_0 = 0$, is

$$x'(t) = Ax(t) + \int\limits_0^t k(t - s)x(s)ds + F(t), \, x(0) = x_0.$$

One can obtain VPF for this equation in a similar way. On the other hand, we can also use Laplace transform technique to obtain VPF. We merely state the corresponding result.

Theorem 2.1.3: *Suppose that $k, F \in L^1(R_+)$ and $z(t)$, a differential resolvent satisfies the adjoint equation*

$$z'(t) = Az(t) + \int\limits_0^t k(t - s)z(s)ds, \; z(0) = I.$$

Suppose further that

$$\det[sI - A - \widehat{k}(s)] \neq 0 \, for \, \operatorname{Re} s > 0,$$

where $\widehat{k}(s)$ is the Laplace transform of $k(t)$. Then the solution $x(t)$ of the foregoing IDE is of the form

$$x(t) = z(t)x_0 + \int\limits_0^t z(t - s)F(s)ds.$$

We shall next extend the ideas to matrix linear IDE

$$y'(t) = A(t)y(t) + y(t)B(t) + \int\limits_{t_0}^t [k_1(t, s)y(s) + y(s)k_2(t, s)]ds + F(t)$$

$$y(t_0) = y_0, \tag{2.1.10}$$

where $A(t)$, $B(t)$, $k_1(t, s)$, $k_2(t, s)$ are continuous $n \times n$ matrices defined on R_+, R^2 respectively, $y \in R^{n \times n}$ and $F \in C[R_+, R^{n \times n}]$.

The perturbation term $F(t)$ can be replaced by a more general term of the $F(t, y(t))$, where $F \in C[R_+ \times R^{n \times n}, R^{n \times n}]$.

Before obtaining VPF for (2.1.10), we shall establish a method for finding an equivalent linear matrix differential equation. The next result is in this direction which extends Theorem 2.1.2.

Theorem 2.1.4: *Assume that there exists $n \times n$ continuous matrix functions $L(t, s)$, $R(t, s)$ on R_+^2 such that $L_s(t, s)$, $R_s(t, s)$ exist, are continuous and satisfy the system*

$$k_1(t, s) + L_s(t, s) + L(t, s)A(s) + A(s)R(t, s)$$

$$+ \int_{s}^{t} [L(t,u)k_1(u,s) + k_1(u,s)R(t,u)]du = 0 \qquad (2.1.11)$$

$$k_2(t,s) + R_s(t,s) + B(s)R(t,s) + L(t,s)B(s)$$

$$+ \int_{s}^{t} [L(t,u)k_2(u,s) + k_2(u,s)R(t,u)]du = 0,$$

$$L(t,t) = R(t,t) = I, \qquad (2.1.12)$$

where $A(t)$, $B(t)$, $k_1(t,s)$, $k_2(t,s)$ are as given in (2.1.10). Then (2.1.10) is equivalent to the initial value problem for matrix linear differential equation

$$y'(t) = [A(t) - L(t,t)]y(t) + y(t)[B(t) - L(t,t)]$$

$$+ L(t,t_0)y_0 + y_0 R(t,t_0) + F(t) + \int_{t_0}^{t} [L(t,s)F(s) + F(s)R(t,s)]ds, \quad (2.1.13)$$

$$y(t_0) = y_0.$$

Proof: Let $y(t)$ be any solution of (2.1.10) existing on $[t_0, \infty)$. Set

$$Q(s) = L(t,s)y(s) + y(s)R(t,s) \qquad (2.1.14)$$

where $L(t,s)$ and $R(t,s)$ commute with matrix $y(t)$ for $t \geq t_0$. Differentiate $Q(s)$ and substitute for $y'(s)$ from (2.1.10) to get

$$Q'(t) = L_s(t,s)y(s) + L(t,s)y'(s) + y'(s)R(t,s) + y(s)R_s(t,s)$$

$$= [L_s(t,s) + L(t,s)A(s)]y(s) + L(t,s)\int_{t_0}^{s} k_1(s,u)y(u)du$$

$$+ y(s)[R_s(t,s) + B(s)R(t,s)] + \int_{t_0}^{s} y(u)k_2(s,u)R(t,s)du$$

$$+ L(t,s)y(s)B(s) + \int_{t_0}^{s} L(t,s)y(u)k_2(s,u)du.$$

Integrate between t_0 and t, use Fubini's theorem and apply commutative property of $y(s)$ with $L(t,s)$, $R(t,s)$ to get

$$Q(t) - Q(t_0) = L(t,t)y(t) + Y(t)R(t,t) - L(t,t_0)y_0 - y_0 R(t,t_0)$$

$$= \int_{t_0}^{t} \{L_s(t,s) + L(t,s)A(s) + A(s)R(t,s)$$

$$+ \int\limits_{s}^{t} [L(t,u)k_1(u,s) + k_1(u,s)R(t,u)]ds\}y(s)ds$$

$$+ \int\limits_{t_0}^{t} y(s)\{R_s(t,s) + L(t,s)B(s) + B(s)R(t,s)$$

$$+ \int\limits_{s}^{t} L(t,u)k_2(u,s) + k_2(u,s)R(t,u)du\}ds$$

$$+ \int\limits_{t_0}^{t} [L(t,s)F(s) + F(s)R(t,s)]ds.$$

In view of the relations (2.1.11) and (2.1.12), we obtain

$$= - \int\limits_{t_0}^{t} k_1(t,s)y(s)ds - \int\limits_{t_0}^{t} y(s)k_2(t,s)ds + \int\limits_{t_0}^{t} [L(t,s)F(s) + F(s)R(t,s)]ds$$

$$= - y'(t) + A(t)Y(t) + y(t)B(t) + F(t) + \int\limits_{t_0}^{t} [L(t,s)F(s) + F(s)R(t,s)]ds.$$

Hence $y(t)$ satisfies (2.1.13).

To prove the converse, let $y(t)$ be any solution of (2.1.13) existing on $[t_0, \infty)$. Define

$$z(t) = y'(t) - A(t)y(t) - y(t)B(t) - \int\limits_{t_0}^{t} [k_1(t,s)y(s) + y(s)k_2(t,s)]ds - F(t).$$

We prove that $z(t) \equiv 0$, so that $y(t)$ verifies (2.1.10). For this purpose, substitute $y'(t)$ from (2.1.13) to yield

$$z(t) = - [L(t,t)y(t) - L(t,t_0)y_0] - [y(t)R(t,t) - y_0R(t,t_0)]$$

$$- \int\limits_{t_0}^{t} [k_1(t,s)y(s) + y(s)k_2(t,s)]ds$$

$$+ \int\limits_{t_0}^{t} [L(t,s)F(s) + F(s)R(t,s)]ds.$$

Substitute $k_1(t,s)$ and $k_2(t,s)$ from the system of equations (2.1.11) and (2.1.12) to arrive at

$$z(t) = - [L(t,t)y(t) - L(t,t_0)y_0 - \int\limits_{t_0}^{t} L_s(t,s)y(s)ds]$$

$$+ \int\limits_{t_0}^{t} [L(t,s)A(s)y(s) + A(s)y(s)R(t,s) + \int\limits_{t_0}^{s} L(t,s)k_1(s,u)y(u)du$$

$$+ \int\limits_{t_0}^{s} k_1(s,u)y(u)R(t,s)du]ds$$

$$- [y(t)R(t,t) - y_0 R(t,t_0) - \int\limits_{t_0}^{t} y(s)Rs(t,s)ds]$$

$$+ \int\limits_{t_0}^{t} [y(s)L(t,s)B(s) + y(s)B(s)R(t,s) + \int\limits_{t_0}^{s} L(t,s)y(u)k_2(s,u)du$$

$$+ \int\limits_{t_0}^{s} k_1(s,u)y(u)R(t,s)du]ds + \int\limits_{t_0}^{t} [L(t,s)F(s) + F(s)R(t,s)]ds.$$

We note that

$$\frac{d}{ds}(L(t,s)y(s)) = L_s(t,s)y(s) + L(t,s)y'(s),$$

which after integration, yields

$$L(t,t)y(t) = L(t,t_0)y_0 = \int\limits_{t_0}^{t} [L_s(t,s)y(s) + L(t,s)y'(s)]ds.$$

Similarly,

$$y(t)R(t,t) - y_0 R(t,t_0) = \int\limits_{t_0}^{t} [y(s)R_s(t,s) + y'(s)R(t,s)]ds.$$

Employing these results in the preceding step, we get

$$z(t) = \int\limits_{t_0}^{t} L(t,s) \left[-y'(s) + A(s)y(s) + y(s)B(s) + \int\limits_{t_0}^{s} k_1(s,u)y(u)du \right.$$

$$\left. + \int\limits_{t_0}^{s} y(u)k_2(s,u)ds + F(s) \right] ds$$

$$+ \int\limits_{t_0}^{t} \left[-y'(s) + A(s)y(s) + y(s)B(s) + \int\limits_{t_0}^{s} k_1(s,u)y(u)du \right.$$

$$\left. + \int\limits_{t_0}^{s} y(u)k_2(s,u)du + F(s) \right] R(t,s)ds$$

$$= -\int_{t_0}^{t} [L(t,s)z(s) + z(s)R(t,s)]ds.$$

But this implies that $z(t) \equiv 0$ because of uniqueness of solutions of matrix Volterra integral equations. The proof is complete.

Once we have developed the equivalence between linear matrix integrodifferential equation (2.1.10) and the matrix linear differential equation (2.1.13), it is immediate to obtain the VPF for (2.1.10). We shall therefore state such a result without proof.

Theorem 2.1.5: *Let $y(t)$ be a matrix-valued solution of (2.1.10) existing for $t \geq t_0$ such that $(t_0) = y_0$, $y_0 \in R^{n \times n}$, let $v(t,t_0)$, $w(t,t_0)$ be the matrix solution of the linear matrix differential equations*

$$v' = [A(t) - L(t,t)]v, \quad v(t_0) = I,$$

$$w' = w[B(t) - R(t,t)], \quad w(t_0) = I,$$

respectively, when the matrixes A, B. L, R are as defined in Theorem 2.1.3. Then $y(t)$ verifies the integral equation

$$y(t) = v(t,t_0)y_0 w(t,t_0) + \int_{t_0}^{t} v(t,t_0)v^{-1}(s,t_0)H(s,y(s))$$

$$w^{-1}(s,t_0)w(t,t_0)ds, \quad t \geq t_0,$$

where

$$H(s,y(s)) = L(s,t_0)y_0 + y_0 R(s,t_0) + F(s)$$

$$+ \int_{t_0}^{s} [L(s,u)F(u) + F(u)R(s,u)]du.$$

2.2 NONLINEAR VARIATION OF PARAMETERS

We wish to develop, in this section, VPF for nonlinear integrodifferential equations (IDE). For this purpose, we need to discuss continuous dependence and differentiability of solutions with respect to initial data. Since such results are merely stated in Chapter 1, we shall prove them here in this general set up.

Consider the initial value problem for nonlinear IDE

$$x'(t) = f(t,x(t)) + \int_{t_0}^{t} g(t,s,x(s))ds, \quad x(t_0) = x_0. \qquad (2.2.1)$$

We shall assume the following conditions:

(H_1) $f \in C[R_+ \times R^n, R^n]$ and $g \in C[R_+ \times R_+ \times R^n, R^n]$;

(H_2) f_x, g_x exist and are continuous on $R_+ \times R^n$, $R_+ \times R_+ \times R^n$ respectively;

(H_3) $x(t, t_0, x_0)$ is the unique solution of (2.2.1) existing on $J = [t_0, t_0 + T]$ and denote

$$H(t, t_0, x_0) = f_x(t, x(t, t_0, x_0)),$$

$$G(t, s, t_0, x_0) = g_x(t, s, x(t, t_0, x_0)).$$

We prove the following result.

Theorem 2.2.1: *Let the hypothesis (H_1), (H_2) and (H_3) hold. Then*

(a) $\Phi(t, t_0, x_0) = \frac{\partial x}{\partial x_0}(t, t_0, x_0)$ *exists and is the solution of*

$$y'(t) = H(t, t_0, x_0)y(t) + \int_{t_0}^{t} G(t, s, t_0, x_0)y(s)ds \qquad (2.2.2)$$

such that $\Phi(t_0, t_0, x_0) = I$;

(b) $\Psi(t, t_0, x_0) = \frac{\partial x}{\partial t_0}(t, t_0, x_0)$ *exists and is the solution of*

$$z'(t) = H(t, t_0, x_0)z(t) + \int_{t_0}^{t} G(t, s, t_0, x_0)z(s)ds - g(t, t_0, x_0) \qquad (2.2.3)$$

such that $\Psi(t_0, t_0, x_0) = -f(t_0, x_0)$;

(c)   *the functions $\Phi(t, t_0, x_0)$, $\Psi(t, t_0, x_0)$ satisfy the relations*

$$\Psi(t, t_0, x_0) + \Phi(t, t_0, x_0)f(t_0, x_0) + \int_{t_0}^{t} R(t, \sigma, t_0, x_0)g(\sigma, t_0, x_0)d\sigma = 0, \qquad (2.2.4)$$

where $R(t, s, t_0, x_0)$ is the solution of the initial value problem

$$\frac{\partial R}{\partial s}(t, s, t_0, x_0) + R(t, s; t_0, x_0)H(s, t_0, x_0)$$

$$+ \int_{s}^{t} R(t, \sigma; t_0, x_0)G(\sigma, s, t_0, x_0)d\sigma = 0, \qquad (2.2.5)$$

$$R(t, t; t_0, x_0) = I \text{ on } t_0 \leq s \leq t \text{ and}$$

$$R(t, t_0, t_0, x_0) = \Phi(t, t_0, x_0).$$

Proof: The assumptions on f and g guarantee the existence of solutions $x(t, t_0, x_0)$ of (2.2.1). This solution is unique with respect to (t, t_0, x_0) on the interval of existence. Consequently the functions H and G are continuous in (t, t_0, x_0) and (t, s, t_0, x_0) respectively. This fact, in turn, implies that the solutions of (2.2.2) and (2.2.3) exist and are unique on the interval on which $x(t, t_0, x_0)$ is defined.

To prove (a), let $e_k = (e_k^1, \dots, e_k^n)$ be a vector satisfying

$$e_k^j = 0 \text{ if } j \neq k, \epsilon_k^k = 1.$$

Denote $x(t, h) = x(t, t_0, x_0 + e_k h)$ where h is small. Then $x(t, h)$ is defined on J and

$$\lim_{h \to 0} x(t, h) = x(t, t_0, x_0),$$

uniformly on J. Set $x(t) = x(t, t_0, x_0)$. It follows that

$$[x(t, h) - x(t)]' = \int_0^1 f_x[t, sx(t, h) + (1 - s)x(t)]ds[x(t, h) - x(t)]$$

$$+ \int_{t_0}^t \int_0^1 g_x[t, s, \sigma x(s, h) + (1 - \sigma)x(s)]d\sigma[x(s, h) - x(s)]ds.$$

Here we have used the Mean Value Theorem. See Lakshmikantham and Leela [1]. Define

$$x_h(t) = \frac{[x(t,h) - x(t)]}{h}.$$

Clearly the existence of $\frac{\partial x(t, t_0, x_0)}{\partial x_0}$, which we want to prove, is equivalent to the existence of the limit of $x_h(t)$ as $h \to 0$. Here $x(t_0, h) = x_0 + e_k h$ and $x_h(t_0) = e_k$. Hence, we claim that $x_h(t)$ is the solution of the initial value problem

$$y'(t) = H(t, t_0, x_0, h)y(t) + \int_{t_0}^t G(t, s, x_0, x_0, h)y(s)ds, \qquad (2.2.6)$$

$$y(t_0) = e_k,$$

where we have set

$$H(t, t_0, x_0, h) = \int_0^1 f_x[t, sx(t, h) + (1 - s)x(t)]ds$$

and

$$G(t, s, t_0, x_0, h) = \int_0^1 g_x[t, s, \sigma x(s, h) + (1 - \sigma)x(s)]d\sigma.$$

Further, we obtain that

$$\lim_{h \to 0} H(t, t_0, x_0, h) = H(t, t_0, x_0),$$

$$\lim_{h \to 0} G(t, s, t_0, x_0, h) = G(t, s, t_0, x_0),$$

uniformly on J and $J \times J$ respectively. Now, treat (2.2.6) as a set of initial value problems in which H and G are continuous and h is small. Further, solutions of (2.2.6) are unique. Hence $x_h(t)$ is a continuous function of h for fixed (t, t_0, x_0)

implying that $\lim_{h \to 0} x_h(t)$ exists and consequently is the solution of (2.2.2) on J such that

$$\frac{\partial x(t_0, t_0, x_0)}{\partial x_0} = I.$$

We also note that under the given hypothesis and the equation (2.2.2) $\frac{\partial x(t, t_0, x_0)}{\partial x_0}$ is also continuous with respect to (t, t_0, x_0).

(b)   Here, we set $x(t, h) = x(t, t_0 + h, x_0)$ and get that

$$[x(t, h) - x(t)]' = \int_0^1 f_x[t, sx(t, h) + (1 - s)x(t)]ds\, x[x(t, h) - x(t)]$$

$$+ \int_{t_0+h}^t \int_0^1 g_x[t, s, \sigma x(s, h) + (1 - \sigma)x(s)]d\sigma$$

$$\times [x(s, h) - x(s)]ds - \int_{t_0}^{t+h} g(t, s, x(s))ds.$$

Define, as in (a), $x_h(t) = \frac{x(t, h) - x(t)}{h}$, $h \neq 0$. We then easily establish that $x_h(t)$ is the solution of the initial value problem

$$y'(t) = H(t, t_0, x_0, h)y(t) + \int_{t_0+h}^t G(t, s, t_0, x_0, h)y(s)ds - L(t, t_0, x_0, h),$$

such that $y(t_0 + h) = a(h)$, where

$$L(t, t_0, x_0, h) = \frac{1}{h} \int_{t_0}^{t_0+h} g(t, s, x(s))ds$$

and

$$a(h) = -\frac{1}{h} \int_{t_0}^{t_0+h} f(s, x(s))ds - \frac{1}{h} \int_{t_0}^{t_0+h} \int_{t_0}^s g(s, \sigma, x(\sigma))d\sigma ds.$$

Observe that

$$\lim_{h \to 0} L(t, t_0, x_0, h) = g(t, t_0, x_0),$$

and

$$\lim_{h \to 0} a(h) = -f(t_0, x_0).$$

Now we repeat the argument made in the proof of the part (a) and conclude that $\frac{\partial x(t, t_0, x_0)}{\partial t_0}$ exists, is continuous in (t, t_0, x_0) and is the solution of (2.2.3).

(c)   Note that $\Phi(t, t_0, x_0)$ and $\Psi(t, t_0, x_0)$ are the solutions of the initial value problem (2.2.2) and (2.2.3) respectively. At this stage, apply Theorem 2.1.1 to yield the relations (2.2.4). The proof is complete.

Remark 2.2.1: If in (2.2.1) $g(t, s, x) \equiv 0$, then (2.2.1) reduces to equation (1.1.11) and hence Theorem 2.2.1 includes Theorem 1.2.4.

The main result of this section is to establish the VPF for nonlinear IDE

$$y'(t) = f(t, y(t)) + \int_{t_0}^{t} g(t, s, y(s))ds + F(t, s, (Sy)(t)), \qquad (2.2.7)$$

$$y(t_0) = x_0.$$

We now proceed to establish this result.

Theorem 2.2.2: *Assume that the hypothesis of Theorem 2.2.1 holds. Let $F \in C[R_+ \times R^n \times R^n, R^n]$ and*

$$(Sy)(t) = \int_{t_0}^{t} k(t, s, y(s))ds \text{ with } k \in C[R_+ \times R_+ \times R^n, R^n].$$

If $x(t, t_0, x_0)$ is the solution of (2.2.1) existing on J, then any solution $y(t, t_0, x_0)$ of (2.2.7) existing on J satisfies the integral equation

$$y(t, t_0, x_0) = x(t, t_0, x_0) + \int_{t_0}^{t} \Phi(t, s, y(s))F(s, y(s), (Sy)(s))ds$$

$$+ \int_{t_0}^{t} \int_{s}^{t} [\Phi(t, \sigma, y(\sigma)) - R(t, \sigma, s, y(s))]g(\sigma, s, y(s))d\sigma ds \qquad (2.2.8)$$

for $t \geq t_0$, where $y(t) = y(t, t_0, x_0)$ and $R(t, s, t_0, x_0)$ is the solution of the initial value problem (2.2.5).

Proof: Define $p(s) = x(t, s, y(s))$ and $y(s) = y(s, t_0, x_0)$. Then

$$p'(s) = \psi(t, s, y(s)) + \Phi(t, s, y(s))y'(s).$$

Substitute y' from (2.2.7), integrate between t_0 and t and use Fubini's theorem to get,

$$p(t) - p(t_0) - \int_{t_0}^{t} \Phi(t, s, y(s))F(s, y(s), (Sy)(s))ds$$

$$= \int_{t_0}^{t} [\Psi(t, s, y(s)) + \Phi(t, s, y(s))f(s, y(s))$$

$$+ \int_{s}^{t} \Phi(t, \sigma, y(\sigma))g(\sigma, s, y(s))d\sigma]ds.$$

The relation (2.2.4) together with the fact that $x(t, t, y(t)) = y(t, t_0, x_0)$ leads us to the conclusion (2.2.8).

Remark 2.2.2: (i) Let $f(t, x) = A(t)x$ and $g(t, s, x) = B(t, s)x$ in (2.2.1). It is then clear that

$$x(t, t_0, x_0) = R(t, t_0)x_0$$

and

$$\Phi(t, t_0, x_0) = \frac{\partial x(t, t_0, x_0)}{\partial x_0} = R(t, t_0),$$

where $R(t, s)$ is the solution of (2.2.5) such that $R(t, t) = I$. The VPF (2.2.8) then reduces to

$$y(t, t_0, x_0) = R(t, t_0)x_0 + \int_{t_0}^{t} R(t, s)F(s, y(s), (Sy)(s))ds, t \geq t_0,$$

since $\Phi(t, \sigma, y(\sigma)) = R(t, \sigma, s, y(s))$ is independent of $y(s)$ and s.
(ii) If $g(t, s, x) \equiv 0$, then the relation (2.2.8) reduces to the VPF proved in Theorem 1.2.9.
(iii) If $x(t, t_0, x_0)$, $x(t, t_0, y_0)$ are any two solutions of (2.2.1) existing on $[t_0, \infty)$, then we have

$$x(t, t_0, x_0) - x(t, t_0, y_0) = \left[\int_{0}^{1} \frac{\partial x}{\partial x_0}(t, t_0, sx_0 + (1 - s)y_0)ds \right] \cdot (x_0 - y_0).$$

(iv) In the special case, when (2.2.1) is an ordinary differential system, one can utilize the ideas of Section 1.2 to obtain various VPF for perturbed IDS (2.2.7). It is clear that such VPFs do not depend on the type of perturbed terms. Although, one can formulate several variant VPFs in this case, we do not intend to discuss them since these can be obtained in a straight forward manner.

2.3 MONOTONE ITERATIVE TECHNIQUE

We shall discuss, in the present section, the existence of extremal solutions of the initial value problem (IVP) for nonlinear integrodifferential equation (IDE) by employing monotone iterative technique.

Consider the nonlinear IDE of Volterra type

$$u' = f(t, u, Tu), \quad t \in J, u(0) = u_0, \tag{2.3.1}$$

where $J = [0, a]$, $a > 0$, $u_0 \in R$, $f \in C[J \times R \times R, R]$ and

$$(Tu)(t) = \int_{0}^{t} k(t, s)u(s)ds, t \in J; \tag{2.3.2}$$

$u \in C[J, R]$ and $k \in C[J \times J, R_+]$.

The following lemma is crucial in the subsequent discussion.

Lemma 2.3.1: *Assume that $p \in C^1[J, R]$ is such that*

$$p' \leq -Mp - NTp, \, t \in J, \, p(0) \leq 0, \tag{2.3.3}$$

where $M > 0$ and $N \geq 0$ are constants such that

$$Nk_0 a(e^{Ma} - 1) \leq M, \tag{2.3.4}$$

where $k_0 = \max\limits_{J \times J} k(t, s)$. Then $p(t) \leq 0$ for $t \in J$.

Proof: Set $v(t) = p(t)e^{Mt}, \, t \in J$. Then (2.3.3) reduces to

$$v'(t) \leq -N \int_0^t k^*(t, s)v(s)ds, t \in J \tag{2.3.5}$$

$$v(0) \leq 0;$$

where $k^*(t, s) = k(t, s)e^{M(t-s)}$. It is enough to show that

$$v(t) \leq 0, t \in J. \tag{2.3.6}$$

Assume that (2.3.6) is not true. Then there exists a t_0, $0 < t_0 < a$ such that $v(t_0) > 0$, let $\min\{v(t): 0 \leq t \leq t_0\} = -b$. Then $b \geq 0$.
(*i*)   If $b = 0$, then $v(t) \geq 0$ for $0 \leq t \leq t_0$ and (2.3.5) then means that $v'(t) \leq 0$ for $0 \leq t \leq t_0$. As a consequence $v(t_0) \leq v(0) \leq 0$, which contradicts the fact that $v(t_0) > 0$.
(*ii*) If $b > 0$, then there exists a t_1, $0 \leq t_1 < t_0$ such that $v(t_1) = -b < 0$. Hence there exists a t_2 with $t_1 < t_2 < t_0$ such that $v(t_2) = 0$. Employing the mean value theorem, we get the existence of t_3, $t_1 < t_3 < t_2$ and

$$v'(t_3) = \frac{v(t_2) - v(t_1)}{t_2 - t_1} = \frac{b}{t_2 - t_1} > \frac{b}{a}. \tag{2.3.7}$$

Now in view of (2.3.5) and (2.3.4), we obtain

$$v'(t_3) \leq -N \int_0^{t_3} k^*(t_3, s)v(s)ds$$

$$\leq Nb \int_0^{t_3} k^*(t_3, s)ds$$

$$\leq Nbk_0 \int_0^{t_3} e^{M(t_3 - s)}ds$$

$$= \frac{Nbk_0}{M}(e^{Mt_3} - 1)$$

$$\leq \tfrac{Nbk_0}{M}(e^{Ma} - 1) \leq \tfrac{b}{a},$$

which contradicts (2.3.7). Hence (2.3.6) holds. The proof is complete.

We list below some needed hypotheses for convenience.

(H_1) There exists $v_0, w_0 \in C^1[J, R]$ such that $v_0(t) \leq w_0(t)$, $t \in J$ and

$$v_0' \leq f(t, v_0, Tv_0),\, t \in J,\, v_0(0) \leq u_0;$$
$$w_0' \geq f(t, w_0, Tw_0),\,\, t \in J,\, w_0(0) \geq u_0.$$

(H_2) There exists constants $M, N, M > 0, N \geq 0$ such that

$$f(t, u, v) - f(t, \overline{u}, \overline{v}) \geq -M(u - \overline{u}) - N(Tv - T\overline{v})$$

for $t \in J, v_0(t) \leq \overline{u} \leq u \leq w_0(t)$ and

$$(Tv_0)(t) \leq (T\overline{v})(t) \leq (Tv)(t) \leq (Tw_0)(t),$$

and $Nk_0a(e^{Ma} - 1) < M$.

We are now in a position to prove the following result.

Theorem 2.3.1: *Suppose that conditions (H_1) and (H_2) hold. Then there exist monotone sequences $\{v_n\}$ and $\{w_n\}$ which converge uniformly and monotonically on J to the minimal and maximal solutions ρ, r in $[v_0, w_0]$ respectively. That is, if $u \in C^1[J, R]$ is any solution of the IVP (2.3.1) such that $v_0(t) \leq u(t) \leq w_0(t)$ for $t \in J$, then*

$$v_0(t) \leq v_1(t) \leq \ldots \leq v_n(t) \leq \ldots \leq \rho(t) \leq u(t) \leq r(t)$$

$$\leq \ldots \leq w_n(t) \leq \ldots \leq w_1(t) \leq w_0(t), t \in J. \qquad (2.3.8)$$

Proof: For $h \in [v_0, w_0]$, consider the IVP of a linear integrodifferential equation

$$u' + Mu = -NTu + g(t),\, u(0) = u_0, \qquad (2.3.9)$$

for $t \in J$, where $g(t) = f(t, h(t), (Th)(t)) + Mh(t) + N(Th)(t)$. Using the VPF, we obtain, for $t \in J$,

$$u(t) = e^{-Mt}\left[u_0 + \int_0^t \{g(s) - N(Tu)(s)\}e^{Ms}ds\right]. \qquad (2.3.10)$$

To establish the existence of a unique solution (2.3.10), set $F: C[J, R] \to C[J, R]$ defined by

$$(Fu)(t) = e^{-Mt}\left[u_0 + \int_0^t \{g(s) - N(Tu)(s)\}e^{Ms}ds\right].$$

Under the given conditions, it is easy to show that

$$| Fu - Fv |_0 \leq Nk_0 a^2 | u - v |_0$$

for $u, v \in C[J, R]$ and where $| \cdot |_0$ denotes the norm in $C[J, R]$. From (2.3.4), it is clear that $Nk_0 a^2 < 1$. Hence by the connection principle, we conclude that F has a unique fixed point u in $C[J, R]$ and that this u is the unique solution of (2.3.9).

Now we set $u = Ah$ where $u = u(t)$ is the unique solution of (2.3.9). The operator A maps $[v_0, w_0]$ into $C[J, R]$. We now show that
(a) $v_0 \leq Av_0$, $Aw_0 \leq w_0$, and
(b) A is nondecreasing in $[v_0, w_0]$.

To prove (a), set $v_1 = Av_0$ and $p = v_0 - v_1$. From (2.3.9), we have

$$v_1' + Mv_1 = - NTv_1 + f(t, v_0, Tv_0) + Mv_0 + NTv_0,$$

$$v_1(0) = u_0.$$

In view of (H_1), it follows that

$$p' \leq - Mp - NTp, \quad p(0) \leq 0.$$

Lemma 2.3.1 concludes that $p(t) \leq 0$, $t \in J$, which thereby implies $v_0 \leq Av_0$. We can similarly show that $Aw_0 \leq w_0$ which proves (a).

To prove (b), let $h_1, h_2 \in [v_0, w_0]$ such that $h_1 \leq h_2$ and let $p = u_1 - u_2$ where $u_1 = Ah_1$ and $u_2 = Ah_2$. From (2.3.9), the hypothesis (H_1) and (H_2) we obtain

$$p' = u_1' - u_2'$$

$$= - Mu_1 - N(Tu_1) + f(t, h_1, Th_1) + Mh_1 + N(Th_1)$$

$$+ Mu_2 + N(Tu_2) - f(t, h_2, Th_2) + Mh_2 - N(Th_2)$$

$$\leq - M(u_1 - u_2) - N(Tu_1 - Tu_2) + M(h_2 - h_1)$$

$$+ N(Th_1 - Th_2) + M(h_1 - h_2) + N(Th_1 - Th_2)$$

$$= - Mp - N(Tp).$$

Further $p(0) = 0$ and hence by Lemma 2.3.1, we conclude that $p(t) \leq 0$ for $t \in J$ proving that $Ah_1 \leq Ah_2$ and thus (b) is proved.

Define $v_n = Av_{n-1}$ and $w_n = Aw_{n-1}$ for $n = 1, 2, \ldots$. By (a) and (b) proved above, we obtain, for $t \in J$,

$$v_0(t) \leq v_1(t) \leq \ldots \leq v_n(t) \leq \ldots \leq w_n(t) \leq \ldots \leq w_1(t) \leq w_0(t).$$

Employing standard argument, it follows that there exists functions $\rho, r \in C^1[J, R]$ such that

$$\lim_{n \to \infty} v_n = \rho \text{ and } \lim_{n \to \infty} w_n = r$$

uniformly and monotonically on J. It is easy to show that ρ and r are solutions of (2.3.1) in view of the fact that v_n and w_n satisfy

$$v'_n = f(t, v_{n-1}, Tv_{n-1}) - M(v_n - v_{n-1}) - N(Tv_n - Tv_{n-1}),$$
$$v_n(0) = u_0,$$

and

$$w'_n = f(t, w_{n-1}, Tw_{n-1}) - M(w_n - w_{n-1}) - N(Tw_n - Tw_{n-1})$$

$$w_n(0) = u_0.$$

To prove that ρ and r are minimal and maximal solutions of (2.3.1) respectively, we need to show that, if u is any solution of (2.3.1) such that $u \in [v_0, w_0]$ on J, then

$$v_0 \leq \rho \leq u \leq r \leq w_0 \text{ on } J. \tag{2.3.12}$$

To do this, assume that for some $k > 2$, $v_{k-1}(t) \leq u(t) \leq w_{k-1}(t)$ for $t \in J$ and set $p = v_k - u$. Employing (2.3.9) and (H_2), we obtain

$$(v_k)' = -(Mv_k) - MTv_k + f(t, v_k, Tv_{k-1}) + M(v_{k-1}) + N(Tv_{k-1}),$$

$$u' = f(t, u, Tu).$$

Hence

$$p' = (v_k)' - u' = f(t, v_k, Tv_{k-1}) - f(t, u, Tu)$$

$$+ M(v_{k-1} - v_k) + N(Tv_{k-1} - Tv_k),$$

$$\leq M(u - v_{k-1}) + N(Tu - Tu_{k-1}) + M(v_{k-1} - v_k) + N(Tv_{k-1} - Tv_k)$$

$$= -M(v_k - u) - N(Tv_k - Tu) = -Mp - NTp.$$

and
$$p(0) = 0.$$

Employing Lemma 2.3.1, we obtain that $p(t) \leq 0$, $t \in J$ implying $v_k(t) \leq u(t)$ for $t \in J$. Similarly, we can show that $u(t) \leq w_k(t)$. By induction, we conclude that for all n, $v_n(t) \leq u(t) \leq w_n(t)$, $t \in J$. By taking the limit as $n \to \infty$, we conclude that

$$\rho(t) \leq u(t) \leq r(t), \ t \in J.$$

Hence (2.3.8) holds. The proof is complete.

2.4 STABILITY CRITERIA

This section deals with the study of stability behavior of linear and nonlinear integrodifferential equations. Various stability properties will be discussed. We first consider the linear integrodifferential system.

Consider the system of convolution type given by

$$x'(t) = Ax(t) + \int\limits_0^t k(t-s)x(s) \tag{2.4.1}$$

where A is an $n \times n$ constant matrix and $k(t)$ is $n \times n$ continuous matrix on R_+^2. Let $x(t, t_0, x_0)$ be any solution of (2.4.1) with the initial function $\phi(t)$ on $0 \le t \le t_0$, for $t_0 > 0$. Then (2.4.1) takes the form

$$x'(t) = Ax(t) + \int\limits_{t_0}^t k(t-s)x(s)ds + \int\limits_0^{t_0} k(t-s)\phi(s)ds. \tag{2.4.2}$$

By recalling VPF given in Theorem 2.1.3, we get

$$x(t) = z(t-t_0)\phi(t_0) + \int\limits_{t_0}^t z(t-s)\left[\int\limits_0^t k(s-\sigma)\phi(\sigma)d\sigma\right]ds. \tag{2.4.3}$$

Replacing t by $t + t_0$, we obtain

$$x(t+t_0) = z(t)\phi(t_0) + \int\limits_{t_0}^t z(t-\sigma)\left[\int\limits_0^{t_0} k(\sigma+t_0-u)\phi(u)ds\right]d\sigma, t \ge t_0 \tag{2.4.4}$$

Set

$$p(t) = \int\limits_0^\infty \left|\int\limits_0^t z(t-\sigma)k(\sigma+u)d\sigma\right| du, t \ge 0, \tag{2.4.5}$$

and assume that $p(t)$ exists for $t \ge 0$.

We define stability of trivial solution of (2.4.1) as follows.

Definition 2.4.1: The solution $x(t) \equiv 0$ of (2.4.1) is said to be stable if given $\epsilon > 0$ and $t_0 \in R_+$, there exists a $\delta = \delta(t_0, \epsilon) > 0$ such that $|\phi|_{t_0} = \max\limits_{0 \le s \le t_0} |\phi(s)| < \delta$ implies $|x(t)| < \epsilon, t \ge t_0$.

The definitions of other stability notions can be formulated similarly. Let us begin by proving the following stability result.

Theorem 2.4.1: *Suppose that $k \in L^1(R_+^2)$. Then $x(t) \equiv 0$ of (2.4.1) is*

(i) *uniformly stable if and only if the two functions $z(t)$ and $p(t)$ are uniformly bounded on R_+;*

(ii) *uniformly asymptotically stable if and only if it is uniformly stable and both*
       $z(t)$ *and* $p(t)$ *tend to zero as* $t \to \infty$.

Proof: Let the trivial solution of (2.4.1) be uniformly stable. Then there exists a constant B such that for ay (t_0, ϕ) with $t_0 \geq 0$, $|\phi|_{t_0} \leq 1$, we have

$$|x(t + t_0, t_0, \phi)| \leq B, \ t \geq 0.$$

Let $t_0 = 0$. Then (2.4.3) yields

$$|x(t, 0, \phi)| = |z(t)\phi(0)| \leq B, t \geq 0$$

when $|\phi(0)| \leq 1$, which implies $|z(t)| \leq B$ for all $t \geq 0$.
   Similarly, if $y(t) = x(t + t_0, t_0, \phi) - z(t)\phi(t_0)$, then $|y(t)| \leq 2B$. Hence, by (2.4.4),

$$y(t) = \int_0^t z(t-s) \left[\int_0^{t_0} k(\sigma + t_0 - u)\phi(u)du \right] d\sigma$$

$$= \int_0^t z(t-s) \left[\int_0^{t_0} k(\sigma + u)\phi(t_0 - u)du \right] d\sigma$$

$$= \int_0^{t_0} \left[\int_0^t z(t - \sigma)k(\sigma + u)d\sigma \right] \phi(t_0 - u)du.$$

This shows that for all $t, t_0 \geq 0$,

$$\int_0^{t_0} \left| \int_0^t z(t - \sigma)k(\sigma + u)d\sigma \right| du \leq 2B$$

which implies that $p(t) \leq 2B$ for $t \geq 0$.
   To prove converse statement, assume that $|z(t)| \leq M$, $p(t) \leq M$ for some fixed constant M. Then (2.4.4) gives

$$|x(t + t_0, t_0, \phi)| \leq |z(t)\phi(t_0)| + \left| \int_0^{t_0} \left[\int_0^t z(t - s)k(s + u)ds \right] \phi(t_0 - u)du \right|$$

$$\leq M|\phi(t_0)| + M|\phi|_{t_0} \leq 2A|\phi|_{t_0}.$$

Hence $x \equiv 0$ of (2.4.1) is uniformly stable.
   In Theorem 2.1.2 we have shown that linear integrodifferential systems are equivalent to linear differential systems under certain conditions. This fact is employed below to obtain stability properties of integrodifferential systems. Let us consider the system

$$x'(t) = A(t)x(t) + \int\limits_0^t k(t,s)x(s)ds + F(t), x(0) = x_0 \qquad (2.4.6)$$

where $A(t)$ is an $n \times n$ matrix continuous for $0 \le t < \infty$, $k(t,s)$ is an $n \times n$ matrix continuous for $0 \le s \le t < \infty$, and $F(t)$ is an n-vector continuous function on R_+.

By Theorem 2.1.2, we know that the system (2.4.6) is equivalent to

$$y'(t) = B(t)y(t) + \int\limits_0^t \Phi(t,s)y(s)ds + H(t), \quad y(0) = x_0 \qquad (2.4.7)$$

where

$$B(t) = A(t) - L(t,t),$$

$$\Phi(t,s) = k(t,s) + \frac{\partial L(t,s)}{\partial s} + L(t,s)A(s) + \int\limits_s^t L(t,u)k(u,s)du,$$

$$H(t) = F(t) + L(t,0)x_0 + \int\limits_0^t L(t,s)F(s)ds,$$

$L(t,s)$ being an $n \times n$ continuously differentiable matrix function for $0 \le s \le t < \infty$.

We are now in a position to prove the following result.

Theorem 2.4.2: *Let $B(t)$ be an $n \times n$ continuous matrix which commutes with its integral and let M and α be positive real numbers. Assume that the relation*

$$\left| \exp\left(\int\limits_s^t B(\tau)d\tau \right) \right| \le Me^{-\alpha(t-s)}, \ 0 \le s \le t < \infty \qquad (2.4.8)$$

holds. Then every solution $y(t)$ of (2.4.7) with $y(0) = x_0$ satisfies

$$|y(t)| \le M|x_0|e^{-\alpha t} + M\int\limits_0^t e^{-\alpha(t-\tau)}|H(\tau)|d\tau$$

$$+ M\int\limits_0^t \left[\int\limits_s^t e^{-\alpha(t-\tau)}|\Phi(t,s)|d\tau \right] |y(s)|ds. \qquad (2.4.9)$$

Proof: We multiply both sides of (2.4.7) by $\exp[-\int\limits_0^t B(\tau)d\tau]$, rearrange the terms, to obtain

$$\frac{d}{dt}\left[\exp\left(-\int\limits_0^t B(\tau)d\tau \right) y(t) \right] = \exp\left[-\int\limits_0^t B(\tau)d\tau \right] \left[H(t) + \int\limits_0^t \Phi(t,s)y(s)ds \right].$$

Integrate between 0 and t, to get

$$\exp\left[\left(-\int_0^t B(\tau)d\tau\right)y(t)\right] = x_0 + \int_0^t \exp\left(-\int_0^s B(\tau)d\tau\right)H(s)ds$$

$$+ \int_0^t \left[\exp\left(-\int_0^s B(\tau)d\tau\right)\right]\left[\int_0^s \Phi(s,u)y(u)du\right]ds.$$

Change the order of integration on the right-hand side in the above expression and use (2.4.8) to derive (2.4.9).

Remark 2.4.1: A constant matrix commutes with its integral. Hence a constant matrix B is admissible in the above proof.

Using the estimate of Theorem 2.4.2, we can prove the exponential stability of (2.4.1).

Theorem 2.4.3: *Let $L(t,s)$ be a continuously differentiable $n \times n$ matrix satisfying the following conditions on $0 \leq s \leq t < \infty$*
(a)   the hypothesis of Theorem 2.4.2 hold,
(b) $|L(t,s)| \leq L_0 \exp(-\gamma(t-s))$,
(c) $\displaystyle\sup_{0 \leq s \leq < \infty} \int_s^t e^{\alpha(\tau-s)}|\Phi(\tau,s)|d\tau \leq \alpha_0$ where L_0, γ, α_0 are positive real
       numbers and $\alpha < \gamma$.
(d) $F(t) \equiv 0$, $F(t)$ being defined in (2.4.6).
If $\alpha - M\alpha_0 > 0$, then every solution $x(t)$ of (2.4.1) tends to zero exponentially as $t \to \infty$.

Proof: It is enough to show that every solution $y(t)$ of (2.4.7) tends to zero exponentially as $t \to \infty$. In view of assumptions (a), (b) and (d), and the inequality (2.4.9), we obtain

$$e^{\alpha t}|y(t)| \leq M|x_0| + ML_0|x_0|\int_0^t e^{-(\gamma-\alpha)\tau}d\tau$$

$$+ M\int_0^t \left[\int_s^t e^{\alpha\tau}|\Phi(\tau,s)|d\tau\right]|y(s)|ds.$$

Because of the assumption (c), we get

$$e^{\alpha t}|y(t)| \leq M|x_0| + \frac{ML_0|x_0|}{(\gamma-\alpha)} + \int_0^t M\alpha_0 e^{\alpha s}|y(s)|ds.$$

Now an application of Gronwall inequality yields the estimate

$$|y(t)| \leq M|x_0|\left[1 + \frac{L_0}{(\gamma-\alpha)}\right]e^{-(\alpha-M\alpha_0)t}, \quad t \geq 0.$$

Since $\alpha - M\alpha_0 > 0$, the desired conclusion follows.

Remark 2.4.2: If $F(t) \equiv 0$ in (2.4.6), then the above theorem asserts the exponentially asymptotic stability of the trivial solution of (2.4.6). In case $F(t) \neq 0$, solutions of (2.4.6) tend to zero as $t \to \infty$ provided $F(t)$ is integrable. To prove this statement, we need to employ VPF and Theorem 2.4.3.

We shall next consider stability properties of nonlinear integrodifferential equation

$$x'(t) = f(t, x(t)) + \int_{t_0}^{t} g(t, s, x(s))ds, \quad x(t_0) = x_0, \tag{2.4.10}$$

for $t \geq t_0 \geq 0$, when $f \in C[R_+ \times R^n, R^n]$ and $g \in C[R_+ \times R_+ \times R^n, R^n]$. Assume that f_x, g_x exist and are continuous, and $f(t, 0) \equiv 0$, $g(t, s, 0) \equiv 0$. Let $x(t, t_0, x_0)$ be the unique solution of (2.4.10), existing for $t \geq t_0 \geq 0$. Consider the related variational equations

$$y'(t) = f_x(t, x(t))y(t) + \int_{t_0}^{t} g_x(t, s, x(s))y(s)ds \tag{2.4.11}$$

and

$$z'(t) = f_x(t, 0)z(t) + \int_{t_0}^{t} g_x(t, s, 0)z(s)ds. \tag{2.4.12}$$

We need the following definition before we proceed further.

Definition 2.4.1: The trivial solution $x \equiv 0$ of (2.4.10) is said to possess stability properties in variation if the trivial solution $x \equiv 0$ of (2.4.12) possesses the corresponding stability property.

Relative to the stability properties of the trivial solution of (2.4.10), we can prove the following result.

Theorem 2.4.4: *Suppose that the assumptions in Theorem 2.1.2 hold. Then stability properties of the function*

$$p(t, t_0, |x_0|) = |\psi(t, t_0)| |x_0| + \int_{t_0}^{t} |\psi(t, s)| |L(t, t_0)| |x_0| ds \tag{2.4.13}$$

where $\psi(t, s)$ is the fundamental matrix solution of $v' = B(t)v$ with $\Psi(t_0, t_0) = I$, imply the corresponding stability properties of (2.4.10).

Proof: Set $A(t) = f_x(t, 0)$, $k(t, s) = g_x(t, s, 0)$ and $F(t) \equiv 0$, it is enough to consider the IVP

$$v'(t) = B(t)v(t) + L(t, t_0)x_0 \tag{2.4.14}$$

in order to study the corresponding variational equation (2.4.12). By VPF, we get from (2.4.14)

$$v(t) = \Psi(t, t_0)x_0 + \int_{t_0}^{t} \Psi(t, s)L(s, t_0)x_0 ds.$$

Hence, it follows that

$$|v(t)| \leq p(t, t_0, |x_0|),$$

where $p(t, t_0, |x_0|)$ is given by (2.4.13).

Theorem 2.4.4 establishes the fact that the stability properties of (2.4.10) are related to the stability properties of the variational system (2.4.11). Observe that any solution $v(t)$ of (2.4.11) such that $v(t_0) = x_0$ is given by

$$v(t) = \frac{\partial x(t, t_0, x_0)}{\partial x_0} x_0.$$

Further, any solution $x(t, t_0, x_0)$ of (2.4.10) has the representation,

$$x(t, t_0, x_0) = \left[\int_0^1 \frac{\partial x(t, t_0, x_0, s)}{\partial x_0} \right] x_0.$$

Hence, when $|v(t)| < \sigma, t \geq t_0$. For $|x_0| < \delta(\sigma)$, we have $|x(t, t_0, x_0)| \leq \sigma$.

In view of the foregoing considerations, we can now study the system (2.4.11) having the form

$$y'(t) = A(t)y(t) + \int_{t_0}^{t} k(t, s)y(s)ds + F(t, y(t)), \; y(t_0) = x_0 \qquad (2.4.15)$$

where

$$F(t, y(t)) = [f_x(t, x(t)) - f_x(t, 0)]y(t)$$

$$+ \int_{t_0}^{t} [g_x(t, s, x(s)) - g_x(t, s, 0)]y(s)ds. \qquad (2.4.16)$$

We know, in view of Theorem 2.1.2 that (2.4.15) is equivalent to

$$v'(t) = B(t)v(t) + L(t, t_0)x_0 + H(t, v), \; v(t_0) = x_0 \qquad (2.4.17)$$

where

$$H(t, v) = F(t, v) + \int_{t_0}^{t} L(t, s)F(s, v(s))ds. \qquad (2.4.18)$$

Assume that

$$|f_x(t, x) - f_x(t, 0)| \leq \lambda(t), \; \lambda \in C[R^+, R^+] \qquad (2.4.19)$$

and

$$|g_x(t, s, x) - g_x(t, s, 0)| \leq \eta(t, s), \qquad (2.4.20)$$

whenever $|x| < \rho$ for some $\rho > 0$. At this stage, set functions $\gamma(t,s)$ and $d(t,s)$ satisfying inequalities

$$\gamma(t,s) \geq \eta(t,s) + \lambda(s) \mid L(t,s) \mid + \int_s^t \mid L(t,\sigma) \mid \eta(\sigma,\delta)d\sigma$$

$$d(t,s) \geq \mid \psi(t,s) \mid \left[\lambda(s) + \int_s^t \mid \psi(t,\sigma) \mid \gamma(\sigma,s)d\sigma \right]. \qquad (2.4.21)$$

Consider the integral equation

$$r(t) = p(t,t_0, \mid x_0 \mid) + \int_{t_0}^t d(t,s)r(s)ds \qquad (2.4.22)$$

having solution $r(t)$ given by

$$r(t) = p(t,t_0, \mid x_0 \mid) - \int_{t_0}^t r(t,s)p(s,t_0, \mid x_0 \mid)ds,$$

where $r(t,s)$ is the resolvent kernel, satisfying

$$r(t,s) = -d(t,s) + \int_s^t r(t,\sigma)d(\sigma,s)d\sigma.$$

We are ready to prove the following result.

Theorem 2.4.5: *Suppose that the hypothesis of Theorem 2.1.2 holds together with (2.4.19) and (2.4.20). Then any solution $v(t)$ of (2.4.17) satisfies the pointwise estimation*

$$\mid v(t) \mid \leq r(t), \ t \geq t_0$$

provided $\mid x(t) \mid \leq \rho$ for all $t \geq t_0$ where $r(t) = r(t,t_0, \mid x_0 \mid)$ is the solution of (2.4.22).

Proof: Let $v(t)$ be any solution of (2.4.17). Then by the VPF, we have

$$v(t) = \psi(t,t_0)x_0 + \int_{t_0}^t \psi(t,s)[L(s,t_0)x_0 + H(s,v(s))]ds.$$

since

$$\mid F(t,v) \mid \leq \mid f_x(t,x(t)) - f_x(t,0) \mid \mid v(t) \mid$$

$$+ \int_{t_0}^t \mid g_x(t,s,x(s)) - g_x(t,s,0) \mid \mid v(s) \mid ds,$$

we obtain by using Fubini's theorem, (2.4.18), (2.4.19), (2.4.20)

$$| H(t, v) | \leq \lambda(t) \, | \, v(t) | + \int_{t_0}^{t} r(t, s) | v(s) | \, ds.$$

Use Fubini's theorem again to get

$$\int_{t_0}^{t} | \psi(t, s) H(s, v(s)) | \, ds \leq \int_{t_0}^{t} | \psi(t, s) | \left[\lambda(s) + \int_{s}^{t} | \psi(t, \sigma) | r(\sigma, s) d\sigma \right] | v(s) | \, ds.$$

It follows from (2.4.18), (2.4.14) and (2.4.21) that

$$| v(t) | \leq p(t, t_0, | x_0 |) + \int_{t_0}^{t} d(t, s) | v(s) | \, ds, t \geq t_0.$$

Hence $v(t) \leq r(t, t_0, | x_0 |)$ where $r(t, t_0, | x_0 |)$ is the solution of (2.4.22). The proof is complete.

Remark 2.4.3: We note that the equations (2.4.16), (2.4.17) are equivalent and (2.4.16) is a restatement of (2.4.12). Theorem 2.4.5 claims that the stability properties of (2.4.22) imply the corresponding stability properties of (2.4.11).

In order to deal directly with (2.4.10), we note that it can be written as

$$x'(t) = A(t)x(t) + \int_{t_0}^{t} k(t, s)x(s)ds + F_1(t, x(t)), x(t_0) = x_0,$$

where

$$F_1(t, x(t)) = [f(t, x(t)) - f_x(t, 0)]x(t)$$

$$+ \int_{t_0}^{t} [g(t, s, x(s)) - g_x(t, s, 0)]x(s)ds$$

or

$$F_1(t, x(t)) = \left[\int_{0}^{1} [f_x(t, \theta x) - f_x(t, 0)]d\theta \right] x$$

$$+ \left[\int_{t_0}^{t} \int_{0}^{1} [g_x(t, s, x(s)\theta) - g_x(t, s, 0)]d\theta \right] x(s)ds$$

$| x(t, t_0, x_0) | \leq M | x_0 |$, for $t \geq t_0$, whenever $| x_0 | < \delta$ and $t_0 \geq 0$.

We have discussed, so far, applications of VPF for linear integrodifferential systems. Finally we present below one result showing the boundedness of solutions of nonlinear integrodifferential equations by employing nonlinear VPF proved in Theorem 2.2.2.

Theorem 2.6.5: *Assume that*
(a)   *the assumptions of Theorems 2.2.1, 2.2.2 hold and $f(t, 0) = 0$, $g(t, s, 0) = 0$;*

(b) *there exists positive numbers α, M and ρ such that if $|x_0| \leq \rho$*

$$|R(t, s, t_0, x_0)| \leq Me^{-\alpha(t-s)},$$

$$|\Phi(t, t_0, x_0)| \leq Me^{\alpha(t-t_0)}, \ t \geq s \geq t_0.$$

(c) $|g(t, s, x)| \leq k|x|e^{-2\alpha t}$ *whenever* $|x| \leq \rho$ *and*

$$|F(t, x, (Sy)(t))| \leq \lambda(t)|x| + \int_{t_0}^{t} \beta(t, s)|y(s)| \, ds$$

where $\lambda \in C[R_+, R_+]$, $\beta \in C[R_+ \times R_+, R_+]$ and $|x|$, $|y(s)| \leq \rho$.
Then any solution $y(t, t_0, x_0)$ of (2.2.7) with $|x_0|$ small enough satisfies the estimate

$$|y(t, t_0, x_0)| \leq M|x_0|R_0(t, t_0)e^{-\alpha(t-t_0)}, t \geq t_0$$

where $R_0(t, s)$ is the solution of the matrix integrodifferential equation

$$\frac{\partial R_0(t,s)}{\partial s} + R_0(t, s)\lambda_0(s) + \int_{s}^{t} R_0(t, \sigma)\gamma(\sigma, s)d\sigma = 0,$$

$$R(t, t) = I,$$

such that $R_0(t, t_0)e^{-\alpha(t-t_0)} \leq N$, $t \geq t_0$, where $\lambda_0(s) = M\lambda(s) + \frac{2Mk}{\alpha}e^{-\alpha(t_0+s)}$ and $\gamma(t, s) = \beta(t, s)e^{\alpha(t-s)}$.

Proof: Let $x(t, t_0, x_0)$ and $y(t, t_0, y_0)$ be solutions of (2.2.1), (2.2.7) respectively, existing on $[t_0, \infty)$. Then in view of (b) and Remark 2.2.2(iii), it follows that

$$|x(t, t_0, x_0)| \leq M|x_0|e^{-\alpha(t-t_0)}, \ t \geq t_0.$$

For $t \geq t_0$, for which $|y(t)| \leq \rho$, the assumptions (b), (c) and Theorem 2.2.2, yield

$$|y(t)| \leq M|x_0|e^{-\alpha(t-t_0)} + M\int_{t_0}^{t}\lambda(s)e^{-\alpha(t-s)}|y(s)| \, ds$$

$$+ \int_{t_0}^{t} Me^{-\alpha(t-s)}\int_{t_0}^{s}\beta(s, \sigma)|y(\sigma)| \, d\sigma ds$$

$$+ \int_{t_0}^{t}\int_{s}^{t} 2Mke^{-\alpha(t-\sigma)}e^{-2\alpha\sigma}|y(s)| \, d\sigma ds.$$

Using Fubini's theorem and defining $v(t) = |y(t)|e^{\alpha t}$, we get

$$v(t) \leq M \mid x_0 \mid e^{\alpha t_0} + \int_{t_0}^{t} \lambda_0(\eta)v(\eta)d\eta + \int_{t_0}^{t} \int_{\eta}^{t} e^{\alpha(\sigma-\eta)}\beta(\sigma,\eta)v(\sigma d\sigma d\eta.$$

Hence

$$y(t) \leq M \mid x_0 \mid e^{-\alpha(t-t_0)}R_0(t,t_0), t \geq t_0$$

$$\leq \rho$$

whenever $\mid x_0 \mid \, < \min\{\rho, \frac{\rho}{MN}\}$. The proof is complete.

In case $\beta(t,s) \equiv 0$, then $R_0(t,t_0)e^{-\alpha(t-t_0)} \leq N$ which implies that

$$\frac{\exp\left(\int_{t_0}^{t}\lambda(s)ds\right)}{\exp[\alpha(t-t_0)]} \leq L < \infty.$$

2.5 CONTROLLABILITY

The present section investigates the controllability problem relative to linear and nonlinear integrodifferential equations. The application of linear VPF given in Theorem 2.1.1, and use of Schauder fixed point theorem provides the desired control function. We shall use the same notation for a norm even in a Banach space. We consider the system represented by a linear integrodifferential equation of the type

$$x'(t) = A(t)x(t) + \int_{t_0}^{t} k(t,s)x(s)ds + B(t)u(t), t \in J = [t_0, t_1], \quad (2.5.1)$$

$$x(t_0) = x_0,$$

where A and B are $n \times n$ and $n \times m$ continuous matrices on J, $k(t,s)$ is an $n \times n$ continuous matrix on $J \times J$, $x \in R^n$, $u \in R^m$, $x_0 \in R^n$.

The Definition 1.10.1 of complete controllability as given before can be extended for the equation (2.5.1) with minor modifications.

Recalling Theorem 2.1.1, we know that solution $x(t)$ of (2.5.1) is given by

$$x(t) = R(t,t_0)x_0 + \int_{t_0}^{t} R(t,s)B(s)u(s)ds, x(t_0) = x_0 \quad (2.5.2)$$

where $R(t,s)$ is as defined in (2.1.5). We have the following result which is an analog of Theorem 1.10.1. We state it without proof.

Theorem 2.5.1: *Consider the control problem (2.5.1) whose solution $x(t)$ is given by (2.5.2). The system (2.5.1) is completely controllable if and only if the controllability matrix*

$$\Phi(t_0, t_1) = \int_{t_0}^{t_1} R(t_0, \tau) B(\tau) B^T(\tau) R^T(t_0, \tau) d\tau \qquad (2.5.3)$$

is nonsingular, where $R(t, s)$ is the resolvent matrix. In this case, one of the controls which steers the state (2.5.2) to a preassigned x_1 at time t_1 is given by

$$u(t) = -B^T(t) R^T(t_0, t) \Phi^{-1}(t_0, t_1)[x_0 - \Phi(t_0, t_1) x_1]. \qquad (2.5.4)$$

Before we proceed further, let us have an illustrative example.

Example 2.5.1: Consider the integrodifferential system

$$x' = \begin{bmatrix} -1 & 0 & 0 \\ 0 & -1 & 0 \\ 0 & 0 & -1 \end{bmatrix} x + \int_{t_0}^{t} \begin{bmatrix} e^{t-s} & 0 & 0 \\ 0 & e^{t-s} & 0 \\ 0 & 0 & e^{t-s} \end{bmatrix} x(s) ds$$

$$+ \begin{bmatrix} 4 & 3 \\ 2 & 1 \\ -2 & 0 \end{bmatrix} u(t), \quad x(t_0) = x_0. \qquad (2.5.5)$$

The resolvent kernel of (2.5.5) is given by

$$R(t, s) = \begin{bmatrix} \alpha & 0 & 0 \\ 0 & \alpha & 0 \\ 0 & 0 & \alpha \end{bmatrix}, \alpha = 1 + \tfrac{1}{2} e^{-t} - \tfrac{1}{2} e^{t-2s}.$$

The controllability matrix given by (2.5.3) is

$$\Phi(t_1, t_0) = \int_{t_0}^{t_1} \beta^2 \begin{bmatrix} 25 & 11 & -8 \\ 7 & 5 & -4 \\ -8 & -4 & -4 \end{bmatrix} d\tau$$

where $\beta = 1 + \tfrac{1}{2} e^{-t_0} - \tfrac{1}{2} e^{t_0 - 2\tau}$.

Here $|\Phi(t_1, t_0)| \neq 0$. This is true for all t_0, t_1. Thus the system is controllable. It is important to observe that the system of differential equations given by

$$x' = \begin{bmatrix} -1 & 0 & 0 \\ 0 & -1 & 0 \\ 0 & 0 & -1 \end{bmatrix} x + \begin{bmatrix} 4 & 3 \\ 2 & 1 \\ -2 & 0 \end{bmatrix} u, \quad x(t_0) = x_0$$

is not controllable.

Let us now consider the control process described by a nonlinear integrodifferential equation

$$x' = A(t, x, u) x + \int_{t_0}^{t} k(t, s, x(s), u) x(s) ds + B(t, x, u) u + f(t, x, u),$$

$$x(t_0) = x_0 \text{ and } t \in J, \tag{2.5.6}$$

where $x, f \in R^n$, $u \in R^m$, $A(t, x, u), k(t, s, x, u)$ are $n \times n$ matrices and $B(t, x, u)$ is an $n \times m$ matrix. The functions $A(t, x, u)$, $B(t, x, u)$, $k(t, s, x, u)$ are all continuous with respect to their arguments.

Our aim is to provide sufficient conditions to ensure that the system state can be transferred from $x(t_0) = x_0$ to x_1 in the allotted time $t_1 - t_0$.

Since the control problem (2.5.6) is nonlinear, we first consider the related control problem represented by the linear integrodifferential equation

$$x'(t) = A(t, y(t), v(t))x(t) + \int_{t_0}^{t} k(t, s, y(s), v(s))x(s)ds$$

$$+ B(t, y(t), v(t))u(t) + f(t, y(t), v(t)), \tag{2.5.7}$$

$$x(t_0) = x_0,$$

where $y(t)$, $v(t)$ are continuous functions of appropriate dimensions. Observe that $A(t, y(t), v(t))$, $k(t, s, y(t), v(t))$, $B(t, y(t), v(t))$ and $f(t, y(t), v(t))$ are functions of time.

In view of (2.1.5), (2.5.2), the solution of (2.5.7) is given by

$$x(t) = R(t, t_0, y(t), v(t)) + \int_{t_0}^{t} R(t, s, y(s), v(s))B(s, y(s), v(s))u(s)ds$$

$$+ \int_{t_0}^{t} R(t, s, y(s), v(s))f(s, y(s), v(s))ds,$$

$$R(t, t, y(t), v(t)) = I \tag{2.5.8}$$

where $R(t, t_0, y(t), v(t))$ is the corresponding resolvent function.

Theorem 2.5.2: *Consider the control problem (2.5.7) whose solution $x(t)$ is given by (2.5.8). The system (2.5.7) is completely controllable if and only if the controllability matrix*

$$\Phi(t_0, t, y(t), v(t)) = \int_{t_0}^{t_1} R(t_0, \tau, y(\tau), v(\tau))B(y(\tau), v(\tau), \tau)$$

$$\times B^T(y(\tau), v(\tau), \tau)R^T(t_0, \tau, y(\tau), v(\tau))d\tau \tag{2.5.9}$$

is nonsingular, where $R(t, s, y(t), v(t))$ is the resolvent of (2.5.7). In this case, one of the controls which steers the state (2.5.8) to a preassigned x_1 at time t_1 is given by

$$u(t, t_0, x_0, t_1, x_1, y(t), v(t))$$

$$= -B^T(y(t), v(t), t)R^T(t_0, t, y(t), v(t))\Phi^{-1}(t_0, t_1, y(t), v(t))$$

$$\times \left[x_0 - R^{-1}(t_1, t_0, y(t), v(t))x_1 + \int_{t_0}^{t} R(t_0, \tau, y(\tau), v(\tau))f(\tau, y(\tau), v(\tau))d\tau \right] \quad (2.5.10)$$

Proof: The proof is similar to Theorem 1.10.1 and hence omitted.

If the vectors y, v are chosen to agree with x, u, then these vectors are also solutions for system (2.5.6), and the controllability of system (2.5.6) is guaranteed. Hence this is an existence problem of a fixed point for (2.5.8) and (2.5.10). Notice that if there exist a set of fixed points for (2.5.8) and (2.5.10), this solution is also a fixed point for (2.5.10) and the relation (2.5.11), given by

$$x(t) = R(t, t_0, y(t), v(t)) \left[x_0 + \int_{t_0}^{t} R(t_0, s, y(s), v(s))B(s, y(s), v(s))v(s)ds \right.$$

$$\left. + \int_{t_0}^{t} R(t_0, s, y(s), v(s))f(s, y(s), v(s))ds \right]$$

where $R(t, t, y(t), v(t)) = I$ and

$$R(t, s, y(t), v(t)) = R(t, t_0, y(t), v(t))R(t_0, s, y(t), v(t)). \quad (2.5.11)$$

Let $C^{n+m}(J)$ denote the Banach space of $(n+m)$-dimensional continuous functions on J. Consider

$$G: C^{n+m}(J) \to C^{n+m}(J)$$

defined by $G(y(t), v(t)) = (x(t), u(t))$.

The operator G is clearly continuous on $C^{n+m}(J)$, and if there exists a closed bounded convex set Ω of $C^{n+m}(J)$ such that $(x, u) = G(y, v) \in \Omega$ for any $(y, v) \in \Omega$, then from (2.5.10) and (2.5.11), $G(\Omega)$ is bounded. Hence by Schauder's fixed point theorem, we can conclude that there exist at least one fixed point of G. Hence we have the following theorem.

Theorem 2.5.3: *If there exists a closed bounded convex subset Ω of $C^{n+m}(J)$ such that the operator G defined above is invariant for Ω, then the control system (2.5.6) satisfying condition (2.5.9) is completely controllable on J.*

Define

$$\Omega = \{(y, v) \in \Omega, \ |y(t)| \le \rho(t), \ |v(t)| \le \gamma(t)\}.$$

We have the following result for nonlinear systems.

Theorem 2.5.4: *Assume that*
(i) $A(t, x, u)| \le M_1; M_1 > 0,$
(ii) $|B(t, x, u)| \le M_2; M_2 > 0,$
(iii) $|f(t, x, u)| \le \psi(t, |x|, |u|); \psi(t, \rho, \gamma)$ *is a continuous function with*

respect to its arguments and nondecreasing for any $\rho > 0$, $\gamma > 0$;

(iv) $|R(t, t_0, y, v)| \leq \exp\left(\int_{t_0}^{t} |A(s, y, v)| \, ds \leq \exp(M_1(t_1 - t_0))\right) = M_3 > 0$;

(v) $|R(t_0, x, y, v)| \, |B(s, y, v)| \leq \exp(M_1(t_1 - t_0))M_2$; $t_0 \leq s \leq t_1$;

(vi) $|B^T(t_0, t, y, v)R^T(t_0, t, y, v)\Phi^{-1}(t_0, t, y, v)\{R^{-1}(t_1, t_0, y(v)) - x_0\}|$
$\leq N_1 > 0$;

(vii) $|B^T(t_0, t, y, v)R^T(t_0, t, y, v)\Phi^{-1}(t_0, t, y, v)| \, |R(t_0, \tau, y, v)| \leq N_2$,
$t_0 \leq \tau \leq t_1$;

$(viii)$ $\exp(2M_1(t_1 - t_0)) = M_5$; $\exp(2M_1(t_1 - t_0))M_2 = M_4$;

(ix) $M_3 + M_4 \int_{t_0}^{t} \gamma(s)ds + Ms \int_{t_0}^{t} \psi(s, \rho(s), \gamma(s))ds \leq \rho(t)$; $\rho(t) > 0$,

(x) $N_1 + N_2 \int_{t_0}^{t} \psi(s, \rho(s), \gamma(s))ds \leq \gamma(s)$; $\gamma(s) > 0$.

Then the control system (2.5.6) is completely controllable on J.

Proof: By (i)-(iii) and (2.5.11), we have

$$|x(t)| \leq |R(t, t_0, y, v)|\{|x_0| + \int_{t_0}^{t} |R(t_0, s, y, v)| \, |B(s, y, v)| \, |v(s)| \, ds$$

$$+ \int_{t_0}^{t} |R(t_0, s, y, v)| \, |f(s, y(s), v(s))| \, ds\}.$$

Now by (iii), (iv), (v) and $(viii)$, we have

$$|x(t)| \leq \exp(M_1(t_1 - t_0))\{|x_0| + \exp(M_1(t_1 - t_0))M_2 \int_{t_0}^{t} |v(s)| \, ds$$

$$+ \exp(M_1(t_1 - t_0))\int_{t_0}^{t} \psi(s, |y(s)|, |v(s)|)ds\}$$

$$\leq M_3 + M_4 \int_{t_0}^{t} \gamma(s)ds + Ms \int_{t_0}^{t} \psi(s, \rho(s), \gamma(s))ds; \text{ on } \Omega. \qquad (2.5.12)$$

From (2.5.10) using (vi) and (vii), we have

$$|u(t)| \leq N_1 + N_2 \int_{t_0}^{t} \psi(s, \rho(s), \gamma(s))ds.$$

By using (ix), (x) and Theorem 2.5.3, the system (2.5.6) is completely controllable on J. The proof is complete.

Corollary 2.5.1: *If f in (2.5.6) does not contain u by making the following changes in the conditions of Theorem 2.5.4,*
$(iii)^*$ $\quad |f(t, x, u)| \leq \psi(t, |x|)$;

$(ix)^*$ $M_3 + M_4 \int\limits_{t_0}^{t} \gamma(s)ds + Ms\int\limits_{t_0}^{t} \psi(s,\ |\ \rho(s)\ |\ )ds \le \rho(t);$

$(x)^*$ $N_1 + N_2 \int\limits_{t_0}^{t_1} \psi(s,\rho(s))ds \le \gamma(s);$

we have the same conclusions as in Theorem 2.5.4.

Corollary 2.5.2: *Let f in (2.5.6) does not contain u and condition (2.5.9) and (i), (ii) of Theorem 2.5.4, and $(iii)^*$ of Corollary 2.5.1 be satisfied and in addition*

(xi) $N_3 + N_4 \int\limits_{t_0}^{t_1} \psi(s,\rho(s))ds + Ms\psi(s,\rho(s)) \le \rho(t),$ *where $N_3 > 0$, $N_3 =$*

       $M_4 N_1$, $N_4 = M_4 N_2;$

(xii) $M_3 \le \rho(t_0)$, $M_3 > 0;$

hold, then if, there exist a nonnegative solution $\rho(t)$ for (xi) and (xii), the system (2.5.6) is completely controllable on J.

Proof: Differentiating both sides on $(ix)^*$ with respect to t and substituting $(x)^*$ implies (xi) and (xii) which completes the proof.

Corollary 2.5.3: *If f in (2.5.6) does not contain x, by making the following changes in the conditions of Theorem 2.5.4*

$(iii)^{**}$ $|\ f(u,t)\ |\ \psi(t,\ |\ u\ |\ );$

$(ix)^{**}$ $M_3 + M_4 \int\limits_{t_0}^{t}\gamma(s)ds + M_5 \int\limits_{t_0}^{t} \psi\Big(s,\gamma(s)\Big)ds \le \rho(t);$

$(x)^{**}$ $N_1 + N_2 \int\limits_{t_0}^{t} \psi(s,\gamma(s))ds \le \gamma(s).$

We have the same conclusion as Theorem 2.5.4.

Noting that if $(x)^{**}$ has a solution $\gamma(s)$, then we can always find a $\rho(t)$ satisfying $(ix)^{**}$ and hence we have the following corollary.

Corollary 2.5.4: *If f in (2.5.6) does not contain x, condition (2.5.9) and (i), (ii) of Theorem 2.5.4 and $(iii)^*$ of Corollary 2.5.1 satisfied in addition. If $(x)^*$ of Corollary 2.5.1 has at least one nonnegative solution $\gamma(s)$ for $N_1 > 0$ then the system (2.5.6) is completely controllable on J.*

2.6 NOTES

Theorem 2.1.1 is taken from Lakshmikantham and Rao [1], which is originally given in Miller [1]. For Theorems 2.1.2 and 2.1.3, see Lakshmikantham and Rao [1], while Theorems 2.1.4 and 2.1.5 are new. See Miller [2] and Corduneanu [2] for VPF for linear Volterra integral equations and applications.

The nonlinear VPF for integrodifferential equations is due to Hu, Lakshmikantham, Rao [1] which is included in Section 2.2. Also refer to earlier results in this direction to Brauer [3], Bernfeld and Lord [1] and Beesack [1].

The contents of Section 2.3 are new. Similar result using monotone technique in Banach space appear in Guo [1]. For general theory of monotone iterative technique refer to the monograph by Ladde, Lakshmikantham and Vatsala [1]. See also Uvah and Vatsala [1], Lakshmikantham and Zhang [1] for recent results.

The stability results of Section 2.4 have been adapted from Lakshmikantham and Rao [1]. For several other results on stability properties of integro-differential equations, refer to Miller [1], Levin and Nohel [1], Nohel [1], Grossman and Miller [2], Rao and Srinivas [1] and Zhang and Muqui [1].

The application to controllability of nonlinear integro-differential equations included in Section 2.5 is new. It extends the results of Yomomoto [1]. See also Balchandran and Lalitha [1], Dauer [1] and Kartsatos [1].

3. Differential Equations with Delay

3.0 INTRODUCTION

The future state of the physical system depends on many circumstances, not only on the presents state but also on its past history. The simplest type of such dynamic systems are called difference differential equations. More general differential equations with delay are also called functional differential equations (FDE) and we shall use this nomenclature as well.

The differential equations with delay occur in a variety of forms and as a result several representations of the variation of parameters formulae are available. In order to avoid monotony, we shall give the VPF for linear as well as nonlinear case, for neutral functional differential equations which include as special cases, various types of such equations. The present chapter also includes applications of VPF to stability analysis and asymptotic behavior.

In Section 3.1, we prove VPF for linear neutral FDE and point out few special cases of this general form. In order to prove the nonlinear VPF, we need to establish that under appropriate conditions, solutions of nonlinear neutral FDE depend continuously on the initial conditions and that they are differentiable with respect to initial data. We obtain such results in Section 3.2 and then develop VPF for nonlinear case.

By using Lyapunov-like functions as well as functions, one may generalize the results of VPFs which is achieved in Section 3.3.

Among several applications available in the literature, we chose the study of asymptotically invariant sets in Section 3.4, while Section 3.5 includes the results on asymptotic equivalence property for neutral FDEs. The VPF in terms of Lyapunov-like functions and functions which provides the generalized comparison principle, is developed in Section 3.6. We offer simple examples to illustrate the ideas involved whenever necessary.

3.1 LINEAR VARIATION OF PARAMETERS

Linear differential equations with delay occur in various forms in applications. Corresponding to each type, it is possible to develop the variation of parameters formula and study the qualitative properties of solutions of such equations.

The simplest type of linear differential system with delay in which rate of change of the system depends on the influence of its hereditary effect is given by

$$x'(t) = A(x(t) + Bx(t - r)). \tag{3.1.1}$$

The related perturbed system is of the form

$$y'(t) = Ay(t) + By(t - r) + h(t), \tag{3.1.2}$$

where A and B are $n \times n$ constant matrices, $t \in R_+$, $x, y \in R^n$ and $h \in C[R_+, R^n]$. Such equations are called difference-differential equations. If $r = 0$, then (3.1.1) reduces to ordinary differential equation without delay. If $r > 0$, then (3.1.1) is a retarded difference-differential equation and if $r < 0$, then it is called advanced differential-difference equation.

In certain situations, the rate of change of the system depends on the present state as well as the rate of change of the past state. The linear systems involving such phenomena appear in the form

$$[x(t) - x(t-r)]' = Ax(t) + Bx(t-r), \tag{3.1.3}$$

and the related perturbed system is given by

$$[y(t) - y(t-r)]' = Ay(t) + By(t-r) + h(t). \tag{3.1.4}$$

The equations of this type are known as neutral difference-differential equations.

In the foregoing equations the matrices A and B may depend on t. In that case, the matrices $A(t)$ and $B(t)$ need to be continuous on R_+. Moreover, in place of one delay term in (3.1.1) or (3.1.3) one may consider linear equations in which n delay terms occur. The system then takes the form

$$x'(t) = A(t)x(t) + \sum_{i=1}^{n} B_i(t)x(t-r_i).$$

In some situations, the constant delay term r may be replaced by a function $r(t)$ appropriately defined so that difference-differential equation makes sense.

We now define an initial value problem (IVP) for (3.1.1). Given an initial function $\phi_0(t)$ defined on $[t_0 - r, t_0]$, $t_0 \in R_+$, a function $x(t)$ is said to be a solution of (3.1.1), if it is defined and continuous on $[t_0 - r, t_0 + A]$, $A > 0$, such that $x(t) = \phi_0(t)$ on $[t_0 - r, t_0]$, $x(t)$ is differentiable on $[t_0, t_0 + A]$ and satisfies (3.1.1) on $[t_0, t_0 + A]$.

More general linear systems then (3.1.1) may be described by the equations

$$x'(t) = L(t, x_t), \tag{3.1.5}$$

$$y'(t) = L(t, y_t) + h(t), \tag{3.1.6}$$

where x_t is defined as follows: if x is a function defined on some interval $[t_0 - r, t_0 + A]$, $A > 0$, $t_0 \in R_+$, and $r > 0$ is the delay term, then for each $t \in [t_0, t_0 + A]$

(i) x_t is the graph of x on $[t - r, t]$ shifted to the interval $[-r, 0]$, i.e.,
      $x_t(s) = x(t+s), t - r \leq s \leq t, t \geq t_0$.
(ii) x_t is the graph of x on $[t_0 - r, t]$.

In the case of (ii), L is a functional of Volterra type which is determined by t and the values of $x(s)$, $t_0 - r \leq s \leq t$. The equations (3.1.5), (3.1.6) are known as functional differential equations (FDEs). We shall consider FDE's in which x_t has the meaning described by (i). In such equations, we can have bounded delays or unbounded delays. In what follows, we restrict our discussion to bounded delays only.

In what follows, assume that $(t_0, \phi) \in R_+ \times C$ where $C = C[[-r, 0], R^n]$. Let the initial condition for the equations (3.1.5) or (3.1.6) be given by $x_{t_0} = \phi$. For $\phi \in C$, assume that $|\phi|_0 = \sup\limits_{s \in [-r, 0]} |\phi(s)|$.

Assume that the operator $L(t, \phi)$ in (3.1.6) be linear in ϕ and that $h \in \mathcal{L}_1^{loc}([t_0, \infty), R^n)$, the space of functions mapping $[t_0, \infty) \to R^n$ which are Lebesgue integrable on each compact set of $[t_0, \infty)$.

It is clear that the system (3.1.3) is a natural generalization of the system (3.1.1). In a similar way, we can consider the generalization of the system (3.1.5) to neutral FDE in the form

$$\frac{d}{dt} D(x, x_t) = L(t, x_t),$$ (3.1.7)

and its perturbation

$$\frac{d}{dt} D(t, y_t) = L(t, y_t) + h(t),$$ (3.1.8)

where $D \in C[R_+ \times C, R^n]$. In order to guarantee that the coefficient of $x(t)$ occurring in $D(t, x_t)$ is not zero, we introduce the following definition.

Definition 3.1.1: Let $D \in C[R_+ \times C, R^n]$ and $D(t, \phi)$ be linear in ϕ. Then the Rietz Representation Theorem shows that there exists an $n \times n$ matrix function $\mu(t, \theta)$ of bounded variation in θ for each t, such that $\mu(t, \theta) = 0$ for $\theta \geq 0$, $\mu(t, \theta) = \mu(t, -r)$ for $\theta \leq -r$, $\mu(t, \theta)$ is continuous from the left on $(-r, 0)$ and

$$D(t, \phi) = \int\limits_{-r}^{0} [d\theta \mu(t, \theta)] \phi(\theta), t \geq 0.$$

Further, there exists $m \in \mathcal{L}_1^{loc}((-\infty, \infty), R)$ such that

$$|D(t, \phi)| \leq m(t) |\phi|,$$

for $t \in (-\infty, \infty)$, $\phi \in C$. Clearly $\operatorname*{var}\limits_{[-r, 0]} \mu(t, \theta) \leq m(t)$.

For $\theta_0 \in C$, if

$$\det[\mu(t_0, \theta_0^+) - \mu(t_0, \theta_0^-)] \neq 0$$

then we say that $D(t, \phi)$ is atomic at θ_0 for t_0. If for every $t \geq 0$, $D(t, \phi)$ is atomic at θ_0 for t, then we say that $D(t, \phi)$ is atomic at θ_0 for R_+.

As a linear example, consider

$$D(t, \phi) = a(t)\phi(0) + b(t)\phi(-r),$$

where $a(t)$ and $b(t)$ are $n \times n$ continuous matrices for $t \in R_+$. Suppose that $\det[a(t)] \neq 0$ for $t \in R_+$. Define

$$\mu(t,\theta) = \begin{cases} a(t), & \text{if } \theta = 0 \\ 0, & \text{if } -r < \theta < 0 \\ b(t), & \text{if } \theta = r. \end{cases}$$

Then

$$D(t,\phi) = \int_{-r}^{0} d_\theta\mu(t,\theta)]\phi(\theta).$$

Here $\det[\mu(t,0^+) - \mu(t,0^-)] = \det[a(t)] \neq 0$, $t \in R_+$. Hence $D(t,\phi)$ is atomic at $\theta = 0$, $t \in R_+$.

In the equation (3.1.7) we have assumed that $L(t,\phi)$ is a linear functional in ϕ. Hence, following the foregoing discussion, there exists an $n \times n$ matrix function $\eta(t,\theta)$ such that

$$L(t,\phi) = \int_{-r}^{0} [d_\theta\eta(t,\theta)]\phi(\theta), \quad t \in R_+, \phi \in C.$$

Furthermore, there exists a function, $m \in \mathcal{L}_1^{loc}((-\infty,\infty), R)$ such that

$$|L(t,\phi)| \leq m(t)|\phi|, \quad t \in R, \phi \in C.$$

Hence the equation (3.1.7) takes the form

$$\frac{d}{dt}[D(t,x_t)] = \int_{-r}^{0} [d_\theta\eta(t,\theta)]x_t(\theta). \tag{3.1.9}$$

At this stage, we observe that

(*i*)　　the equation (3.1.1) is a special case of (3.1.9), if we assume that $D(t,\phi) = \phi(0)$ and

$$\eta(t,\theta) = \begin{cases} -A(t) - B(t), & \theta \leq -r \\ -A(t), & -r < \theta < 0 \\ 0, & \theta = 0; \end{cases}$$

(*ii*)　　the equation (3.1.3) is a special case of (3.1.8) if

$$D(t,\phi) = \phi(0) - \phi(-r)$$

and that $\eta(t,\theta)$ in this case is the same as in (*i*) above;

(*iii*)　　the equation (3.1.5) is a special case of (3.1.9) if

$$D(t,\phi) = \phi(0).$$

The above discussion leads to the fact that the equation (3.1.7) or equivalently equation (3.1.9) represents a general form of FDE. Our main aim in this section is to develop VPF for linear FDEs. Hence it is enough to obtain such a result for the equation (3.1.8) so that it will cover all particular cases cited above. Further, it will serve as a model for several other FDEs which occur in applications. Hence we shall concentrate our attention on neutral FDEs (3.1.7) and (3.1.8).

We are now concerned with the systems

$$\frac{d}{dt}D(x_t) = L(t, x_t), \, x_{t_0} = \phi_0, \tag{3.1.10}$$

and

$$\frac{d}{dt}D(y_t) = L(t, y_t) + h(t), \, y_{t_0} = \phi_0, \tag{3.1.11}$$

where $D \in C[\mathcal{C}, R^n]$, $L \in C[R_+ \times \mathcal{C}, R^n]$, $h \in \mathcal{L}_1^{loc}[[t_0, \infty), R^n]$, $t_0 \in R_+$ and

$$\left.\begin{aligned} D(\phi) &= \phi(0) - \int_{-r}^{0} [d\mu(\theta)]\phi(\theta) \\ L(t, \phi) &= \int_{-r}^{0} [d\eta(\theta)]\phi(\theta). \end{aligned}\right\} \tag{3.1.12}$$

Here $\phi \in \mathcal{C}$ and $t \geq t_0$, the matrices μ and η being defined above. The system (3.1.11) is equivalent to

$$D(y_t) = \phi_0(0) + \int_{t_0}^{t} L(s, y_s)ds + \int_{t_0}^{t} h(s)ds, t \geq t_0$$

$$x_{t_0} = \phi_0;$$

or equivalently,

$$D(y_t) = \phi_0(t_0) + \int_{t_0}^{t}\left[\int_{-r}^{0} d_\theta\eta(s, \theta)y_s(0)\right]ds + \int_{t_0}^{t} h(s)ds, \tag{3.1.13}$$

$$x_{t_0} = \phi_0.$$

We assume that under the given hypotheses, there exists a unique function $y_t(t_0, \phi_0)$ defined and continuous for $t \geq t_0$ which satisfies the system (3.1.11) and that it is pointwise bounded for $t \geq t_0$. We need the following lemma.

Lemma 3.1.1: *Let $L_1([t_0, \infty), R^n)$ denote the space of equivalence classes in $\mathcal{L}_1([t_0, \infty), R^n)$. Suppose $T: L_1([t_0, \infty), R^n) \to R^n$ is a continuous linear operator. Then there exists a unique $n \times n$ matrix function $v(\theta)$, $t_0 \leq \theta < \infty$, which is integrable and essentially bounded such that*

$$Th = \int_{t_0}^{\infty} v\theta)h(\theta)d\theta$$

where $h \in \mathcal{L}_1^{loc}([t_0, \infty), R^n)$.

We now prove the main result of this section.

Theorem 3.1.1: *Let $h \in \mathcal{L}_1^{loc}([t_0, \infty), R^n)$, L satisfy the hypothesis given above and $y_t(t_0, \phi_0)$ be the solution of the system (3.1.11). Then*

$$y(t_0, \phi_0)(t) = x(t_0, \phi_0)(t) + \int_{t_0}^{t} U(t, s) h(s) ds, t \geq t_0$$

$$x_{t_0} = \phi_0, \qquad\qquad (3.1.14)$$

where $U(t, s)$ verifies the equation

$$D(U(t, s)) = \begin{cases} \int_{0}^{t} L(u, U_u(\,\cdot\,, s)) du + I, & a.e. \ in \ s, \ t \geq s \\ 0 & s - r \leq t < s, \end{cases} \qquad (3.1.15)$$

where $U_t(\,\cdot\,, s) = U(t + \theta, s)$, $-r \leq \theta \leq 0$. The function $U(t, s)$ is called the fundamental matrix.

Proof: Under the given hypotheses and Lemma 3.1.1, we claim that there exists a unique matrix function $v(t, t_0, \,\cdot\,) \in L_\infty([t_0, t], R^{n^2})$ $t \geq t_0$, which is independent of t_0 such that

$$y_t(t_0, 0) = \int_{t_0}^{t} v(t, t_0, s) h(s) ds, t \geq t_0. \qquad (3.1.16)$$

Define

$$U(t, s) = \begin{cases} v(t, t_0, s), & \text{for } s \leq t \\ 0, & \text{for } s - r \leq t < s. \end{cases}$$

To establish (3.1.14), it is enough to prove that $\int_{t_0}^{t} U(t, s) h(s) ds$ is a particular integral of (3.1.11).

From (3.1.13) and using Fubini's theorem, we obtain

$$\int_{t_0}^{t} D(v_t(t, s)) h(s) ds = \int_{t_0}^{t} \left(\int_{-r}^{0} [d_\theta \eta(s, \theta)] \int_{t_0}^{s+\theta} V(s + \theta, u) h(u) du \right) ds + \int_{t_0}^{t} h(s) ds$$

$$= \int_{t_0}^{t} \left(\int_{-r}^{0} [d_\theta \eta(s, \theta)] \int_{t_0}^{s} V(s + \theta, u) h(u) du \right) ds + \int_{t_0}^{t} h(s) ds$$

$$= \int_{t_0}^{t} \left[\int_{u}^{t} \left(\int_{-r}^{0} [d_\theta \eta(s, \theta)] V(s + \theta, u) ds \right) \right] h(u) du + \int_{t_0}^{t} h(s) ds$$

$$= \int_{t_0}^{t} \left(\int_{s}^{t} \left\{ \int_{-r}^{0} d_\theta \eta(u, \theta) u(u + \theta, s) \right\} du \right) h(s) ds + \int_{t_0}^{t} h(s) ds$$

$$= \int_{t_0}^{t} \left[\int_{s}^{t} \left\{ \int_{-r}^{0} d_\theta \eta(u, \theta) u(u + \theta, s) \right\} du + I \right] h(s) ds.$$

Hence, it follows that

$$D(V_t(t, s)) = \int_{s}^{t} \left(\int_{-r}^{0} [d_\theta \eta(u, \theta)] u(u + \theta, s) \right) du + I.$$

Consequently, $U(t, s)$ verifies the relation (3.1.15). One can take (3.1.15) as the defining relation for $U(t, s)$ for $t \geq s$ with $U(t, s) = 0$ for $t < s$. Thus the VPF (3.1.14) for the neutral FDE (3.1.11) is proved.

Corollary 3.1.1: *It $L(t, \phi) \equiv L(\phi)$ is independent of t in system* (3.1.11), *then*

$$U(t, s) = V(t - s, 0) = V(t - s); \ (say)$$

and hence VPF (3.1.14) *takes the form*

$$y(t_0, \phi_0)(t) = x(t_0, \phi_0)(t) + \int_{t_0}^{t} V(t - s) h(s) ds$$

$$x_{t_0} = \phi_0.$$

Remarks 3.1.1: (i) The integral equation (3.1.15) is equivalent to

$$\frac{\partial}{\partial t} [D(U(t, s)] = L(t, V_t(\cdot, s)), \ t \geq s$$

$$U(t, s) = \begin{cases} 0, & s - r \leq t \leq s; \\ I, & t = s. \end{cases}$$

In the above discussion, we have defined the matrix $U(t, s)$ as fundamental matrix. It is important to note the initial data for such a matrix.
(ii) The VPF (3.1.14) may also be written as

$$y_t(t_0, \phi_0) = x_t(t_0, \phi_0) + \int_{t_0}^{t} V_t(\cdot, s) h(s) ds, \ t \geq t_0.$$

It is to be understood that the above integral is in Euclidean space and the equality here is for each θ, $-r \leq \theta \leq 0$.
(iii) At this stage, define an operator T such that

$$T(t, t_0) \phi = x_t(t_0, \phi).$$

Then T is linear and continuous. Further, since $U(t,s)$ has a continuous first order derivative in t for $t > s + r$, we write

$$V_t(\,\cdot\,,s) = T(t,s)x_{t_0},$$

$$x_{t_0}(\theta) = \begin{cases} 0, & -r \le \theta < 0; \\ I, & \theta = 0. \end{cases}$$

With this notation, the VPF (3.1.14) takes the form

$$y_t(t_0,\phi_0) = T(t,t_0)\phi_0 + \int_{t_0}^{t} T(t,s)x_{t_0}h(s)ds, t \ge t_0 \qquad (3.1.17)$$

where the integral (3.1.17) is interpreted as

$$y_t(t_0,\phi_0)(\theta) = T(t,t_0)\phi_0(\theta) + \int_{t_0}^{t} T(t,s)x_{t_0}(\theta)h(s)ds \qquad (3.1.18)$$

for $t \ge t_0$ and $-r \le \theta \le 0$.

(*iv*) We have already pointed out that the equation (3.1.11) is quite general and it includes, as special cases, equation (3.1.2), (3.1.4) and (3.1.6). Hence (3.1.14) also works as a VPF for all these special cases after introducing appropriate modifications.

At this stage, we observe that the equation (3.1.2) is constantly in use in applications. The VPF for this equation has been obtained by following different techniques, like Laplace transforms and adjoint equation method. We provide below an independent proof of the VPF for the equation (3.1.2) by adjoint equation method. Assume in (3.1.2), $y_{t_0} = \phi_0$ for initial condition.

The formal adjoint system for (3.1.1) is given by

$$z'(s) = -A^T(s)y^T(s) - B^T(s+r)y^T(s+r). \qquad (3.1.19)$$

Let $z(s,t)$, an $n \times n$ continuous matrix function, be such that

$$z(t,t) = \begin{cases} I, \\ 0, & t < s \le t + r \end{cases} \qquad (3.1.20)$$

and as a function of s, it verifies the adjoint system (3.1.19), yielding

$$\frac{\partial z}{\partial s}(s,t) = -z(s,t)A(s) - z(s+r,t)B(s+r), s \le t. \qquad (3.1.21)$$

Multiply by $z(s,t)$ on either side of (3.1.2) to get

$$z(s,t)y'(s) = z(s,t)A(s)y(s) + z(s,t)B(s)y(s-r) + z(s,t)h(s).$$

Integration between t_0 and t, gives

$$z(t,t)y(t) - z(t_0,t)\phi_0(t_0) - \int\limits_{t_0}^{t} \frac{dz(s,t)}{ds} y(s)ds$$

$$= \int\limits_{t_0}^{t} z(s,t)A(s)y(s)ds + \int\limits_{t_0-r}^{t} z(s+r,t)B(s+r)\phi_0(s)ds$$

$$+ \int\limits_{t_0}^{t} z(s+r,t)B(s+r)y(s)ds$$

$$- \int\limits_{t-r}^{t} z(s+r,t)B(s+r)y(s)ds + \int\limits_{t_0}^{t} z(s,t)h(s)ds.$$

We now make use of the relations (3.1.20) and (3.1.21) and finally get

$$y(t) = z(t_0,t)\phi_0(t_0) + \int\limits_{t_0-r}^{t_0} z(s+r,t)B(s+r)\phi_0(s)ds$$

$$+ \int\limits_{t_0}^{t} z(s,t)h(s)ds. \tag{3.1.22}$$

The foregoing discussion leads to the following result.

Theorem 3.1.2: *Consider the difference-differential equation* (3.1.2) *where A and B are $n \times n$ continuous matrices defined on R_+ and $h \in C[R_+, R^n]$. Let $z(s,t)$ be an $n \times n$ continuous matrix function satisfying* (3.1.20) *which is a fundamental solution of* (3.1.19). *Then the solution of* (3.1.2) *is given by* (3.1.22).

3.2 NONLINEAR VARIATION OF PARAMETERS

This section deals with the development of VPF for nonlinear functional differential equations. As pointed out in the previous section, it is convenient to consider below nonlinear FDE of neutral type so as to cover several special cases. In order to obtain such a form one needs, as a prerequisite, to show that the solution of the unperturbed system is unique and that it depends continuously on initial data as well as, differentiable with respect to initial conditions. We shall therefore discuss such results for neutral FDE before proceeding further.

We are concerned with the systems

$$\frac{d}{dt}(Dx_t) = f(t, x_t), \quad x_\sigma = \phi, \, t \geq \sigma \tag{3.2.1}$$

and

$$\frac{d}{dt}(Dy_t) = f(t, y_t) + g(t, y_t), \quad y_\sigma = \phi, \, t \geq \sigma \tag{3.2.2}$$

where $D \in C[\mathcal{C}, R^n]$ is a continuous linear operator defined in (3.1.12), $f, g \in C[\Omega, R^n]$ $\Omega \subset R_+ \times \mathcal{C}$, Ω open, f being nonlinear with first Frechet derivative with respect to $\phi \in \mathcal{C}$.

For every solution $x_t(\sigma, \phi)$ of (3.2.1), we define a nonautonomous linear FDE of neutral type

$$\tfrac{d}{dt}(Dz_t) = f_x(t, x_t(s, y_s(\sigma, \phi)))z_t, \ t \ge s \ge \sigma \ge 0 \tag{3.2.3}$$

which is called the linear variational equation of (3.2.1) with respect to the solution $x_t(s, y_s(\sigma, \phi))$. Note that $z_t(\sigma, \phi)$ is the unique solution of (3.1.3) for $\phi \in \mathcal{C}$. Define, as in the previous section, the operator $T(t, \sigma) : \mathcal{C} \to \mathcal{C}, t \ge \sigma$ by the relation

$$T(t, \sigma)\phi = z_t(\sigma, \phi). \tag{3.2.4}$$

Clearly $T(t, \sigma)$ is strongly continuous for $\sigma \le t < \infty$.

We write the equation (3.2.3) in the form

$$\tfrac{d}{dt}(Dz_t) = L(t, z_t)$$

and consider the $n \times n$ matrix solution of the equation

$$D(x_t) = I + \int_\sigma^t L(s, x_s)ds, t \ge \sigma,$$

$$x_\sigma(\theta) = \begin{cases} \theta, & -r \le \theta < 0, \\ I, & \theta = 0; \end{cases} \tag{3.2.5}$$

$$x_t = T(t, \sigma)\phi, \ x_t(\theta) = [T(t, \sigma)x_\sigma](\theta), \ -r \le \theta \le 0.$$

This solution is of bounded variation on compact sets, and continuous from the right side.

Consider the difference equation

$$Dx_t = 0, \ t \ge 0$$

and let $c_D = \{\phi \in \mathcal{C}, D\theta = 0\}$. Then this equation defines a strongly continuous semigroup of linear transformations $T_D(t) : c_D \to c_D, t \ge 0$, where

$$T_D(t)\psi = x_t(\psi), \ t \ge 0, \ \psi \in c_D$$

and $x_t(\psi)$ is the solution of $Dx_t = 0$ through $(0, \psi)$ for $t \ge 0$.

We need the following definition.

Definition 3.1.1: The operator D is said to be uniformly stable if there exist constants $k \ge 1, \alpha > 0$ such that

$$|T_D(t)\phi| \le ke^{-\alpha t} |\phi|, \phi \in c_D, t \ge \sigma \ge 0.$$

Lemma 3.1.1: *If in* (3.2.1), *the continuous operator* D *is uniformly stable and* f *maps bounded subsets of* $R_+ \times C$ *into bounded sets in* R^n, *then*

$$T(t, \sigma)\phi = x_t(\sigma, \phi) = T_D(t)(\psi\phi) + T_1(t, \sigma)\phi$$

where $\psi: c \to c_D$ *is continuous linear operator defined by* $\psi\phi = \phi - \Phi D(\phi)$ *with* $\Phi = (\phi, \ldots, \phi_n)$, $\phi_i \in C$ $i = 1, \ldots, n$ *such that* $D(\Phi) = I$ *and* $T_D(t)$ *is an* α *contraction with* $\alpha < 1$ *and* $T_1(t, \sigma)$ *is compact.*

Lemma 3.1.2: *If* A *is a convex bounded closed subset of a Banach space and if* $f: A \to A$ *is an* α-*contraction with* $\alpha < 1$, *then* f *has a fixed point.*

Below we assume that the initial function ϕ in (3.1.1) is differentiable with respect to θ and it satisfies the compatibility condition $D\dot\phi = f(\sigma, \phi)$.

We prove now a result leading to differentiability with respect to initial conditions.

Theorem 3.2.1: *If* $x_t(s, y_s(\sigma, \phi))$ *and* $y_s(\sigma, \phi)$ *are differentiable solutions of* (3.1.1) *and* (3.1.2) *respectively with respect to* t *and* s, *then*

$$\frac{d}{dt}x_t(s, y_s(\sigma, \phi)) \text{ and } T(t, s: y_s(\sigma, \phi))x_\sigma g(t, y_t))$$

are solutions of the variational equation (3.2.3) *that coincide with* $x_\sigma g(t, y_t(\sigma, \phi))$ *at* $t = s$.

Proof: That $T(t, s: y_s(\sigma, \phi))x_\sigma$ is a solution of (3.2.3) is immediate. So is $T(t, s, y_s(\sigma, \phi))x_\sigma g(t, y_t)$, since g is constant with respect to θ, and for $t = s$

$$T(t, s, y_s(\sigma, \phi))x_\sigma g(s, y_s(\sigma, \phi)) = x_\sigma g(s, y_s(\sigma, \phi)). \tag{3.2.6}$$

Since D is linear, we get

$$\frac{d}{dt}D\left(\frac{d}{ds}x_t(s, y_s(\sigma, \phi))\right)$$

$$= \frac{d}{dt}\frac{d}{ds}D(x_t(s)y_s(\sigma, \phi)))$$

$$= \frac{d}{ds}\frac{d}{dt}D(x_t(s, y_s(\sigma, \phi)))$$

$$= \frac{d}{ds}f(t, x_t(s, y_s(\sigma, \phi)))$$

$$= f_y(t, x_t(s, y_s(\sigma, \phi)))\frac{d}{ds}[x_t(s, y_s(\sigma, \phi))]$$

which shows that $\frac{d}{ds}x_t(s, y_s(\sigma, \phi))$ is a solution of (3.2.3).

Now define the following functions:

$$H(u, v) = Dx_u(v),$$

$$h(s) = (s, y_s(\sigma, \phi)),$$

and

$$l(s) = (sh(s)) = H(s, h(s)).$$

As a result, we get

$$\left(\frac{d}{ds}\ell(s)\right)_{s=s_0} = \frac{\partial H}{\partial u}(s_0, h(s_0)) + \frac{\partial H}{\partial v}(s_0, h(s_0)) \cdot \frac{d}{ds}h(s_0)$$

$$= \frac{\partial H}{\partial u}(s_0, h(s_0)) + \frac{dH}{ds}(s_0, h(s)) \mid_{s=s_0}$$

$$= \frac{dH}{ds}(s, h(s_0)) \mid_{s=s_0} + \frac{dH}{ds}(s_0, h(s)) \mid_{s=s_0}.$$

Therefore

$$\frac{d}{ds}Dx_s(s, y_s(\sigma, \phi)) = \left[\frac{d}{dt}Dx_t(s, y_s(\sigma, \phi))\right]_{t=s}$$

$$+ \left[\frac{d}{dt}Dx_s(t, y_t(\sigma, \phi))\right]_{t=s}$$

From (3.2.2), we have

$$\frac{d}{ds}Dx_s(s, y_s(\sigma, \phi)) = \frac{d}{ds}Dy_s(\sigma, \phi)$$

$$= f(s, y_s(\sigma, \phi)) + g(s, y_s(\sigma, \phi))$$

and from (3.2.1), we obtain

$$\left[\frac{d}{dt}Dx_t(s, y_s(\sigma, \phi))\right]_{t=s} = f(s, y_s(\sigma, \phi)).$$

Then

$$\left[\frac{d}{dt}Dx_s(t, y_t(\sigma, \phi))\right]_{t=s} = g(s, y_s(\sigma, \phi)).$$

From the linearity of D, we see that

$$\left[D\left(\frac{d}{ds}x_t(s, y_s(\sigma, \phi))\right)\right]_{t=s} = g(s, y_s(\sigma, \phi)). \tag{3.2.7}$$

We now claim that

$$\left[\frac{d}{dt}x_t(s, y_s(\sigma, \phi))\right]_{t=s}(\theta) = x_\sigma(\theta)g(s, y_s(\sigma, \phi)); \quad -r \le \theta \le 0. \tag{3.2.8}$$

To establish this claim, consider

$$h(\theta) = \left[\frac{d}{ds}x_t(s, y_s(\sigma, \phi))\right]_{t=s}(\theta), \ -r \le \theta \le 0.$$

Note that for each θ, and t sufficiently close to s, such that $t + \theta < s$, we have

$$x(t + \theta, y_s(\sigma, \phi)) = y(t + \theta, \sigma, \phi).$$

Since $x_s(s, y_s(\sigma, \phi)) = y_s(\sigma, \phi)$, from (3.2.2), we obtain

$$\frac{d}{ds}x(t + \theta, y_s(\sigma, \phi)) = \frac{d}{ds}y(t + \theta, \sigma, \phi) = 0$$

for every t and, in particular, for $t = s$.

Thus $h(\theta) = 0$ for $\theta \in [-r, 0)$. From (3.2.1) and (3.2.7), we have at $\theta = 0$,

$$\left[\frac{d}{ds}x_s(s, y_s(\sigma, \phi))\right]_{t=s}(0) = D\left(\left[\frac{d}{ds}x_t(s, y_s(\sigma, \phi))\right]_{t=s}\right.$$

$$\left. + \int_{-r}^{0}[d\mu(\theta)]\left[\frac{d}{ds}x_s(s, y_s(\sigma, \phi))\right]_{t=s}(\theta)\right).$$

The integral term above is zero since $\theta < 0$, $h(\theta) = 0$ and $\int_{-r}^{0}[d\mu(\theta)]\phi(\theta)$ is nonatomic at zero. Hence we conclude

$$\left[\frac{d}{ds}x_t(s, y_s(\sigma, \phi))\right]_{t=s}(0) = g(s, y_s(\sigma, \phi)), \theta = 0.$$

Therefore, for each θ in $[-r, 0]$, it follows that

$$\left\{\frac{d}{ds}[x_t(s, y_s(\sigma, \phi))]\right\}_{t=s} = x_\sigma g(s, y_s(\sigma, \phi)).$$

Since the solutions of (3.2.3) are unique and the relations (3.2.6), (3.2.8) hold, we finally arrive at

$$\frac{d}{ds}x_t(s, y_s(\sigma, \phi)) = T(t, s; y_s(\sigma, \phi))x_\sigma g(s, y_s(\sigma, \phi)) \qquad (3.2.9)$$

and the proof is complete.

We now prove the VPF for nonlinear FDE (3.2.2).

Theorem 3.2.2: *Under the conditions specified for the systems (3.2.1) and (3.2.2), a relationship between solutions is given by*

$$y_t(\sigma, \phi) = x_t(\sigma, \phi) + \int_{\sigma}^{t}T(t, s; y_s(\sigma, \phi))x_\sigma g(s, y_s(\sigma, \phi))ds. \qquad (3.2.10)$$

Proof: We use below the conclusions of Theorem 3.2.1 and show that $y_t(\sigma, \phi)$ given by (3.2.10) satisfies (3.2.2). From (3.2.10), we have

$$\frac{d}{dt}[D(y_t(\sigma,\phi))] = \frac{d}{dt}D\left[(x_t(\sigma,\phi)) + \int_\sigma^t T(t,s;y_s(\sigma,\phi))x_\sigma g(s,y_s)ds\right].$$

From (3.2.1) and since D is linear, we get

$$\frac{d}{dt}[D(y_t(\sigma,\phi))] = f(t,x_t) + \frac{d}{dt}\int_\sigma^t D[T(t,s;y_s(\sigma,\phi))x_\sigma g(s,y_s)]ds$$

$$= f(t,x_t) + D[T(t,t,y_t(\sigma,\phi))x_\sigma g(t,y_t)]$$

$$+ \int_\sigma^t \frac{d}{dt}D[T(t,s,y_\sigma(\sigma,\phi))x_\sigma g(s,y_s)]ds$$

$$= f(t,x_t) + g(t,x_t) + \int_\sigma^t \frac{d}{dt}D[T(t,s;y_s(\sigma,\phi))x_\sigma g(s,y_s)]ds,$$

since $D(x_\sigma) = 1$. Now from (3.2.9) and (3.2.3) and Theorem 3.2.1, we see that

$$\frac{d}{dt}D[T(t,s:y_s(\sigma,\phi)x_\sigma g(s,y_s))]$$

$$= f_x(t,x_t(s,y_s(\sigma,\phi)))T(t,s,y_s(\sigma,\phi)x_\sigma g(s,y_s)$$

$$= f_x(t,x_t(s,y_s(\sigma,\phi)))\frac{d}{ds}x_t(s,y_s(\sigma,\phi)$$

$$= \frac{d}{ds}f(t,x_t(s,y_s(\sigma,\phi)).$$

Thus

$$\frac{d}{dt}D(y_t(\sigma,\phi)) = f(t,x_t(\sigma,\phi)) + g(t,y_t) + \int_\sigma^t \frac{d}{ds}f(t,x_t(s,y_s(\sigma,\phi))ds$$

$$= f(t,x_t(\sigma,\phi)) + g(t,y_t) + f(t,y_t(\sigma,\phi)) - f(t,x_t(\sigma,\phi))$$

$$= f(t,y_t(\sigma,\phi)) + g(t,y_t).$$

The proof is therefore complete.

Remark 3.1.2: (i) As a special consideration of the above discussion, let us choose $D(\phi) = \phi(0)$, $\phi \in C$ in (3.1.12), then the equations (3.2.1), (3.2.2) and (3.2.3) reduce to the form

$$x'(t) = f(t,x_t), x_\sigma(\theta) = \phi(\theta), \theta \in [-r,0],$$

$$y'(t) = f(t,y_t) + g(t,y_t), y_\sigma(\theta) = \phi(\theta),$$

for $t \geq \sigma$ and

$$z'(t) = f_x(t, x_t(s, y_s(\sigma, \phi)))z_t, t \geq s \geq \sigma \geq 0.$$

$$\equiv L(t, z_t)$$

respectively. For this situation, the matrix $x(t)$, $t \geq -r$ satisfies the equation

$$x(t) = I + \int_\sigma^t L(s, x_s)ds.$$

The representation of the VPF given by (3.2.10) holds in this case.

(ii) In case $D(\phi) = \phi(0) - \phi(-r)$, $\phi \in C$, then the equations (3.2.1), (3.2.2) take the form

$$\frac{d}{dt}(x'(t) - x(t-r)) = f(t, x(t), x(t-r))$$

$$\frac{d}{dt}(y'(t) - y(t-r)) = f(t, y(t), y(t-r)) + g(t, y(t), y(t-r))$$

$$x_\sigma(\theta) = y_\theta(\theta) = \phi(\theta), \quad -r \leq \theta \leq 0, \phi \in C.$$

The corresponding variational equation is then given by

$$\frac{d}{dt}[\psi(t) - \psi(t-r)] = f_x(t, x(\sigma, \phi)(t), x(\sigma, \phi)(t-r))\psi(t)$$

$$+ f_u(t, x(\sigma, \phi)(t), x(\sigma, \phi)(t-r))\psi(t-r).$$

The equation corresponding to (3.2.5) verified by the fundamental matrix $x(t)$ is then

$$\frac{d}{dt}[x(t) - x(t-r)] = A(t)x(t) + B(t)x(t-r),$$

$$x_\sigma(\theta) = \begin{cases} 0, & -r \leq \theta < 0 \\ I, & r = 0, \end{cases}$$

where the matrices $A(t)$ and $B(t)$ are f_x and f_u given in the earlier step. The related VPF then appears in the form

$$y(\sigma, \phi)(t) = x(\sigma, \phi)(t) + \int_{t_0}^t T(t, s; y_s(\sigma, \phi))x_\sigma g(s, y(s), y(s-r))ds,$$

where $x_t(\theta) = [T(t, \sigma)x_\sigma](\theta)$, $\theta \in [-r, 0]$.

(iii) Of particular interest are the difference differential equations of the form

$$x'(t) = f(t, x(t), x(t-r))$$

$$y'(t) = f(t, y(t), y(t-r)) + g(t, y(t), y(t-r)), \quad t \geq \sigma.$$

$$x_\sigma(\theta) = y_\sigma(\theta) = \phi(\theta), \ \theta \in [-r, 0]. \tag{3.2.11}$$

The representation of VPF for the above system is obtained immediately from the neutral difference differential equation discussed in (ii) above.

(iv) *Finally*, we remark that, as an analog of Theorem 1.2.7, one may consider systems

$$x' = f_1(t, x), \ x(t_0) = \phi(0) \tag{3.2.12}$$

$$y' = f_2(t, y_t), \ y_t(\theta) = \phi(\theta), \ -r \leq \theta \leq 0 \tag{3.2.13}$$

where $f_s \in C[R_+ \times C, R^n]$ and $f_1 \in C[R_+ \times R^n, R^n]$, $\frac{\partial f_1}{\partial x}(t, x)$ exists and is continuous on $R_+ \times R^n$.

Let $x(t, t_0, x_0)$ be the solution of (3.2.12) and $y(t_0, \phi)(t)$ be any solution of (3.2.13) existing for $t \geq t_0$, then $y(t_0, \phi)$ satisfies the integral equation

$$y_{t_0} = \phi,$$

$$y(t_0, \phi)(t) = x(t, t_0, \phi(0)) + \int_{t_0}^{t} \Phi(t, s, y(s))[f_2(s, y_s) - f_1(s, y(s))]ds \ t \geq t_0,$$

where $\Phi(t, t_0, x_0 = \frac{\partial x(t, t_0, x_0)}{\partial x_0}$ and $y(t) = y(t_0, \phi)(t)$.

3.3 VPF IN TERMS OF LYAPUNOV-LIKE FUNCTIONALS

In this section, we establish a generalized VPF for FDEs. The applications of this result will be given in a subsequent section. We consider below FDEs

$$x'(t) = f(t, x_t), \tag{3.3.1}$$

$$y'(t) = f(t, y_t) + g(t, y_t), \tag{3.3.2}$$

for $t \geq \sigma, \sigma, t \in R_+$ with initial conditions

$$x_\sigma(\theta) = y_\sigma(\theta) = \phi(\theta), \ \theta \in [-r, 0].$$

Let $\Omega \subset R_+ \times C$, Ω is open, $C = C[[-r, 0], R^n]$ and $f, g \in C[\Omega, R^n]$. The systems (3.3.1) and (3.3.2) are special cases of systems (3.2.1) and (3.2.2) respectively and are obtained by letting $D(\phi) = \phi(0), \phi \in C$. Assume the hypothesis

(H_1) $f(t, \phi)$ has a continuous Fréchet derivative $f_x(t, \phi)$ with respect to ϕ in Ω and $g \in C[\Omega, R^n]$.

The continuity of $f_x(t, \phi)$ in Ω implies that $f(t, \phi)$ is locally Lipschitzian in ϕ in each compact subset of Ω and hence solutions $x_t(\sigma, \phi)$ of (3.3.1) are unique and continuous in the three arguments (t, σ, ϕ). Corresponding to each solution $x_t(\sigma, \phi)$ of (3.3.1), we have a linear FDE

$$z'(x) = [f_x(t, x_t(\sigma, \phi))]z_t \tag{3.3.3}$$

which is the variational equation of (3.3.1) with respect to $x_t(\sigma, \phi)$. Let $T(t, \sigma : \phi)$, $t \geq \sigma$ denote the family of linear operators associated with equation (3.3.3) such that $T(t, \sigma : \phi) : C \to C$ and $T(t, \sigma : \phi) = z_t(\sigma, \phi)$.   Further $T(t, \sigma : \phi)$ is strongly continuous for $0 \leq \sigma \leq t < \infty$.

We need the following lemma.

Lemma 3.3.1: *For any $(\sigma, \phi) \in \Omega$ and each $t \in [\sigma, \infty)$, the Fréchet derivative of the unique solution $x_t(\sigma, \phi)$ of (3.3.1) exists and equals $T(t, \sigma : \phi)$. Further,*

$$\frac{\partial}{\partial \sigma}[x_t(\sigma, \phi)] - T(t, \sigma : \phi)\xi$$

whenever $\phi \in \widetilde{C}$, where

$$\widetilde{C} = \{\phi \in C, \phi'(\theta) \text{ exists, is bounded and piecewise continuous on } [-r, 0]\}$$

and

$$\xi(\theta) = \begin{cases} -f(t, \phi), & \theta = 0 \\ -\phi'(\theta) & -r \leq \theta < 0. \end{cases}$$

Also, the Fréchet derivative of $x_t(s, \phi)$ with respect to ϕ is given by

$$[x_t(\sigma, \phi)]' = T(t, \sigma : \phi)\psi, \psi(\theta) = \phi'(\theta), r \leq \theta \leq 0.$$

In fact, a result similar to Lemma 3.3.1 is proved in Theorem 3.2.1 under a general setting. Hence the proof is omitted here.

We assume that following hypothesis.

(H_2) $V \in C[R_+ \times C_\rho, R^n], C_\rho = \{\phi \in C, |\phi|_0 < \rho, \rho > 0\}, |\phi|_0 = \sup_{-r \leq s \leq 0} |\phi|, V(t, \phi)$ has a continuous Fréchet derivative with respect to ϕ in C_ρ denoted by $V'(t, \phi)$.

The main result of this section now follows.

Theorem 3.3.1: *Let f, g and V satisfy the conditions given in (H_1) and (H_2). Let $x_t(\sigma, \phi)$ denote the solution of (3.1.1) existing for $t \geq \sigma$. Let $y_t(\sigma, \phi)$ denote any solution of (3.2.2). Then*

$$V(t, y_t(\sigma, \phi) = V(t, x_t(\sigma, \phi))$$

$$+ \int_\sigma^t V'(t, x_t(s, y_s))[T(t, s) : y_s)g_s](0)ds, \tag{3.3.4}$$

so long as $(\sigma, \phi) \in [0, \infty) \times \widetilde{C}$ and t belongs to the interval of existence of solutions $y_t(\sigma, \phi)$ and where

$$g_s(\theta) = \begin{cases} 0, & -r \leq \theta < 0, \\ g(s, y_s), & \theta = 0. \end{cases}$$

Proof: Let t belong to the interval of existence of $y_t(\sigma, \phi)$ and $x_t(s, y_s)$ be any solution of (3.1.1) through (s, y_s). Employing Lemma 3.3.1, hypotheses (H_1) and (H_2), we obtain

$$\frac{d}{ds} V(t, x_t(s, y_s)) = V'(t, x_t(s, y_s)) \frac{d}{ds}(x_t(x, y_s))$$

$$= V'(x_t(s, y_s))[T(t, s: y_s)\xi_s + T(t, s: y_s \psi_s]$$

$$= V'(t, x_t(s, y_s))[T(t, s: y_s)g_s](0).$$

Integrating this relation between σ and t, we obtain

$$V(t, y_t(\sigma, \phi)) - V(t, x_t(\sigma, \phi))$$

$$= \int\limits_{\sigma}^{t} V'(t, x_t(s, y_s))[T(t, s: y_s)g_s](0)ds,$$

yielding (3.3.4). The proof is complete.

Remarks 3.3.1: (i) The VPF (3.3.4) is obtained by using Lyapunov-like functionals when $V(t, \phi) = |\phi|_0$, $\phi \in \widetilde{C}$, we have a special case of (3.3.4), namely

$$|y_t(\sigma, \phi)|_0 = |x_t(\sigma, \phi)|_0 + \int\limits_{\sigma}^{t} [T(t, s: y_s)g_s](0).$$

(ii) Under the same hypothesis, we can obtain the following variant of the VPF

$$\frac{d}{ds} V(t, x_t(s, y_s(\sigma, \phi)) - x_t(\sigma, \phi))$$

$$= V'(t, x_t(s, y_s(\sigma, \phi)) - x_t(\sigma, \phi))[T(t, s: y_s(\sigma, \phi))g_s](0).$$

Integration between σ and t yields

$$V(t, y_t(\sigma, \phi) - x_t(\sigma, \phi))$$
$$= V(t, 0) + \int\limits_{\sigma}^{t} V'(t, x_t(s, y_s(\sigma, \phi)) - x_t(\sigma, \phi)) [T(t, s: y_s(\sigma, \phi))g_s](0)ds.$$

(iii) Suppose that $x_t(\sigma, \phi)$ and $x_t(\sigma, \psi)$ are solutions of (3.3.1) through (σ, ϕ) and (σ, ψ) respectively, existing for $t \geq t_0$, such that ϕ and ψ belong to a convex subset of C. Then for $t \geq \sigma$,

$$V(t, x_t(\sigma, \psi)) - V(t, x_t(\sigma, \phi))$$

$$\int\limits_{0}^{1} V'(t, x_t(\sigma, \phi + p(\psi - \phi))[T(t, s: \phi + p(\psi - \phi))(\psi - \phi)](0)dp.$$

Since ϕ and ψ belong to a convex subset of C, $x_t(\sigma, \phi + p(\psi - \phi))$ is defined for $0 \leq p \leq 1$. Now

$$\frac{d}{dp} V(t, x_t(\sigma, \phi + p(\psi - \phi)))$$

$$= V'(t, x_t(\sigma, \phi + p(\psi - \phi)))[T(t, \sigma : \phi + p(\psi - \phi))(\psi - \phi)](0).$$

Integrating this relation between 0 and 1, we obtain the desired result.

One can consider the solutions of the FDEs (3.3.1), (3.3.2) as elements of R^n for all $t \geq \sigma \geq 0$. In such a case, the qualitative properties can be studied by employing Lyapunov functions in place of functionals. The derivative of a Lyapunov function along a solution of the FDE is then a functional. Below we extend the generalized comparison principle, using Lyapunov functions. We assume the following hypothesis:

(H_3) $V \in C[[-r, \infty) \times S_\rho, R^n]$, $S_\rho = \{x \in R^n, \ |x| < \rho, \rho > 0\}$. For $\phi \in C_\rho$, define

$$D^+ V(t, \phi(0), \phi) = \limsup_{h \to 0^+} \frac{1}{h}[V(t + h, \phi(0) + hf(t, \phi)) - V(t, \phi(0))].$$

$V(t, x)$ possesses a continuous derivative with respect to $x \in S_\rho$ for each $t \in R_+$.

Based on these assumptions, we obtain the following result.

Theorem 3.3.2: *Let f, g and V satisfy the conditions given in (H_1) and (H_3). Let $x(t, \sigma, \phi)$ denote the solution of (3.3.1) existing for $t \geq \sigma$. Let $y(t, t_0, \phi)$ denote any solution of (3.3.2). Then*

$$V(t, y(t, \sigma, \phi)) = V(t, x(t, \sigma, \phi))$$

$$+ \int_\sigma^t V'(t, x(t, s, y_s))[T(t, s : y_s)g_s](0)ds \qquad (3.3.5)$$

so long as $(\sigma, \phi) \in [0, \infty) \times C_\rho$ and t is in the interval in which $y(t, \sigma, \phi)$ exists.

The proof of this theorem follows the argument of Theorem 3.3.1.

Remark 3.3.2: When the delay $r = 0$, the relation (3.3.5) reduces to the special case of Theorem 1.3.1 which in itself includes as a special case the VPF due to Alekseev, namely Theorem 1.2.7.

When $V(t, x) = x$, then the relation (3.3.5) takes the form

$$y(t, \sigma, \phi) = x(t, \sigma, \phi) + \int_\sigma^t [T(t, s : y_s)g_s](0)ds$$

which may also be written as

$$y_t(\sigma, \phi) = x_t(\sigma, \phi) + \int\limits_{\sigma}^{t} T(t, s: y_s(\sigma, \phi)) x_\sigma g(s, y_s(\sigma, \phi)) ds, \ \ t \geq \sigma$$

where x_t is the fundamental matrix solution of the variational equation (3.3.3).

The equation (3.3.2) includes as a special case, the equation (3.2.11), the difference differential equation. The VPF (3.3.5) may be modified suitably so that we get the VPF for (3.2.11).

3.4 ASYMPTOTICALLY SELF-INVARIANT SETS

We consider in this section, asymptotically invariant sets and stability properties of such sets. For this purpose, we restrict ourselves to the difference differential equations. Even though we discuss such results for linear equations, one can extend these conclusions to nonlinear difference differential equations as well as various types of FDEs.

Consider the linear system

$$x'(t) = A(t)x(t) + B(t)x(t - \tau), \ x_{t_0} = \phi_0 \tag{3.4.1}$$

and the perturbed system

$$y'(t) = A(t)y(t) + B(t)y(t - \tau) + w(t), \ y_{t_0} = \phi_0 \tag{3.4.2}$$

where $A(t)$ and $B(t)$ are $n \times n$ continuous matrices and $w \in C[R_+, R^n]$. Let $x(t_0, \phi_0)(t) \in R^n$ for $t \geq t_0$ represent the solution of (3.4.1) passing through (t_0, ϕ_0). We need the following definitions.

Definition 3.4.1: A function $\lambda \in C[R_+, R_+]$ is said to belong to a class L if $\lambda(t)$ is decreasing in t and $\lim\limits_{t \to \infty} \lambda(t) = 0$.

Definition 3.4.2: The set $\phi = 0$ is said to be asymptotically self-invariant (ASI) with respect to (3.4.1) if there exists a $q \in L$ such that

$$| x(t_0, 0)(t) | \leq q(t_0), \ t \geq t_0.$$

Assume that the set $\phi = 0$ is ASI with respect to the system (3.4.1).

Definition 3.4.3: The ASI set $\phi = 0$ of (3.4.1) is said to be
(i)   equi-stable, if, for each $t_0 \in R^+$,

$$| x(t_0, \phi_0)(t) | \leq k(t_0, \tau) | \phi_0 |_0 + q(t_0), \ t \geq t_0,$$

where $k \in C[R^+ \times R^+, R^+]$ and $q \in L$;
(ii)   equi-exponentially asymptotically stable, if

$$| x(t_0, \phi_0)(t) | \leq k(t_0, \tau) | \phi_0 |_0 e^{-\alpha(t-t_0)} | H(t, t_0), \ t \geq t_0$$

where $k \in C[R^+ \times R^+, R^+]$, $\alpha > 0$, $H \in C[R^+ \times R^+, R^+]$, $H(t,t) = 0$, $H(t,t_0) \leq p(t_0)$, $p \in L$ and

$$\lim_{t \to \infty} [\sup_{t_0 \geq T_0} H(t,t_0)] = 0$$

for some positive number T_0.

Clearly (ii) implies (i). In case $q(t_0) \equiv 0$ and $H(t,t_0) \equiv 0$, then Definition 3.4.3 reduces to equi-stability and equi-asymptotic stability of the trivial solution of (3.4.1).

Following the conclusions of Theorem 3.1.2, we define the function $Y(s,t)$ satisfying the initial conditions

$$Y(t,t) = I, \text{(unit matrix)}$$

$$Y(s,t) \equiv 0, t < s \leq t + \tau$$

and $Y(s,t)$, as a function of s, satisfies the adjoint system

$$\frac{\partial Y(s,t)}{\partial s} = -Y(s,t)A(s) - Y(s+\tau,t)B(s+\tau), s < t. \tag{3.4.3}$$

Assume that

$$|B(t)| \leq B_0, t_0 - \tau \leq t \leq t_0,$$

$$|Y(s,t)| \leq N \exp(\beta s - \alpha(t-s)), N > 1, \alpha > 0, \beta \geq 0. \tag{3.4.4}$$

Employing the VPF, we find that any solution of (3.4.2) satisfies

$$y(t_0, \phi_0)(t) = Y(t_0,t)\phi_0(0) + \int_{t_0-\tau}^{t_0} Y(s+\tau,t)B(s+\tau)\phi_0(s)ds$$

$$+ \int_{t_0}^{t} Y(s,t)w(s)ds. \tag{3.4.5}$$

In view of the assumptions (3.4.4), we then get

$$|y(t_0, \phi_0)(t)| \leq N e^{\beta t_0}[1 + \frac{B_0 e^{(\alpha+\beta)\tau}}{\alpha+\beta}] \, | \, \phi_0 \, |_0 e^{-\alpha(t-t_0)}$$

$$+ \int_{t_0}^{t} e^{-\alpha(t-s)} N e^{\beta s} \, |w(s)| \, ds$$

$$\leq k(t_0, \tau) |\phi_0|_0 e^{-\alpha(t-t_0)} + \int_{t_0}^{t} e^{-\alpha(t-s)}\gamma(s)ds, \, t \geq t_0, \tag{3.4.6}$$

where

$$k(t_0, \tau) = N e^{\beta t_0 + \alpha \tau} [1 + \tfrac{B_0 e^{(\alpha+\beta)\tau}}{\alpha+\beta}] \text{ and } \gamma(t) = N e^{\alpha \tau + \beta t} \mid w(t) \mid .$$

Setting

$$H(t, t_0) = \int_{t_0}^{t} e^{-\alpha(t-s)} \gamma(s) ds$$

and assuming that $w(t)$ is such that

$$\int_{t}^{t+1} \gamma(s) ds \to 0 \text{ as } t \to \infty, \tag{3.4.7}$$

it follows that $\lim\limits_{t \to \infty} [\sup\limits_{t_0 \geq 1} H(t, t_0)] = 0$, $H(t, t_0) \leq p(t_0)$, $t \geq t_0$, $p \in L$. Hence we conclude from (3.4.6) that the set $\phi = 0$ is ASI with respect to (3.4.2). Further, if $y(t_0, \phi_0)$ and $y(t_0, \psi_0)$ are two solutions of (3.4.2), then on the basis of assumptions (3.4.4) we can obtain that

$$\mid y(t_0, \phi_0)(t) - y(t_0, \psi_0)(t) \mid \leq k(t_0, \tau) \mid \phi_0 - \psi_0 \mid_0 e^{-\alpha(t-t_0)}, \ t \geq t_0. \tag{3.4.8}$$

The foregoing discussion can be summarized in the following result.

Theorem 3.4.1: *Suppose that conditions (3.4.4) and (3.4.7) are satisfied. Then the set $\phi = 0$ is ASI, relative to the linear difference differential equation (3.4.2) and it is equi-exponentially asymptotically stable. Further, for any two solutions $y(t_0, \phi_0)$, $y(t_0, \psi_0)$, the relation (3.4.8) is valid.*

3.5 ASYMPTOTIC EQUIVALENCE

We study in this section, the asymptotic equivalence between perturbed system and unperturbed system of neutral FDEs. The technique employed for this purpose is the VPF for nonlinear neutral FDEs proved in Section 3.2.

Our concern here is with the systems

$$\tfrac{d}{dt} D x_t = f(t, x_t), \ x_\sigma = \phi, \ t \geq \sigma; \tag{3.5.1}$$

and

$$\tfrac{d}{dt} D y_t = f(t, y_t) + g(t, y_t), \ y_\sigma = \phi, \ t \geq \sigma, \tag{3.5.2}$$

where D, f, g have the same meaning as in Section 3.2.

First of all, we need to establish some notation and definitions.

We denote by $C([-r, 0], R)$ the set of all continuous functions defined on $[-r, 0]$ with real values and by $C_+ = C[-r, 0], R_+)$.

Let

$$u' = w(t, u_t). \tag{3.5.3}$$

be a functional differential equation where $w \in C([\sigma, \infty) \times C([-r, 0], R)$.

Definition 3.5.1: For every $\phi \in C = C[-r,0], E^n)$ the Euclidean-norm of ϕ, $|\phi|_1$ is such that

$$|\phi|_1(\theta) = |\phi(\theta)|, \quad -r \leq \theta \leq 0.$$

We have that $|\phi|_1 \in C_+$.

Definition 3.5.2: We say that $w(t, \phi)$ is nondecreasing in ϕ for each t fixed in $[\sigma, \infty)$ if given $\phi, \psi \in C([-r,0], R)$ with $|\phi|_1 \leq |\psi|_1$ we have $w(t, \phi) \leq w(t, \psi)$.

Definition 3.5.3: If $\rho(t, \sigma; \phi)$ is a solution of (3.5.3) defined on $[\sigma, \sigma + a]$ with $a > 0$ and if any other solution $u(t, \sigma; \phi)$ of (3.5.3) defined in the same interval $[\sigma, \sigma + a]$ satisfies the inequality

$$u(t, \sigma; \phi) \leq \rho(t, \sigma; \phi), \ t \in [\sigma, \sigma + a]$$

then, $\rho(t, \sigma; \phi)$ is a maximal solution of (3.5.3) on $[\sigma, \sigma + a]$.

We use the following lemmas in subsequent discussions.

Lemma 3.5.1: *If $w \in C([\sigma, \infty) \times C([-r,0], R)$ and $w(t, \phi)$ is nondecreasing in ϕ, for each fixed $t \in [\sigma, \infty)$, then, given an initial function $u_\sigma \in C([-r,0], R)$ there exists $\alpha_1 > 0$ such that (3.5.3) admits a unique maximal solution $u(t, \sigma; u_\sigma)$ defined on $[\sigma, \sigma + \alpha_1]$.*

Lemma 3.5.2: *Let $m \in C([\sigma - r, \sigma + a], R_+)$ satisfy the inequality*

$$m(t) \leq \rho_\sigma(0) + \int_\sigma^t w(s, m_s) ds,$$

where $w \in C[\sigma - r, \sigma + a] \times C_+, R_+)$. Suppose that $w(t, \psi)$ is nondecreasing in ψ for each fixed t, $t \in [\sigma, \sigma + a]$, and that $(t, \sigma; \rho_\sigma)$, with $\rho_\sigma \in C_\rho \{\phi \in C_+: |\phi|_1 < \rho, \rho > 0\}$ is the maximal solution of (3.5.3) existing for $t \geq \sigma$. Then there exists an $x > 0$ such that $m_\sigma \leq \rho_\sigma$ implies that $m(t) \leq u(t, \sigma; u_\sigma)$, $\sigma \leq t \leq \sigma + a$.

Define a nonsingular $n \times n$ real continuous matrix on $[\sigma - r, \infty)$ by $\Delta(t)$ satisfying

$$\Delta_t(\theta) = \Delta(t + \theta), \ (\Delta_t x_t)(\theta) = \Delta_t(\theta) x_t(\theta)$$

where $\theta \in [-r, 0]$.

Suppose that the maximal solution $\rho(t, \sigma, \rho_\sigma)$ of $\rho' = w(t, \rho_t), w(t, \rho_t)$ satisfying the same conditions of Lemma 3.5.2 exists for $t \geq \sigma$ and satisfies $\lim_{t \to \infty} \rho(t) = \rho_\infty < \infty$.

Under these conditions, we have the following result.

Theorem 3.5.1: *Let Ω_1 and D_1 be open sets such that $\Omega_1 \subset C, D_1 \subset \Omega_1$ and $\overline{D}_1 \subset \Omega_1$ and let D_2 be a subset of D_1. If*

(a) $| \Delta_t \gamma_t |_1 \leq \rho_t$ *implies* $\gamma_t \in D_1$, *for* $t \geq \sigma$,

(b) $\| \Delta_t T(t,s;\gamma_s) x_\sigma g(s,\gamma_s) \| \leq w(s,|\Delta_s \gamma_s|)$, *for* $t,s \in [\sigma,\infty)$ *and* $\gamma_1 \in \Omega_1$,

(c) $\| \Delta_t(\sigma,\psi) \| \leq |\phi(0)|$, $t \geq \sigma$, $\psi \in D_2$ *and* ϕ *is the initial function of the solution* $y(t,\sigma;\phi)$ *of* (3.5.2), *then, for every initial* $\phi \in \Omega_1$ *such that* $|\phi|_1 < \rho$, $\rho < \rho_\infty$ *and* $|\phi(0)| \leq \rho_\sigma(0)$ *there exists a solution* $y_t(\sigma,\phi) \in D_1$ *of* (3.5.2), *for* $t \geq \sigma$.

To each such solution there corresponds a solution $x_t^*(\sigma,\phi)$ *of* (3.5.1), $t \geq 0$ *satisfying*

$$\lim_{t \to x} \| \Delta_t(y_t - x_t^*) \| = 0. \tag{3.5.4}$$

Proof: Using (3.2.10) for $\phi \in D_2$, $|\phi|_1 < \rho < \rho_\infty$ and $|\phi(0)| \leq \rho_\sigma(0)$, the solution of (3.5.2) is given by

$$y_t(\sigma,\phi) = x_t(\sigma,\phi) + \int_\sigma^t [T(t,s;x_s(\sigma,\phi)x_\sigma]g(s,y_s(\sigma,\phi))ds$$

for every $t \geq \sigma$.

Hence, for every $\theta \in [-r,0]$,

$$y(t+\theta) = x(t+\theta) + \int_\sigma^t [T(t,s;y_s)x_\sigma](\theta)g(s,y_s)ds$$

and then the hypotheses (b) and (c) yield

$$\| \Delta(t+\theta)y(t+\theta) \| \leq \| \Delta(t+\theta)x(t+\theta) \|$$

$$+ \int_\sigma^t \| \Delta(t+\theta)[T(t,s;y_s)x_\sigma](\theta)g(s,y_s)) \| \, ds$$

$$\leq |\phi(0)| + \int_\sigma^t w(s,|\Delta_s y_s|t_1)ds$$

$$\leq \rho_\sigma(0) + \int_\sigma^t w(s,|\Delta_s y_s|_1)ds.$$

Therefore, by Lemma 3.5.2

$$| \Delta(t+\theta)y(t+\theta) | \leq \rho(t+\theta), t \geq \sigma, \theta \in [-r,0]$$

which implies

$$| \Delta_t y_t(\sigma,\phi) |_1 \leq \rho_t, \ t \geq \sigma,$$

and then, by hypothesis (a),

$$y_t(\sigma,\phi) \in D_1, t \geq \sigma.$$

In order to prove (3.5.4) we need to show that $y_t(\sigma,\phi)$ is bounded-in the future.

For this we show that $y_t(\sigma, \phi)$ is bounded in $[\sigma, t_+)$ and then we can conclude that $t_+ = \infty$.

In fact, suppose that $y_t(\sigma, \phi)$ is unbounded in $[\sigma, t_+)$. Then, there exists a sequence $t_n \to t_+$ and for all positive constants M there is n_0 such that $|y_{t_n}(\sigma, \phi)| \geq M > 0$ for $n \geq n_0$.

On the other hand, we show that there is a constant $K > 0$ such that

$$K \, \| \, y_{t_n}(\sigma, \phi) \, \| \, \leq \, \| \, \Delta_{t_n} y_{t_n}(\sigma, \phi) \, \| \, .$$

It is equivalent to show that there is $K > 0$ such that

$$K \leq \| \, \Delta_{t_n} v_{t_n} \, \| \quad \text{with } v_{t_n} = \frac{y_{t_n}(\sigma, \phi)}{\|x_{t_n}(\sigma, \phi)\|}.$$

Suppose that this is not true. Then there exists a subsequence (t_{n_j}) such that $\| \, \Delta_{t_{n_j}} v_{t_{n_j}} \, \| \to 0$ with $n_j \to \infty$. Hence, there is a subsequence (t_{n_k}) such that for $v_{t_{n_k}} \to v_0$ we have

$$\| \, \Delta_{t_{n_k}} v_{t_{n_k}} \, \| \to \| \, \Delta_{t_-} v_0 \, \| = 0.$$

Since $\Delta(t)$ is a nonsingular matrix we have $v_0 = 0$, against the fact that v_0 is a point on the unitary ball.

Thus, there is a constant $R > 0$ such that

$$\| \, y_t(\sigma, \phi) \, \| \leq R, \ t \in [\sigma, t_+).$$

If $t_+ < \infty$, that is, if the solution $y_t(\sigma, \phi)$ is noncontinuable in $[\sigma, \phi)$ with $\sigma < b < t_+$, we take the closed set

$$[\sigma, b] \times (\overline{B}_R \times \overline{D}_1) \subset [\sigma, t_+) \times \Omega_1$$

where $\overline{B}_R$ is the closed ball with radius R with center at the origin. Hence $t_+ = \infty$. We show now, that the integral

$$\int_\sigma^\infty T(t, s; y_s(\sigma, \phi)) x_\sigma g(s, y_s(\sigma, \phi)) ds \tag{3.5.5}$$

converges uniformly on compact subintervals of $[\sigma, \infty)$.

In fact, since $\lim_{t\to\infty} \rho(t) = \rho_\infty$ and $\rho(t)$ is increasing because its derivative is positive, given $\epsilon > 0$ there is $T \geq \sigma$ such that $T \leq t_1 \leq t_2$, $0 < \rho(t_2) - \rho(t_1) < \epsilon$. Hence, we get

$$\int_{t_1}^{t_2} \| \, \Delta_t T(t, s; y_s(\sigma, \phi)) x_\sigma g(s, y_s(\sigma, \phi)) \, \| \, ds$$

$$\leq \int_{t_1}^{t_2} w(s, |\Delta_s y_s(\sigma, \phi)|_1) ds$$

$$\leq \int\limits_{t_1}^{t_2} w(s, \rho_s)ds = \rho(t_2) - \rho(t_1) < \epsilon,$$

which shows that for t in compact subintervals of $[\sigma, \infty)$ the integral

$$\int\limits_{\sigma}^{x} \Delta_t T(t, s; y_s(\sigma, \phi))x_\sigma g(s, y_s(\sigma, \phi))ds \qquad (3.5.6)$$

converges uniformly. Since $\Delta(t)$ is nonsingular it follows that the integral in (3.5.6) converges uniformly in compact subintervals of $[\sigma, \infty)$.

Thus given a solution $x_t(\sigma, \phi)$ of (3.5.1) we have

$$x_t^*(\sigma, \phi) = x_t(\sigma, \phi) + \int\limits_{\sigma}^{\infty} T(t, s; y_s(\sigma, \phi))x_\sigma g(s, y_s(\sigma, \phi))ds \qquad (3.5.7)$$

is a solution of (3.5.1) which satisfies (3.5.4).

In fact, in view of (3.2.9), we obtain

$$\int\limits_{\sigma}^{\tau} T(t, s; y_s(\sigma, \phi))x_\sigma g(s, y_s(\sigma, \phi))ds = \int\limits_{\sigma}^{\tau} \frac{d}{ds} x_t(s, y_s(\sigma, \phi))ds$$

$$= x_t(\tau, y_\tau(\sigma, \phi)) - x_t(\sigma, y_\sigma(\sigma, \phi)) = x_t(\tau, y_\tau(\sigma, \phi)) - y_t^*(\sigma, \phi), \ \ \sigma \leq \tau.$$

Thus,

$$x_\tau(\tau, y_\tau(\sigma, \phi)) = x_t^*(\sigma, \phi) - \int\limits_{\tau}^{\infty} T(t, s; y_s(\sigma, \phi))x_\sigma g(s, y_s(\sigma, \phi))ds.$$

Therefore, when $\tau \to \infty$ we have $x_t(\tau, y_\tau(\sigma, \phi))$ converging uniformly to $x_t^*(\sigma, \phi)$, since (3.5.5) holds.

On the other hand, from (3.2.9) we have

$$x_t(\tau, y_\tau(\sigma, \phi)) = y_t(\sigma, \phi) + \int\limits_{t}^{\tau} T(t, s; y_s(\sigma, \phi))x_\sigma g(s, y_s(\sigma, \phi))ds$$

for $0 \leq t \leq \tau$, a solution of (3.5.1) in $[\sigma, \tau]$.

By (3.5.5) we can conclude that

$$x_t^*(\sigma, \phi) = \lim_{\tau \to \infty} x_t(\tau, x_\tau(\sigma, \phi))$$

is the uniform limit of solutions of (3.5.1) on $[\sigma, \infty)$ and then it is a solution of (3.5.1) on compact parts of $[\sigma, \infty)$.

Finally from (3.2.10), (3.5.7) and from hypotheses (a) and (b) we have:

$$\| \Delta_t(y_t(\sigma, \phi) - x_t^*(\sigma, \phi)) \|$$

$$\leq \int\limits_{t}^{\infty} \parallel \Delta_t T(t,s;y_s(\sigma,\phi))x_\sigma g(s,y_s(\sigma,\phi)) \parallel ds$$

$$\leq \int\limits_{t}^{\infty} w(s, \mid \Delta_s y_s(\sigma,\phi) \mid_1 ds$$

$$\leq \int\limits_{t}^{\infty} w(s,\rho_s)ds = \rho_\infty - \rho(t) \to 0$$

when $t \to \infty$, which completes the proof.

Before we establish the reciprocal theorem of Theorem 3.5.1, we study some results used in the proof of Theorem 3.5.1, in the following lemma.

Lemma 3.5.3: *Suppose that the hypotheses of Theorem 3.5.1 are satisfied. Let $v \in C([\sigma,\infty), R^n)$. Then,*

(i)   the integral $\int_\sigma^\infty T(t,s;v(s)x_\sigma g(s,v(s))ds$ exists uniformly for t on compact subintervals of $[\sigma,\infty)$;

(ii) for $z \in C([\sigma,\infty), R^n)$ defined by

$$z(t) = x_t(\sigma,\phi) - \int\limits_{t}^{\infty} T(t,s;v(s))x_\sigma g(s,v(s))ds,$$

where $x_t(\sigma,\phi)$ is a solution of (3.5.1), there exists $y \in C([\sigma - r,\infty), R^n)$ defined by

$$y(t) = \begin{cases} z(t)(0), & t \geq \sigma \\ z(\sigma)(t-\sigma), & \sigma - r \leq t \leq \sigma \end{cases}$$

are satisfying $y_t = z(t)$, $t \geq \sigma$ and $(d/dt)Dy_t = f(t,y_t) + g(t,v(t))$ with

$$y_\sigma = \phi - \int\limits_{\sigma}^{x} T(\sigma,s;v(s))x_\sigma g(s,v(s))ds.$$

Theorem 3.5.2: *In the hypotheses of Theorem 3.5.1 suppose that the linear operator D is uniformly stable and suppose that f and g map closed bounded subsets of $R \times C$ into bounded sets of R^n.*

Then given any solutions $x_t(\sigma,\phi)$ of (3.5.1), $t \geq \sigma$ with $\phi \in D_2$, there exist $t_1 \geq \sigma$ and a solution $y_t(\sigma,\phi)$ of (3.5.2), $t \geq t_1$ such that the condition (3.5.4) is verified.

Proof: First of all, observe that the conditions below are equivalent:

(i)   given any real number $\rho_\infty > 0$ there exists a solution of $\rho'(t) = w(t,\rho_t)$ in some interval $[t_1,\infty)$ such that $\lim_{t\to x}\rho(t) = \rho_\infty$;

(ii) $\int^\infty w(s,M)ds < \infty$ for each $M > 0$ real.

Now, let $x_t(\sigma,\phi)$ be a solution of (3.5.1) such that $\phi \in D_2$, $\mid \phi(0) \mid \leq \rho(\sigma)$. Let $\rho(\sigma) < \eta$ for some positive constant η and choose $t_1 \geq \sigma$, sufficiently large such that

$$\int\limits_{t_1}^{\infty} w(s, 2\eta)ds \le \eta.$$

Consider the set

$$B_{2\eta} = \{v \in C([\sigma, \infty), R^n): v(t) = \Delta_t x_t, \ | v(t) |_1 \le 2\eta, t \ge \sigma \ |$$

where $x \in C([\sigma - r, \infty), R^n)$.

It is clear that $B_{2\eta}$ is a convex subset of $C([\sigma, \infty), R^n)$.

Let $S: B_{2\eta} \to C([\sigma, \infty), R^n)$ be the operator defined by

$$(Sv)(t) = \Delta_t \left[x_t(\sigma, \phi) - \int\limits_{t}^{\infty} T(t, s; \Delta_s^{-1}v(s))x_\sigma g(s, \Delta_s^{-1}v(s))ds \right] \quad (3.5.8)$$

with $t \ge t_1 \ge \sigma$ and $x_t(\sigma, \phi)$ a solution of (3.5.1), $\phi \in D_2$ and $| \phi(0) | < \rho(\sigma)$, $\Delta_t^{-1} = (\Delta^{-1})_t$.

We have

(i) $S(B_{2\eta}) \subset B_{2\eta}$.

From hypothesis (b) and (c) we get

$$| (Sv)(t) | \le | \Delta_t x_t(\sigma, \phi) |$$

$$+ \int\limits_{t}^{\infty} | \Delta_t T(t, s; \Delta_s^{-1}v(s))x_\sigma g(s, \Delta_s^{-1}v(s)) | \, ds$$

$$\le \eta + \int\limits_{t}^{x} w(s, | v(s) |_1)ds \le \eta + \int\limits_{t}^{\infty} w(s, 2\eta)ds < 2\eta, \ t \ge t_1.$$

(ii) S is a continuous operator.

Let $(v_n)_{n \in N}$ be a sequence in $B_{2\eta}$ converging uniformly to $v \in B_{2\eta}$. Then for $t \ge t_1$ and $t_2 \ge t_1$ such that for every $\epsilon > 0$

$$\int\limits_{t_2}^{\infty} w(s, 2\eta) < \frac{\epsilon}{4}$$

one can write

$$| (Sv_n)(t) - (Sv)(t) |$$

$$\le \int\limits_{t}^{\infty} | \Delta[T(t, s; \Delta_s^{-1}v_n(s))x_\sigma g(s, \Delta_s^{-1}v_n(s))$$

$$- T(t, s; \Delta_s^{-1}v(s))x_\sigma g(s, \Delta_s^{-1}v(s))] | \, ds$$

$$= \left(\int\limits_{t}^{t_2} + \int\limits_{t_2}^{\infty} \right) | \Delta_t[T(t, s; \Delta_s^{-1}v_n(s))x_\sigma g(s, \Delta_s^{-1}v_n(s))$$

$$- T(t, s; \Delta_s^{-1} v(s)) x_\sigma g(s, \Delta_s^{-1} v(s))] \mid ds.$$

Hence, by hypothesis (b) of Theorem 3.5.1 we have

$$\int_{t_2}^{\infty} \mid \Delta_t [t, s; \Delta_s^{-1} v_n(s)) x_\sigma g(s, \Delta_s^{-1} v_n(s))$$

$$- T(t, s; \Delta_s^{-1} v(s)) x_\sigma g(s, \Delta_s^{-1} v(s))] \mid ds$$

$$\leq \int_{t_2}^{x} w(s, \mid v_n(s)) \mid_1) ds + \int_{t_2}^{\infty} w(s, \mid v(s) \mid_1) ds$$

$$\leq 2 \int_{t_2}^{\infty} w(s, 2\eta) ds < \frac{\epsilon}{2}, \quad t_2 \geq t \geq t_1.$$

Because of the uniform convergence of the sequence (v_n) on compact subintervals of $[\sigma, \infty)$ and to the uniform continuity of T and g in the same intervals, one obtains

$$T(t, s; \Delta_s^{-1} v_n(s)) x_\sigma g(s, \Delta_s^{-1} v_n(s))$$

converges uniformly to $T(t, s; \Delta_s^{-1} v(s)) x_\sigma g(s, \Delta_s^{-1} v(s))$ on compact subintervals of $[\sigma, \infty)$.

Since $M = \sup \mid \Delta_t \mid$, $t \in [t_1, t_2]$, and given $\epsilon > 0$ there exists $n_0 = n_0(\epsilon, t_1, t_2)$ such that for every $n \geq n_0$ it follows

$$\int_{t}^{t_2} \mid \Delta_t [T(t, s; \Delta_s^{-1} v_n(s)) x_\sigma g(s, \Delta_s^{-1} v_n(s))$$

$$- T(t, s; \Delta_s^{-1} v(s)) x_\sigma g(s, \Delta_s^{-1} v(s))] \parallel$$

$$\leq \int_{t}^{t_2} \mid \Delta_t \mid \frac{\epsilon}{2M(t_2-t)} ds \leq \frac{M \epsilon (t_2-t)}{2M(t_2-t)} = \frac{\epsilon}{2}.$$

Thus, $\mid (S v_n)(t) - (S v) \mid < \epsilon$ for every $n \geq n_0(\epsilon)$, $t \in [t_1, t_2]$, which shows that $(S v_n)(t)$ converges uniformly to $(S v)(t)$ on compact subintervals of $[\sigma, \infty)$ and therefore, S is continuous.

By Lemma 3.5.3, the operator $\Delta_t^{-1}(S v)(t)$ is a solution of the differential equation

$$\frac{d}{dt} D(\Delta_t^{-1}(S v)(t)) = f(t, \Delta_t^{-1}(S v)(t)) + g(t, \Delta_t^{-1} v(t)) \qquad (3.5.9)$$

with

$$\Delta_\sigma^{-1}(S v)(\sigma) = \phi - \int_{\sigma}^{\infty} T(t, s; \Delta_s^{-1} v(s)) x_\sigma g(s, \Delta_s^{-1} v(s)) ds. \qquad (3.5.10)$$

By hypotheses, the operator D is uniformly stable and then by Lemma 3.1.1, $\Delta_t^{-1}(Sv)(t)$ is an α-contraction with $\alpha < 1$ plus compact operator. Therefore, by Lemma 3.1.2, $\Delta_t^{-1}(Sv)(t)$ has a fixed point given by

$$\Delta_t^{-1}v(t) = y_t(\sigma, \phi) - \int_t^\infty T(t, s; \Delta_s^{-1}v(s))x_\sigma g(s, \Delta_s^{-1}v(s))ds$$

and satisfying (3.5.9).

From the definition of $B_{2\eta}$, we have $v(t) = \Delta_t y_t$ and

$$\Delta_t^{-1}v(t) = \Delta_t^{-1}\Delta_t y_t = y_t.$$

From (3.5.8) we have that

$$y_t(\sigma, \phi) = x_t(\sigma, \phi) - \int_t^\infty T(t, s; y_s(\sigma, \phi))x_\sigma g(s, y_s(\sigma, \phi))ds$$

is a solution of (3.5.2) and satisfies (3.5.4):

$$\Delta_t[y_t(\sigma, \phi) - x_t(\sigma, \phi)] = \int_t^\infty \Delta_t T(t, s; y_s(\sigma, \phi))x_\sigma g(s, y_s(\sigma, \phi))ds$$

which by (3.5.6) tends to zero when $t \to \infty$, which completes the proof.

The Theorems 3.5.1 and 3.5.2 show that the systems (3.5.1) and (3.5.2) are asymptotically equivalent.

We give below an illustrative example as an application of Theorems 3.5.1 and 3.5.2.

Example 3.5.1: Suppose $r = n = 1$ and $\sigma = 0$. Let D be defined by $D\phi = \phi(0) - K_1\phi(-1)$, with $0 < |K_1| < 1$ so that D is stable. Let $f(\phi) = a \sin \phi(0)$ and $g(t, \phi) = b(t)\phi^2(-1)$ with $a < 0$ and $b(t)$ a nonnegative continuous function satisfying

$$\int_0^\infty (s+2)^3 b(s)ds < \infty.$$

Thus the systems (3.5.1) and (3.5.2) have the form

$$\frac{d}{dt}[x(t) = K_1(t-1)] = a \sin y(t) \qquad (3.5.11)$$

$$\frac{d}{dt}[y(t) - K_1y(t-1)] = a \sin y(t) + b(t)y^2(t-1) \qquad (3.5.12)$$

where g is a continuous function and f is a continuous nonlinear map with a continuous first Fréchet derivative with respect to $\phi \in C$.

Note that we consider differentiable initial function θ with respect to θ and satisfying the compatible condition $\phi(0) - K_1\phi(-1) = a \sin \phi(0)$. It is easy to see that the set of such ϕ is not empty. It is enough to consider differentiable ϕ with $\phi(0) = 0$, $-K_1 = a$ and $\phi(-1) = \sin \phi(0) = 0$.

The variational equation relative to the trivial solution of (3.5.11) is given by

$$\frac{d}{dt}[z(t) - K_1 z(t-1)] = az(t). \tag{3.5.13}$$

The characteristic equation associated to (3.5.13) is

$$\lambda - K_1 \lambda e^{-\lambda} = a$$

whose eigenvalues λ are such that $\text{Re}\,\lambda < 0$, which can be verified by straightforward calculations. Therefore, the solution $x(t)$ of (3.5.11) is uniformly asymptotically stable.

Now we can choose $M > 0$ such that

$$MK \int_0^\infty (s+2)^3 b(s)ds < 1$$

for $K > 0$. Then

$$\rho(t) = M\left[1 - MK \int_0^t (s+2)^3 b(s)ds\right]^{-1}$$

is the unique bounded solution to

$$\rho(t) = K(t+2)^3 b(t)\rho^2(t), \quad \rho(0) = M.$$

If we take $\Delta(t) = |\phi(0)|(t+2)^{-1}$ on $-1 \leq t$ and $w(t, \rho_t) = K(t+2)^3 b(t)\rho_t^2$ and the open set $D_1 \subset R^n$,

$$D_1 = \{\psi \in C/\ |\psi|_1 < M_1(t+2)\}$$

where $M_1 = M[1 - MK \int_0^\infty (s+2)^3 b(s)ds]^{-1}$, the hypotheses are satisfied.

In fact,

(a) let $\gamma_t \in C$ such that $|\Delta_t \gamma_t|_1 \leq \rho_t$. So

$$|\Delta_t \gamma_t|_1 = \frac{1}{t+2}|\gamma_t|_1 \leq \rho_t \Rightarrow |\gamma_t|_1 \leq \rho_1(t+2)$$

$$\Rightarrow |\gamma_t|_1 < M_1(t+2) \Rightarrow \gamma_t \in D_1;$$

(v) since D is stable then for $t, s \in [0, \infty)$ and $\gamma_t \in C$ we get

$$\|\Delta_t T(t, s; \gamma_s)x_\sigma g(s, \gamma_s)\| \leq \frac{1}{t+2}K_2 |g(s, \gamma_s)|$$

$$\leq K_2(t+2)^3 b(s)\left|\frac{1}{t+2}\gamma_s\right|^2$$

$$= K_2(t+2)^3 b(s)|\Delta_s \gamma_s|_1^2$$

$$= K_2(t+2)^3 w(t, |\Delta_t \gamma_t|_1), \quad K_2 > 0, t \geq 0;$$

(c) for $t \geq 0$, $\psi \in C$, and ϕ initial function of the solution $y_t(\sigma, \phi)$ of (3.5.12) we have

$$\| \Delta_t x_t(\sigma, \psi) \| \leq \tfrac{1}{t+2} \| x_t(\sigma, \psi) \| \leq \tfrac{1}{t+2} \mid \phi(0) \mid < \mid \phi(0) \mid$$

since x_t is uniformly asymptotically stable.

Then, for every initial function $\phi \in C$ such that $\mid \phi \mid < \rho_\infty$ and $\mid \phi(0) \mid \leq \rho_\sigma(0)$ the relation (3.5.4) holds.

3.6 GENERALIZED COMPARISON PRINCIPLE

This section is devoted to the study of generalized comparison theorem and stability properties by using VPF in terms of Lyapunov-like functional established in Section 3.3.

We consider the systems

$$x'(t) = f(t, x_t), \ t \geq \sigma \tag{3.6.1}$$

$$y'(t) = f(t, y_t) + g(t, y_t), \tag{3.6.2}$$

$$x_\sigma(\theta) = y_\sigma(\theta) = \phi(\theta), \theta \in [-r, 0],$$

where the functions f, g and the hypotheses (H_1) to (H_3) are the same as given in Section 3.3. Our subsequent discussion is based on the conclusions of Theorems 3.3.1 and 3.3.2 in which we have obtained nonlinear VPFs in terms of Lyapunov-like functionals.

In addition to (H_1) to (H_3) we assume the following hypotheses:

(H_4) $G \in C[R_+ \times R^n, R^n]$, where $G(t, u)$ is quasimonotone nondecreasing in u for each $t \in R_+$. Let $r(t, \sigma, u_0)$ be the maximal solution of the auxiliary differential system

$$u' = G(t, u), \ u(\sigma) = u_0, \ t \geq \sigma, \tag{3.6.3}$$

existing for $t \geq \sigma, \sigma \in R_+$;

(H_5) for $t \geq s \geq \sigma$ and $\phi \in \widetilde{C}$

$$V'(t, x_t(s, \phi))[T(t, s: \phi)g_s](0) \leq G(s, V(t, x_t(s, \phi))), \tag{3.6.4}$$

where $x_t(s, \phi)$ is the solution of (3.6.1) through (s, ϕ).

We are now in a position to prove the following comparison theorem.

Theorem 3.6.1: *Assume that hypotheses (H_1) to (H_3) and (H_4) hold. If $x_t(\sigma, \phi)$ is a solution of (3.6.1) such that*

$$V(t, x_t(\sigma, \phi)) \leq u_0, t \geq \sigma,$$

then for any solution $y_t(\sigma, \phi)$ of (3.6.2), we have

$$V(t, y_t(\sigma, \phi)) \le r(t, \sigma, u_0), \ t \ge \sigma,$$

where $r(t, \sigma, u_0)$ is defined in (H_4).

Proof: From Theorem 3.3.1, and the inequality (3.6.4), we have

$$\frac{d}{ds} V(t, x_t(s, y_s)) = V'(t, x_t(s, y_s))[T(t, s: y_s)g_s](0)$$

$$\le G(s, V(t, x_t(s, y_s))).$$

Set $m(s) = V(t, x_t(s, y_s))$, $m(\sigma) = V(t, x_t(\sigma, \phi))$. Then we get

$$m'(s) \le G(s, m(s)), m(s) \le u_0.$$

Hence it follows that $m(s) \le r(s, \sigma, u_0)$, $t \ge s \ge \sigma$. This inequality leads to the desired conclusion. The proof is complete.

Remarks 3.6.1: (a) This theorem provides us a relationship between the solutions of (3.6.1), the perturbed equation (3.6.2), and the auxiliary equation (3.6.3). This variational comparison theorem plays an important role in the study of qualitative behavior of solutions of (3.6.1) and (3.6.2).
(b) Suppose that f in (3.6.1) is linear, i.e., $f(t, \phi) = L(t, \phi)$, where L is linear in ϕ. Then any solution $x_t(\sigma, \phi)$ of (3.6.1) may be written in the form $T(t, \sigma)\phi$, where T is a linear operator associated with (3.3.3). Hypothesis (H_1) holds. In case $G(t, u) \equiv 0$, then Theorem 3.6.1 yields

$$V(t, y_t(\sigma, \phi)) \le V(t, T(t, \sigma)\phi), \ t \ge \sigma.$$

If $U(t, s) = |\phi|_0$, then it follows that

$$|y_t(\sigma, \phi)|_0 \le |T(t, \sigma)\phi|_0, \ t \ge \sigma.$$

Suppose that $G(t, u) = -\alpha u^2$, $\alpha > 0$. Then we have

$$V(t, y_t(\sigma, \phi)) \le \frac{V(t, T(t, \sigma)\phi)}{1 + \alpha(t - \sigma)V(t, T(t, \sigma)\phi)}, \ t \ge \sigma > 0.$$

In particular, if $V(t, \phi) = |\phi|_0$, then we obtain

$$|y_t(\sigma, \phi)|_0 \le \frac{|T(t, \sigma)\phi|_0}{1 + \alpha(t - \sigma)|T(t, \sigma)\phi|_0}.$$

As an application, we prove a result on practical stability. We define this property as follows.

Definition 3.6.1: The system (3.6.1) is said to be

(i)   practically stable if given (λ, A) with $0 < \lambda < A$ we have $\mid \phi \mid_0 < \lambda$
     implies that $\mid x_t(\sigma, \phi) \mid_0 < A$ for $t \geq \sigma$,
(ii)   practically quasi-stable if given $(\lambda, B, T) > 0$ and some $\sigma \in R_+$, we have
     $\mid \phi \mid_0 < \lambda$ implies $\mid x_t(\sigma, \phi) \mid_0 < B$, $t \geq \sigma + T$;
(iii) strongly practically stable, if (i) and (ii) hold simultaneously.
The other definitions of practical stability can be formulated similarly.

Theorem 3.6.2: *Assume that the hypothesis* (H_1) *in Section 3.3,* (H_4) *and* (H_5)
hold. Suppose that for each $(\sigma, \phi) \in R_+ \times C_B$, $C_B = \{\phi \in C, \mid \phi \mid_0 \leq B\}$,

$$\alpha(\mid \phi \mid_0) \leq \sum_{i=1}^{n} V_i(t, \phi) \leq \beta(\mid \phi \mid_0) \qquad (3.6.5)$$

where α *and* β *are functions belonging to class* $\mathcal{K}$ *and* $a(\lambda) < b(B)$. *Let*
$f(t, 0) = g(t, 0) = 0$ *for* $t \geq \sigma$. *Then, if the system* (3.6.1) *is* (λ, λ) *practically*
stable, the practical stability properties of (3.6.1) *imply the corresponding practical*
stability properties of (3.6.2).

Proof: Suppose that (3.6.3) is strongly practically stable.   Then given
$(\lambda, A, B, T) > 0$ such that $\lambda < A$, $B < A$

$$\sum_{i=1}^{n} u_i(t, \sigma, u_0) < b(A), t \geq \sigma \text{ if } \sum_{i=1}^{n} u_{0i} < a(\lambda) \qquad (3.6.6)$$

and

$$\sum_{i=1}^{n} u_{0i} < a(\lambda) \text{ implies } \sum_{i=1}^{n} u_i(t, \sigma, u_0) < b(B) \qquad (3.6.7)$$

for $t > \sigma + T$.
     Since (3.6.1) is (λ, λ) practically stable, we have

$$\mid x_t(\sigma, \phi) \mid_0 < \lambda, t \geq t_0 \text{ if } \mid \phi \mid_0 < \lambda.$$

We now claim that $\mid \phi \mid_0 < \lambda$ also implies that $\mid y_t(\sigma, \phi) \mid_0 < A$, $t > \sigma$, where
$y_t(\sigma, \phi)$ is any solution of (3.6.2).
     Suppose that this is not true. Then there exists a solution $y_t(\sigma, \phi)$ of (3.6.2) with
$\mid \phi \mid_0 < \lambda$ and a $t_1 > t_0$ such that

$$\mid x_t(\sigma, \phi) \mid_0 \leq A, \sigma \leq t \leq t_1.$$

By Theorem 3.6.1, we have

$$V(t, y_t(t_0, \phi)) \leq r(t, \sigma, V(\sigma, x_t(\sigma, \phi))), \ \sigma \leq t \leq t_1.$$

As a result, we obtain

$$b(A) \leq \sum_{i=1}^{n} V_i(t_1, y_{t_1}(\sigma, \phi)) \leq \sum_{i=1}^{n} r_i(t_1, \sigma, a(x_{t_1}(\sigma, \phi)))$$

$$\leq \sum_{i=1}^{n} r_i(t_1, t_0, a(\lambda)) < b(A).$$

This is a contradiction. Hence we have proved that

$$|\phi|_0 < \lambda \text{ implies } |y_t(\sigma, \phi)|_0 < A, t \geq \sigma.$$

To prove strong practical stability, we have

$$b(|y_t(\sigma, \phi)|_0) \leq \sum_{i=1}^{n} V_i(t, y_t(\sigma, \phi)) \leq \sum_{i=1}^{n} r_i(t, \sigma, V(\sigma, x_t(\sigma, \phi)))$$

for all $t \geq t_0$ if $|x_0| < \lambda$. It then follows that

$$b(|y_t(\sigma, \phi)|_0) \leq \sum_{i=1}^{n} r_i(t, \sigma, a(\lambda)), t \geq \sigma.$$

Now (3.6.7) yields strong practical stability of the system (3.6.2). The proof is complete.

In Theorem 3.3.2, we have established the generalized VPF by employing a Lyapunov function. The derivative of a Lyapunov function along a solution of FDE is then a functional which needs to be estimated by a comparison functional. Below we obtain generalized comparison principle for this situation. For this purpose, we have the following assumptions.

(H_6) $G \in C[R_+ \times R^n \times C_\rho, R^n]$, $G(t, 0, 0) \equiv 0$ and $G(t, u, \psi)$ is nondecreasing in u and ψ for each $(t, u) \in R_+ \times R^n$ and for each $(t, \psi) \in R^+ \times C_\rho$. Let $r(t, \sigma, \psi)$ be the maximal solution of the auxiliary equation

$$u'(t) = G(t, u(t), u_t), u(\theta) = \psi(\theta) \tag{3.6.8}$$

existing for $t \geq \sigma \geq 0$ and $\psi \in C_\rho$.

(H_7)   For $(t, x) \in R_+ \times R^n, t \geq s \geq \sigma \geq 0$,

$$V_x(t, x(t, s, \phi))[T(t, s; \phi)g_s](0) \leq G(t, V(t, x(t, s, \phi)), V_t)$$

where $V_t = V(t + \theta, x(t + \theta, s, \phi)), -r \leq \theta \leq 0$.

Based upon these hypotheses, we prove the following generalized comparison principle.

Theorem 3.6.3: *Assume the hypotheses* (H_1), (H_3), (H_6) *and* (H_7). *If* $x(t, \sigma, \phi)$ *is any solution of* (3.6.1) *such that*

$$V(t + \theta, x(t + \theta, \sigma, \phi)) \leq \psi(\theta), \theta \in [-r, 0], t \geq \sigma,$$

then for any solution $y(t, \sigma, \phi)$ of (3.6.2), we have

$$V(t, y(t, \sigma, \phi)) \leq r(t, \sigma, \psi), \, t \geq \sigma \qquad (3.6.9)$$

where $r(t, \sigma, \psi)$ is as defined in (H_6).

Proof: From the generalized VPF proved in Theorem 3.3.2, we have

$$\frac{d}{ds} V(t, x(t, s, y_s)) = V_x(t, x(t, s, y_s))[T(t, s; y_s)g_s](0)$$

$$\leq G(s, V(t, x(t, s, y_s)), V_t),$$

where $V_t = V(t + \theta, x(t + \theta, s, y_s))$, $-r \leq \theta \leq 0$.
Set $m(s) = V(t, x(t, s, y_s))$, $m_\sigma(\theta) = V(t + \theta, x(t + \theta, \sigma, \phi))$. We then have

$$D^+ m(t) \leq G(s, m(s), m_s), m_\sigma(\theta) \leq \psi(\theta).$$

Hence it follows that $m(s) \leq r(s, \sigma, \psi)$ for $t \geq s \geq \sigma$ which yields the inequality (3.6.9). The proof is complete.

We now apply the conclusions of Theorem 3.3.2 and Theorem 3.6.3 to a stability problem defined below.

Definition 3.6.2: The trivial solution of (3.6.1) is said to be equistable with respect to (3.6.8), if, for each $\epsilon > 0$, $\sigma \in R^+$ there exists a positive function $\delta = \delta(\sigma, \epsilon)$ that is continuous in σ for each $\epsilon > 0$ such that $|\phi|_0 \leq \delta$ implies that $\sum_{i=1}^{n} |r_i(t, \sigma, \psi)| < \epsilon$ for $t \geq \sigma$ where $u_0 \geq V(t, x(t, \sigma, \phi))$ and where $x(t, \sigma, \phi)$ and $r(t, \sigma, \psi)$ are solutions of (3.6.1) and (3.6.8) respectively.

Other definitions of stability can be formulated similarly.
We merely state the following stability result.

Theorem 3.6.4: *Let the hypothesis of Theorem 3.3.2 hold. Suppose further that $f(t, 0) \equiv 0$, $g(t, 0, 0) \equiv 0$ for $(t, \phi) \in [\sigma, \infty) \times C_\rho$,*

$$\alpha(|\phi|_0) \leq \sum_{i=1}^{n} V_i(t, \phi(0), \phi) \leq \beta(|\phi|_0)$$

where α and β are the functions belonging to the class k. Then equistability or asymptotic stability of the trivial solution of (3.6.1) implies the stability or asymptotic stability of the trivial solution of (3.6.2).

Finally, we give a simple but illustrative example in which we study stability property of nonlinear equation by using both the Lyapunov functional and Lyapunov function.

Example 3.6.1: Consider a scalar delay differential equation

$$x'(t) = -a(t)x(t) - b(t)x(t-r), \quad r > 0, t \geq \sigma \geq 0, \quad (3.6.10)$$

where a and b are bounded, continuous functions on R_+. Let $a(t) \geq \mu > 0$ for $t \geq \sigma \geq 0$. The perturbed equation is

$$y'(t) = -a(t)y(t) - b(t)y(t-r) + g(t, y(t), y(t-r)), \quad (3.6.11)$$

where $g(t, 0) \equiv 0$, $t \geq \sigma \geq 0$. The unique solution $x_t(\sigma, \phi)$ of (3.6.10) may be represented as $T(t, \sigma)\phi$, where $T(t, \sigma)$ is a linear operator associated with variational equation. Let

$$[T(t, s)\psi](\theta)[T(t, s)g_s](\theta) \leq \tfrac{1}{2}\lambda(t)\{[T(t, s)\psi](\theta)\}^2 \quad (3.6.12)$$

for $-r \leq \theta \leq 0$, $(t, \psi) \in [\sigma, \infty) \times C_\rho$, where

$$g_s(\theta) = 0 \text{ for } -r \leq \theta < 0 \text{ and } g_s(0) = g(s, \psi(0), \psi(-r)),$$

and $\lambda \in C[R_+, R_+] \cap L_1(0, \infty)$. Let

$$V(t, \psi) = \tfrac{1}{2}\psi^2(0) + \frac{\mu}{2}\int_{-r}^{0} \psi^2(\theta)d\theta.$$

Then

$$\frac{d}{ds}V(t, x_t(s, y_s)) = x_t(s, y_s)(0)[T(t, s)g_s](0)$$

$$+ \mu\int_{-r}^{0} x_t(s, y_s)(\theta)[T(t, s)g_s](\theta)d\theta$$

$$= [T(t, s)y_s](0)[T(t, s)g_s](0$$

$$+ \mu\int_{-r}^{0} [T(t, s)y_s](\theta)[T(t, s)g_s](\theta)d\theta.$$

In view of (3.6.12), it follows that

$$\frac{d}{ds}V(t, x_t(s, y_s)) \leq \lambda(t)V(t, x_t(s, y_s)).$$

The auxiliary equation corresponding to (3.6.3) is $u' = \lambda(t)u$ and hence $r(t, \sigma, u_0) = u_0 \exp(\int_\sigma^t \lambda(s)ds)$ where $u_0 = V(t, x_t(\sigma, \phi))$.

If in (3.6.10), we have $|b(t)| \leq \theta\mu$, $0 < \theta < 1$, $t \in [\sigma, \infty)$, then the trivial solution of (3.6.10) is uniformly asymptotically stable and consequently it is V-uniformly asymptotically stable. Hence we conclude that the trivial solution of (3.6.11) is uniformly asymptotically stable.

The above conclusion, for the equation (3.6.11) is obtained by using Lyapunov functional. Let us now consider (3.6.11) and employ Lyapunov function. Let g in (3.6.11) satisfy the condition

$$[T(t,s)y_t](0)[T(t,s)g_t](0) \leq - \lambda\{[T(t,s)y_t](0)\}^2$$

$$+ \mu\{[T(t,s)y_t](-r)\}^2$$

for $t \in R^+$ and $y(t) \in C[[-r,\infty),R]$ such that $y_t \in S_\rho$, $\lambda > 0$, $\mu > 0$, and $\lambda > \mu$. Let $x(t,\sigma,\phi) = [T(t,\sigma)\phi](0)$ denote the unique solution of (3.6.10), and $V(t,x) = x^2$. Then it is easy to show that the auxiliary equation which corresponds to (3.6.8) is given by

$$u'(t) = - 2\lambda u(t) + 2\mu u(t-r), \ u(\theta) = \phi(\theta), \ t \geq \sigma.$$

The trivial solution $u = 0$ of this equation is uniformly asymptotically stable under the given hypothesis. Now in (3.6.10), if $a < 0$ and $b < 0$ for $t \in [\sigma,\infty)$, then the trivial solution of (3.6.10) is unstable, but it is V-uniformly asymptotically stable. Hence by Theorem 3.6.4, the trivial solution of (3.6.11) is uniformly asymptotically stable.

3.6 NOTES

There exists extensive literature on FDEs. Several applications of VPFs appear in classical monographs by Hale [1], Lakshmikantham and Leela [1], Bellman and Cooke [1], Lakshmikantham, Wen and Zhang [1], and Oguztoreli [1].

Section 3.1 consists of linear VPF for neutral FDE. The method of proof adopted here is taken from Hale [1]. For additional results refer to Hale [1], Bellman and Cooke [1]. The nonlinear VPF for neutral FDE discussed in Section 3.2 is due to Ize and Ventura [1]. The VPF for FDEs and for difference-differential equations have been obtained also in Shanholt [1] and Hastings [1]. The nonlinear VPF for retarded FDE established by Jiongyu [1] is an interesting generalization in this direction. See also Delfour and Mitter [1] and Jiongyu [1] for the results on differentiability of solutions with respect to initial data for FDEs.

The study of VPF in terms of Lyapunov-like functionals as well as function is due to Deo and Torres [1] which is the content of Section 3.3.

Section 3.4 includes an application for asymptotically invariant sets. This work is due to Bernfeld and Lakshmikantham [1]. These results pertain to difference differential equations. Also refer to the papers by Rashbaev [1], Kamala and Lakshmikantham [1] for similar results in this direction.

The application of asymptotic equivalence given in Section 3.5 is taken from Ize and Ventura [1]. For various other applications of stability behavior of solutions of FDEs refer to the work of Ize and Molfetta [1], Ize and Ventura [1], Hale [2], Hale and Mayer [1], Daoyi [1], Li [1], Evans [1, 2] and Jiongyu [1].

The generalized comparison results given in Section 3.6 are due to Deo and Torres [1]. For the study of practical stability refer to Lakshmikantham, Matrosov and Sivasundaram [1]. For several applications to stability analysis, refer to Shendge [1].

4. Difference Equations

4.0 INTRODUCTION

Difference equations appear, as mathematical models, in describing several physical phenomena which are time-dependent. The study of discretization methods for differential equations has also increased the scope of the theory and applications of difference equations. In recent years the investigation of the theory of difference equations has assumed a greater importance as a well-deserved independent discipline.

The present chapter aims in developing the VPFs for linear and nonlinear difference equations of Volterra type which are also known as summary equations. The VPFs obtained are then employed in selected areas of applications.

In Section 4.1, we obtain linear VPF for systems of difference equations. These ideas are then extended to linear difference equations of Volterra type. We also discuss the question when the linear Volterra equation possesses an equivalent representation in terms of difference equations. We then extend such results to matrix difference equations.

Section 4.2 contains the study of VPF for nonlinear difference equations and the difference equations of Volterra type. An example is worked out to demonstrate the results discussed.

Section 4.3 offers a result on periodic solutions of difference equations of Volterra type, while Section 4.4 contains the discrete analog of Sturm-Liouville problem. The aim here is to develop the theory of Green's matrix for the BVP. The stability analysis of difference equations of Volterra type is the content of Section 4.5. The final Section 4.6 provides an application of VPF in numerical methods for nonlinear differential equations.

4.1 LINEAR VARIATION OF PARAMETERS

We begin by consider the linear difference equation of the type

$$\Delta x(n) = A(n)x(t), \qquad (4.1.1)$$

and

$$y(n) = A(n)y(n) + b(n), \qquad (4.1.2)$$

with

$$x(n_0) = y(n_0) = \eta, \qquad (4.1.3)$$

where $A(n)$ is $d \times d$ matrix whose elements $a_{ij}(n)$ are real or complex functions defined on $N_{n_0}^+$

$$N_{n_0}^+ = \{n_0, n_0 + 1, \ldots, n_0 + k, \ldots\}, \; k \in N^+, \; n_0 \in N,$$

$x, y \in R^d$ having components that are functions defined on N_0^+. Further, in (4.1.2) assume that $b(n) \in R^d$ and $\eta \in R^d$ is a given vector. Here the difference operator Δ means $\Delta x(n) = x(n+1) - x(n)$, $n \in N_{n_0}^+$. We aim at developing VPF for the system of equations (4.1.2).

Let $\Phi(n, n_0)$ represent the fundamental solution of the system of equations (4.1.1). The matrix $\Phi(n, n_0)$ verifies the matrix difference equation

$$\Delta x(n) = A(n)x(n), \; x(n_0) = I; \text{(Identity matrix)}.$$

Clearly $\Phi(n, n_0)$ has the following properties:
(*i*) $\Phi(n, s)\Phi(s, t) = \Phi(n, t)$,
(*ii*) $\Phi^{-1}(n, s) = \Phi(s, t)$, whenever Φ^{-1} exists.
Let $x(n, n_0, \sigma)$ denote the solution of the difference equation

$$\Delta x(n) = A(n)x(n), \; x(n_0) = c, \qquad\qquad (4.1.4)$$

where c is any n-vector. The solution $x(n, n_0, c)$ of (4.1.4) is given by

$$x(n, n_0, c) = \Phi(n, n_0)c.$$

To get the solution of (4.1.2), let us assume that the parameter c is a function of $n \in N_{n_0}^+$ and that $x(n, n_0, c_n)$ satisfies the system (4.1.2). By the method of VP we have

$$y(n, n_0, \eta) = x(n, n_0, c_n) = \Phi(n, n_0)c_n. \qquad\qquad (4.1.5)$$

Hence

$$\Delta y(n, n_0, \eta) = \Phi(n + 1, n_0)c_{n+1} - \Phi(n, n_0)c_n$$

$$= A(n)\Phi(n, n_0)c_n + b(n)$$

$$= \Phi(n + 1, n_0)c_n - \Phi(n, n_0)c_n + b_n.$$

It therefore follows that

$$\Phi(n + 1, n_0)[c_{n+1} - c_n] = b_n,$$

or

$$c_{n+1} = c_n + \Phi^{-1}(n + 1, n_0)b_n$$

$$= c_n + \Phi(n_0, n + 1)b_n,$$

which yields

$$c_n + \eta + \sum_{j=n_0}^{n-1} \Phi(n_0, j+1)b(j).$$

Substituting c_n in (4.1.5), we obtain

$$y(n, n_0, \eta) = \Phi(n, n_0)[\eta + \sum_{j=n_0}^{n-1} \Phi(n_0, j+1)b(j)]$$

$$= \Phi(n, n_0)\eta + \sum_{j=n_0}^{n-1} \Phi(n, j+1)b(j). \qquad (4.1.6)$$

The foregoing discussion yielded the following result.

Theorem 4.1.1: *The solution $y(n, n_0, \eta)$ of (4.1.2) and (4.1.3) is given by (4.1.6). This relation is obtained by the method of VP.*

Corollary 4.1.1: *If $A(n)$ in (4.1.2) is an $n \times n$ constant matrix A, then (4.1.6) takes the form*

$$y(n, n_0, \eta) = A^{(n-n_0)}\eta + \sum_{j=n_0}^{n-1} A^{n-j-1}b(j), \quad n \in N_{n_0}^+.$$

Let the matrix $A(n)$ and vector $b(n)$ of (4.1.2) be now defined on $N^{\pm} = 0, \pm 1, \pm 2, \ldots$. Let $k(n)$ denote the fundamental solution of (4.1.1), i.e.,

$$\Delta k(n) = A(n)k(n).$$

Then we have the following result for (4.1.1).

Theorem 4.1.2: *Suppose that $\sum_{j=-\infty}^{n} | k^{-1}(j+1) | < +\infty$ and $| b_j | < M, M > 0,$ $j \in N^{\pm}$. Then*

$$x(n) = \sum_{s=0}^{\infty} k(n)k^{-1}(n-s)b(n-s-1)$$

verifies (4.1.1).

The proof of this theorem is by verification.
The foregoing discussion can be further extended to difference equations of Volterra type given by

$$\Delta x(n) = A(n)x(n) + \sum_{s=n_0}^{n-1} k(n, s)x(s) + F(n), \qquad (4.1.7)$$

$$x(n_0) = \eta;$$

where $A(n), k(n, s)$ are $d \times d$ matrices for each $n, s \in N_{n_0}^+$, $n_0 \leq s \leq n$, and $F: N_{n_0}^+ \to R^d$.

We need the following result in subsequent discussions. Let us present a relation which corresponds to Fubini's theorem, the proof being obtained by mathematical induction.

Lemma 4.1.1: *Let $L(n, s), k(n, s)$ be $d \times d$ matrices defined for s, $n \geq n_0$ such that $L(n, s) = 0$, $k(n, s) = 0$ for $n, s < n_0$. Then the relation*

$$\sum_{s=n_0}^{n-1} L(n, s+1) \sum_{\sigma=n_0}^{s-1} k(s, \sigma) x(\sigma)$$

$$= \sum_{s=n_0}^{n-1} \sum_{\sigma=s+1}^{n-1} L(n, \sigma+1) k(\sigma, s) x(s) \qquad (4.1.8)$$

holds, where $x: N_{n_0}^+ \to R^d$.

We obtain VPF for (4.1.7) in the following result.

Theorem 4.1.3: *Assume that there exists a $d \times d$ matrix $L(n, s)$ defined on $n_{n_0}^+ \times N_{n_0}^+$ satisfying*

$$\Delta L_s(n, s) + L(n, s+1) A(s) + \sum_{\sigma=s+1}^{n-1} L(n, \sigma+1) k(\sigma, s) = 0, L(n, n) = I. \ (4.1.9)$$

Then the unique solution $x(n, n_0, \eta)$ of (4.1.7) is given by

$$x(n, n_0, \eta) = L(n, n_0) \eta + \sum_{s=n_0}^{n-1} L(n, s+1) F(s). \qquad (4.1.10)$$

Proof: Let $x(n) = x(n, n_0, \eta)$ be the solution of (4.1.7). Set $p(s) = L(n, s) x(s)$. Then

$$p(s+1) - p(s) = [L(n, s+1) - L(n, s)] x(s) + L(n, s+1)[x(s+1) - x(s)]$$

$$= \Delta L_s(n, s) x(s) + L(n, s+1) \left[A(s) x(s) + \sum_{\sigma=n_0}^{s} k(s, \sigma) \right] x(\sigma).$$

Summing both sides of the foregoing relation, between n_0 to $n-1$, and employing Lemma 4.1.1, we get

$$L(n, n) x(n) - L(n, n_0) \eta$$

$$= \sum_{s=n_0}^{n-1} \left[\Delta L_s(n, s) + L(n, s+1) A(s) + \sum_{\sigma=s+1}^{n-1} L(n, \sigma+1) k(\sigma, s) \right] x(s)$$

$$+ \sum_{s=n_0}^{n-1} L(n, s+1)F(s).$$

In view of (4.1.8), it follows that

$$x(n) = L(n, n_0)\eta + \sum_{s=n_0}^{n-1} L(n, s+1)F(s)$$

which is (4.1.10). The proof is complete.

Remarks 4.1.1: (i) The matrix $L(n, n_0)$ is called resolvent matrix for the Volterra difference equation (4.1.7) and the equation (4.1.9) is the resolvent equation.
(ii) If the perturbation term in (4.1.7), namely $F(n) = 0$ then we obtain homogeneous difference equation whose solution $x(n, n_0, \eta)$ is given by

$$x(n, n_0, \eta) = L(n, n_0)\eta.$$

(iii) In case th matrix $k(n, s)$ is of convolution type, i.e., $k(n, s) = k(n - s)$, then the resolvent matrix L has the form

$$L(n, s) = L(n - s).$$

In several situations, it is convenient to convert the Volterra difference equations into equivalent ordinary difference equations. This is what we achieve in the subsequent result.

Theorem 4.1.4: *Assume that there exist a $d \times d$ matrix $L(n, s)$ defined on $N_{n_0}^+ \times N_{n_0}^+$ satisfying*

$$k(n, s) + L(n, s+1) - L(n, s) + L(n, s+1)A(s)$$

$$+ \sum_{\sigma=s+1}^{n-1} L(n, \sigma+1)k(\sigma, s) = 0. \qquad (4.1.11)$$

Then the Volterra difference equation (4.1.7) is equivalent to ordinary linear difference equation

$$\Delta y(n) = B(n)y(n) + L(n, n_0)\eta + H(n), \ y(n_0) = \eta \qquad (4.1.12)$$

where

$$B(n) = A(n) - L(n, n) \ and \ H(n) = F(n) + \sum_{s=n_0}^{n-1} L(n, s+1)F(s). \ (4.1.13)$$

Proof: We first prove that every solution of (4.1.7) satisfies (4.1.12). Let $x(n) = x(n, n_0, \eta)$ be the solution of (4.1.7). Set $p(s) = L(n, s)x(s)$. We then have

$$p(s+1) - p(s) = [L(n, s+1) - L(n, s)]x(s) + L(n, s+1)[x(s+1) - x(s)].$$

Substituting $\Delta x(s)$ from (4.1.7), we obtain

$$p(s+1) - p(s) = [L(n, s+1) - L(n, s) + L(n, s+1)A(s)]x(s)$$

$$+ L(n, s+1)\left[\sum_{\sigma=n_0}^{s-1} k(s, \sigma)x(\sigma) + F(s) \right]. \tag{4.1.14}$$

Summing both sides of (4.1.14) from n_0 to $n-1$, we get

$$L(n, n)x(n) - L(n, n_0)\eta$$

$$= \sum_{s=n_0}^{n-1} \left[L(n, s+1) - L(n, s) + L(n, s+1)A(s) + \sum_{\sigma=s+1}^{n-1} L(n, \sigma+1)k(\sigma, s) \right]$$

$$+ \sum_{s=n_0}^{n-1} (n, s+1)F(s).$$

Using Lemma 4.1.1 and the relations (4.1.11), (4.1.13) and (4.1.7), we obtain

$$\Delta x(n) = B(n)x(n) + L(n, n_0)\eta + H(n)$$

which implies that $x(n)$ satisfies (4.1.12).

To prove the converse, i.e., every solution of (4.1.12), is also a solution of (4.1.7), let $y(n) = y(n, n_0, \eta)$ be any solution of (4.1.2) for $n \geq n_0$. Define

$$z(n) = \Delta y(n) - A(n)y(n) - F(n) - \sum_{s=n_0}^{n-1} k(n, s)y(s).$$

In view of (4.1.11). We then obtain

$$z(n) = \Delta y(n) - A(n)y(n) - F(n)$$

$$+ \sum_{s=n_0}^{n-1} \left[L(n, s+1) - L(n, s) + L(n, s+1)A(s) + \sum_{\sigma=s+1}^{n-1} L(n, \sigma+1)k(\sigma, s) \right] y(s).$$

By Lemma 4.1.1, it follows that

$$z(n) = \Delta y(n) - A(n)y(n) - F(n)$$

$$+ \sum_{s=n_0}^{n-1} [L(n, s+1) - L(s, n) + L(n, s+1)A(s)]y(s)$$

$$+ \sum_{s=n_0}^{n-1} L(n, s+1) \sum_{\sigma=n_0}^{s-1} k(s, \sigma)y(\sigma). \tag{4.1.15}$$

Setting $p(s) = L(n, s)y(s)$, we obtain as before

$$L(n, n)y(n) - L(n, n_0)\eta = \sum_{s=n_0}^{n-1}[L(n, s+1) - L(n, s)]y(s)$$

$$+ \sum_{s=n_0}^{n-1}L(s, s+1)\Delta y(s). \tag{4.1.16}$$

Using (4.1.12), (4.1.15), (4.1.16), we find that

$$z(n) = -\sum_{s=n_0}^{n-1}L(n, s+1)\left[\Delta y(s) - A(s)y(s) - F(s) = \sum_{\sigma=n_0}^{n-1}k(s, \sigma)y(\sigma)\right]$$

which in view of the definition of $z(n)$ yields

$$z(n) = -\sum_{s=n_0}^{n-1}L(n, s+1)z(s).$$

Observe that $z(n_0) = 0$ and hence it follows that $z(n) = 0$ for all $n \geq n_0$. The proof is complete.

Corollary 4.1.2: Consider the nonhomogeneous difference equation

$$\Delta x(n) = B(n)x(n) + p(n), \quad x(n_0) = \eta \tag{4.1.17}$$

where $B(n)$ is a $d \times d$ matrix on $N_{n_0}^+$ and $p: N_{n_0}^+ \to R^d$. The solution $x(n) = x(n, n_0, \eta)$ of (4.1.17) is given by the VPF

$$x(n) = \Phi(n, n_0)\eta + \sum_{s=n_0}^{n-1}\Phi(n, s+1)p(s)$$

where $\Phi(n, n_0)$ is the fundamental matrix solution of the difference equation

$$\Delta x(n) = B(n)x(n),$$

and $\Phi(n_0, n_0) = I$.

Finally, we establish below VPF for matrix difference equation given by

$$\Delta x(n) = A(n)x(n) + x(n)B(n) + F(n, x(n))$$

$$x(n_0) = \eta, \tag{4.1.18}$$

the unperturbed system being

$$\Delta y(n) = A(n)y(n) + y(n)B(n), \quad y(n_0) = \eta, \tag{4.1.19}$$

where A and B are $d \times d$ matrices, $F: N_{n_0}^+ \times \mathcal{F} \to \mathcal{F}$, $\mathcal{F}$ being the set of $d \times d$ matrices, $\eta \in \mathcal{F}$ is a preassigned matrix. We prove the following result.

Theorem 4.1.5: *Let $y(n)$ be a matrix valued solution of (4.1.19) for $n \geq n_0$ such that $y(n_0) = \eta$, $\eta \in \mathcal{F}$. Moreover, let $v(n, n_0)$, $w(n, n_0)$ be the matrix solutions of the linear matrix difference equations*

$$\Delta v(n) = A(n)v(n), \ v(n_0) = I \ (I \in \mathcal{F} \text{ is a unit matrix})$$

$$\Delta w(n) = w(n)B(n), \ \ w(n_0) = I \qquad (4.1.20)$$

existing for $n \geq n_0$ respectively. Then $y(n)$, is given by

$$y(n) = v(n, n_0)\eta w(n, n_0) + \sum_{s=n_0}^{n-1} v(n, s+1)F(s, x(s))w(n, s+1) \ \ (4.1.21)$$

verifies the equation (4.1.18) for $n \in N_{n_0}^+$.

Proof: It is easy to see that

$$y(n) = v(n, n_0)\eta w(n, n_0)$$

is the solution of (4.1.19) such that $y(n_0) = \eta$. Employing the method of variation of parameters, we replace η by $\eta(n)$ with the expectation that

$$x(n) = v(n, n_0)\eta(n)w(n, n_0), \ \ n \in N_{n_0}^+ \qquad (4.1.22)$$

is the solution of (4.1.18). Hence it follows from (4.1.20) that

$$\Delta x(n) = \Delta[v(n, n_0)\eta(n)w(n, n_0)]$$

$$= \Delta v(n, n_0)[\eta(n)w(n, n_0)] + v(n, n_0)\Delta \eta(n)w(n, n_0) + [v(n, n_0)\eta(n)]\Delta w(n, n_0).$$

$$= A(n)v(n, n_0)\eta(n)w(n, n_0) + v(n, n_0)\Delta \eta(n)w(n, n_0)$$

$$+ v(n, n_0)\eta(n)w(n, n_0)B(n, n_0).$$

Also from (4.1.18), we have

$$\Delta x(n) = A(n)V(n, n_0)\eta(n)w(n, n_0) + v(n, n_0)\eta(n)w(n, n_0)B(n) + F(n, x(n)).$$

Hence, we conclude that

$$v(n, n_0)\Delta \eta(n)w(n, n_0) = F(n, x(n))$$

or

$$\Delta \eta(n) = \eta(n+1) - \eta(n) = v^{-1}(n, n_0)F(n, x(n))w^{-1}(n, n_0).$$

Summing between η_0 to $n-1$, we obtain

$$\eta(n) = \eta + \sum_{s=n_0}^{n-1} v^{-1}(s+1, n_0) F(s, x(s)) w^{-1}(s+1, n_0).$$

Substituting $\eta(n)$ in (4.1.22) yields the relation (4.1.21) and the proof is complete.

Corollary 4.1.3: *In the case A and B are constant matrices*

$$v(n, n_0) = A^{n-n_0} = v(n-n_0, 0) = v(n-n_0), \quad (say)$$

$$w(n, n_0) = B^{n-n_0} = w(n-n_0, 0) = w(n-n_0), \quad (say)$$

the VPF given by (4.1.21) *takes the form*

$$x(n) = A^{n-n_0} \eta B^{n-n_0} + \sum_{s=n_0}^{n-1} A^{n-s-1} F(s, x(s)) B^{n-s-1}.$$

4.2 NONLINEAR VARIATION OF PARAMETERS

We shall now consider the nonlinear difference equations and nonlinear difference equations of Volterra type and establish the VPF for such equations. As in the foregoing situation, we need to formulate variational equation, the fundamental solution of which plays an important role.

Consider the equation

$$\Delta x(n) = f(n, x(n)), x(n_0) = x_0; \tag{4.2.1}$$

where $f: N_{n_0}^+ \times R^d \to R^d$ and f possesses partial derivations on $N_{n_0}^+ \times R^d$. We prove the following result.

Theorem 4.2.1: *Assume that f is as given in* (4.2.1). *Let $x(n, n_0, x_0)$ denote the solution of* (4.2.1) *and that it exists for $n \geq n_0$. Let*

$$H(n, n_0, x_0) = \frac{\partial f(n, x(n, n_0, x_0))}{\partial x}. \tag{4.2.2}$$

Then

$$\Phi(n, n_0, x_0) = \frac{\partial x(n, n_0, x_0)}{\partial x_0}, \tag{4.2.3}$$

exists and is the solution of

$$\Delta \Phi(n, n_0, x_0) = H(n, n_0, x_0) \Phi(n, n_0, x_0)$$

$$\Phi(n_0, n_0, x_0) = I. \tag{4.2.4}$$

Proof: Let $x(n, n_0, x_0)$ be the solution of (4.2.1) existing on $N_{n_0}^+$. Differentiate (4.2.1) with respect to x_0 to get

$$\Delta\Phi(n, n_0, x_0) = H(n, n_0, x_0)\Phi(n, n_0, x_0)$$

which is (4.2.4).

We are now in a position to prove the VPF for nonlinear difference equation

$$\left.\begin{array}{r}\Delta y(n) = f(n, y(n)) + F(n, y(n)), \\ y(n_0) = x_0.\end{array}\right\} \qquad (4.2.5)$$

Theorem 4.2.2: *Let $f, F: N_{n_0}^+ \times R^d \to R^d$ and let $\frac{\partial f}{\partial x}$ exist and be continuous and invertible on $N_{n_0}^+ \times R^d$. If $x(n, n_0, x_0)$ is the solution of (4.2.1), then any solution of (4.2.5) satisfies the relation*

$$y(n, n_0, x_0) = x\left(n, n_0, x_0 + \sum_{j=n_0}^{n-1}\Psi^{-1}(j+1, n_0, v_j, v_{j+1})F(j, y(j))\right) \quad (4.2.6)$$

where

$$\Psi(n, n_0, v_j, v_{j+1}) = \int_0^1 \Phi(n, n_0, sv_j + (1-s)v_{j-1})ds, \qquad (4.2.7)$$

and

$$\Delta v(n) = \Psi^{-1}(n+1, n_0, v_n, v_{n+1})F(n, y(n)). \qquad (4.2.8)$$

Proof: Set

$$y(n, n_0, x_0) = x(n, n_0, v(n)), \quad v_0 = x_0. \qquad (4.2.9)$$

Then

$$\Delta y(n, n_0, x_0) = y(n+1, n_0, x_0) = y(n, n_0, x_0)$$

$$= x(n+1, n_0, v(n+1)) - x(n, n_0, v(n)) - x(n+1, n_0, v(n)) + x(n+1, n_0, v(n))$$

$$= f(n, x(n, n_0, v(n)) + F(n, x(n, n_0, v(n)))$$

$$= x(n+1, n_0, v(n)) - x(n, n_0, v(n)) + F(n, x(n, n_0, v(n)))$$

from which, we get

$$x(n+1, n_0, v(n+1)) - x(n+1, n_0, v(n)) = F(n, x(n, n_0, v(n))).$$

Applying the mean value theorem, we have

$$\int\limits_0^1 \frac{\partial x(n+1,n_0,sv(n+1)+(1-s)v(n))}{\partial x_0}ds \cdot (v(n+1) - v(n) = F(n,y(n)).$$

Hence by (4.2.3) we obtain

$$\int\limits_0^1 \Phi(n+1,n_0,sv(n+1)+(1-s)v(n))ds \cdot (v(n+1) - v(n)) = F(n,y(n)).$$

In view of (4.2.7), one sees that

$$\Psi(n,n_0,v(n),v(n+1))\Delta v(n) = F(n,y(n)),$$

from which, we get (4.2.8) having the solution

$$v(n) = x_0 + \sum_{j=n_0}^{n-1} \Psi^{-1}(j+1,n_0,v_j,v_{j+1})F(j,y(j))$$

$n \in N_{n_0}^+$.

We now substitute $v(n)$ in (4.2.9) to get (4.2.6). The proof is complete.

Corollary 4.2.1: *Under the hypothesis of Theorem 4.2.2, any solution $y(n,n_0,x_0)$ of (4.2.5) can be written as*

$$y(n,n_0,x_0) =$$
$$x(n,n_0,x_0) + \sum_{j=n_0}^{n-1} \Psi(n,n_0,v(n),x_0)\Psi^{-1}(j+1,n_0,v(j),v(j+1))F(j,y(j)).$$

For the proof, apply the mean value theorem to (4.2.6).

Corollary 4.2.2: *If $f(n,x(n)) = A(n)x(n)$ in (4.2.1) then (4.2.6) reduces to (4.1.6).*

Proof: In this case, $x(n,n_0,x_0) = \Phi(n,n_0)x_0$ and

$$\Phi(n,n_0,x_0) = \Psi(n,n_0) = \Psi(n,n_0,v(n),v(n+1)).$$

Hence

$$\Delta v(n) = \Psi^{-1}(n+1,n_0,v(n),v(n+1))F(n,y(n))$$

and

$$v(n) = x_0 + \sum_{j=n_0}^{n-1} \Psi^{-1}(j+1,n_0,v(j),v(j+1))F(j,y(j)).$$

The claim is now obvious.

Finally, we obtain the VPF for difference equations of Volterra type of the form

$$\Delta x(n) = f\left(n, x(n), \sum_{i=n_0}^{n-1} k(n, i, x(i))\right), \quad x(n_0) = x_0, \qquad (4.2.10)$$

$$\Delta y(n) = \left(f(n, y(n), \sum_{i=n_0}^{n-1} k(n, i, y(i)))\right) + g(n), \, y(n_0) = x_0, \qquad (4.2.11)$$

where $n \in N_{n_0}^+$. Let $k: N_{n_0}^+ \times N_{n_0}^+ \times R^d \to R^d$, $f \in N_{n_0}^+ \times R^d \to R^d$, $g: N_{n_0}^+ \to R^d$ are assumed to be continuous for $n \in N_{n_0}^+$. The given systems are natural analogs in discrete form of integrodifferential equations.

In (4.2.11) the perturbation term $g(n)$ may be replaced by a more general term

$$g\left(n, y(n), \sum_{i=n_0}^{n-1} h(n, i, y(i))\right) \qquad (4.2.12)$$

where $h: N_{n_0}^+ \times N_{n_0}^+ \times R^d \to R^d$ and $g: N_{n_0}^+ \times R^d \times R^d \to R^d$.

Let $x(n) = x(n, n_0, x_0)$ be any solution of (4.2.10). Set $\Delta_n x(n, n_0, w(n)) = x(n + 1, n_0, w(n)) - x(n, n_0, w(n))$, $w(n)$ being real valued function defined on $N_{n_0}^+$. Define

$$\Phi(n, n_0, w(n)) = \Delta_{w(n)} u(n, n_0, w(n))$$

$$= u(n, n_0, w(n + 1)) - u(n, n_0, w(n)).$$

We now prove the following result.

Theorem 4.2.3: *Suppose that $x(n, n_0, x_0)$ is the unique solution of* (4.2.10) *and* $\Phi(j + 1, n_0, p(j))$ $\Phi^{-1}(j + 1, n_0, p(j))$ *exist for $n_0 \le j < n$. Then any solution* $y(n, n_0, x_0)$ *of* (4.2.11) *satisfies the relation*

$$y(n, n_0, \eta) =$$
$$x\left(n, n_0, \eta + \sum_{j=n_0}^{n-1} \Phi^{-1}(j + 1, n_0, p(j))\right) \cdot \Omega(j, n_0, x_0, p(j)); \qquad (4.2.13)$$

for $n \in N_{n_0}^+$, where $p(j)$ is the solution of

$$\Delta p(j) = \Phi^{-1}(j + 1, n_0, p(j)\Omega(j, n_0, \eta, p(j)), p(n_0) = x_0 \qquad (4.2.14)$$

$n_0 \le j < n$ and

$$\Omega(j, n_0, x_0, p(j)) = f\left(j, x(j, n_0, p(j)), \sum_{i=n_0}^{j-1} k(j, i, x(i, n_0, p(i)))\right)$$

$$- f\left(j, x(j, n_0, p(j)), \sum_{i=n_0}^{j-1} k(j, i, x(i, n_0, p(j)))\right) + g(j) \qquad (4.2.15)$$

$n_0 \le j < n$.

Proof: Since $x(n, n_0, x_0)$, $n \in N_{n_0}^+$ is the solution of (4.2.10), by the method of variation of parameters, we can find the solution of (4.2.11) by the relation

$$y(n, n_0, x_0) = x(n, n_0, p(n)), \quad n \in N_{n_0}^+, \quad p(n_0) = x_0 \qquad (4.2.16)$$

where the function $p(n)$ needs to be determined appropriately. For this purpose, it is necessary that

$$\Delta y(j, n_0, x_0) = x(j+1, n_0, p(j+1)) - x(j, n_0, p(j))$$

$$\Delta_{p(j)} x(j+1, n_0, p(j)) \Delta p(j) + \delta_j x(j, n_0, p(j))$$

$$= \Phi(j+1, n_0, p(j)) \Delta p(j) + f\left(j, x(j, p(j)), \sum_{i=n_0}^{j-1} k(j, i, x(i, n_0, p(j)))\right)$$

$$= f\left(j, y(j), \sum_{i=n_0}^{j-1} k(j, i, y(i))\right) + g(j), \quad n_0 \le j < n.$$

From (4.2.16) and the existence of $\Phi^{-1}(j+1, n_0, p(j))$, $n_0 \le j < n$, we obtain

$$\Delta p(j) = \Phi^{-1}(j+1, n_0, p(j)) \left\{ f\left(j, y(j), \sum_{i=n_0}^{j-1} k(j, i, y(i))\right) \right.$$

$$- f\left(j, y(j), \sum_{i=n_0}^{j-1} k(j, i, x(i, n_0, p(j)))\right) + g(j) \right\}$$

$$= \Phi^{-1}(j+1, n_0, p(j)) \Omega(j, n_0 x_0, p(j)), \qquad (4.2.17)$$

for $n_0 \le j < n$ and $p(n_0) = x_0$. Thus we have determined $p(j)$. Further from (4.2.17), we see that

$$p(n) = \Delta p(n-1) + \Delta p(n-2) + \dots + \Delta p(n_0) p(n_0)$$

$$= x_0 + \sum_{j=n_0}^{n-1} \Phi^{-1}(j+1, n_0, p(j)) \Omega(j, n_0, x_0, p(j))$$

for $n \in N_{n_0}^+$. This expression together with (4.2.16) yields (4.2.13).

Remark 4.3.2: The VPF (4.2.13) includes several particular cases when f in (4.2.10) takes the values $A(n)y(n)$ or $f(n, y(n))$, or $A(n) \sum_{i=n_0}^{n-1} B(i)y(i)$, where the matrices A and B are suitable defined.

The following theorem provides sufficient conditions for the existence of the matrix $\Phi(n+1, n_0, p(n))$.

Theorem 4.2.4: *Suppose that $x(n, n_0, x_0)$ is the unique solution of (4.2.10). If f is differentiable with respect to the second and third argument, k is differentiable with respect to the third argument, then*

$$\Phi(n, n_0, p(j)) = \int_0^1 \frac{\partial x(n, n_0, tp(j+1) + (1-t)p(j))}{\partial x_0} \, dt$$

$j, n \in N_{n_0}^+$, $j < n$.

Proof: By using the method of induction, we first show that $\frac{\partial x}{\partial x_0}(n, n_0, x_0)$, $n \in N_{n_0}^+$ exists. Note that since $x(n_0, n_0, x_0) = x_0$,

$$\frac{\partial x}{\partial x_0}(n_0, n_0, x_0) = I.$$

Denote $\sum_{i=n_0}^{n-1} k(n, i, x(i))$ by γ. Then the equation (4.2.10) may be written as

$$x(n+1) = x(n) + f(n, x(n), \gamma), \, n \in N_{n_0}^+;$$

$$x(n_0) = x_0.$$

Hence if $\frac{\partial x(n, n_0, x_0)}{\partial x_0}$ exists, then in view of the hypothesis, differentiation of the above equation with respect to x_0 gives that

$$\frac{\partial x(n+1)}{\partial x_0} = \frac{\partial x(n)}{\partial x_0} + \frac{\partial f}{\partial x}\frac{\partial x}{\partial x_0} + \frac{\partial f}{\partial \gamma} \sum_{i=n_0}^{n-1} \frac{\partial k(n, i, x(i))}{\partial x(i)} \frac{\partial x(i)}{\partial x_0}, \, n = n_0, n_0+1, \ldots$$

and this proves the existence of $\frac{\partial x}{\partial x_0}(n, n_0, x_0)$ for all $x_0 \in N_{n_0}^+$.

Using the mean value theorem, we get

$$\Phi(n, n_0, p(j))\Delta p(j) = x(n, n_0, p(j+1)) - x(n, n_0, p(j))$$

$$= \int_0^1 \frac{\partial x(n, n_0, tp(j+1) + (1-t)p(j)}{\partial x_0} \, dt \cdot \Delta p(j)$$

$j, n \in N_{n_0}^+$, $j \leq n$. The conclusion is now obvious.

We now get an illustrative example. We consider a difference equation of Volterra type (4.2.11) together with the perturbation term of the type (4.2.12). The equation under consideration is

$$\Delta w(n) = \alpha^2 \sum_{i=n_0}^{n-1} \beta_{i(n)} W(i) + 2n + 1 - \alpha^2(2n-3)$$

$$w(n_0) = u_0$$

where $|\alpha| \neq 1$ and

$$\beta_{i(n)} = \begin{cases} 1 & i(n) = n - 1 \geq n_0 \\ -1 & i(n) = n - 2 \geq n_0 \\ 0 & \text{otherwise.} \end{cases}$$

It is clear that

$$\Delta w(n_0) = 2n_0 + 1 - \alpha^2(2n_0 - 3); \quad \Delta^2 w(n_0) = \alpha^2 u_0 + 2(1 - \alpha^2)$$

and hence the unique solution of the above equation is

$$w(n) = c_1 + c_2(\alpha)^{n-n_0} + c_3(-\alpha)^{n-n_0} + n^2 - n_0^2, \qquad (4.2.18)$$

where

$$c_1 = u_0 - \tfrac{\alpha^2}{\alpha^2-1}(4 + u_0 - 4n_0), \ c_2 = \tfrac{\alpha}{2(\alpha-1)}(3\alpha + 1 + u_0 - 2n_0(\alpha + 1))$$

and

$$c_3 = \tfrac{\alpha}{2(\alpha+1)}(-3\alpha + 1 + u_0 + 2n_0(\alpha - 1)).$$

Let $n_0 = 0$. For $\alpha = 2$ and $\alpha = 3$ the above equation reduces to

$$\Delta u(n) = 4\sum_{i=0}^{n-1}\beta_{i(n)}u(i) + 2n + 1 - 4(2n - 3), \ u(0) = u_0, \qquad (4.2.19)$$

and

$$\Delta v(n) = 4\sum_{i=0}^{n-1}\beta_{i(n)}v(i) + 2n + 1 - 4(2n - 3) + 5\sum_{i=0}^{n-1}\beta_{i(n)}v(i) - 5(2n - 3)$$

$$v(0) = u_0. \qquad (4.2.20)$$

The equation (4.2.20) will be treated as a perturbed equation of (4.2.19).

In view of (4.2.18), the unique solution of the above equations (4.2.19) and (4.2.20) can be written as

$$u(n) = c_1 + c_2(2)^n + c_3(-2)^n + n^2, \ v(n) = \hat{c}_1 + \hat{c}_2(3)^n + \hat{c}_3(-3)^n + n^2,$$

where $c_1 = -16/3 - u_0/3$, $c_2 = 7 + u_0$, $c_3 = -5/3 + u_0/3$ and $\hat{c}_1 = -9/2 - u_0/8$, $\hat{c}_2 = 30/4 + 3u_0/4$, $\hat{c}_3 = -3 + 3u_0/8$.

Now consider the VPF (4.2.12) replacing $g(j)$ in (4.2.15) by the function (4.2.12). From (4.2.20), it is clear that

$$g\left(n, u(n), \sum_{i=n_0}^{n-1}h(n, i, v(i))\right) = 5\sum_{i=0}^{n-1}\beta_{i(n)}v(i) = -5(2n - 3).$$

It is easy to verify that $p(1) = u_0 + 15$, $p(2) = \frac{11u_0}{16} + \frac{51}{16}$ and $\Phi^{-1}(n+1, 0, p(n)) = \frac{3}{((3-(-1)^n)2^{n+1})}$, $n \geq 0$. Hence, a simple computation yields

$$v(1, 0, u_0) = u(1, 0, u_0 + \Phi^{-1}(1, 0, p(0))\Omega(0, 0, p(0)))$$

$$= 10 + u_0,$$

and

$$v(2, 0, u_0) = u(2, 0, u_0 + \sum_{j=0}^{1} \Phi^{-1}(j+1, 0, p(j))\Omega(j, 0, u_0, p(j)))$$

$$= 40 + 10u_0.$$

This computation can be extended for all values $n \geq 3$ similarly.

4.3 PERIODIC SOLUTIONS

This section is devoted to the study of periodic solutions of Volterra difference equations. We shall utilize the VPF for Volterra difference equations obtained in Section 4.1. Consider the following system of difference equations of nonconvolution type

$$x(n+1) = A(n)x(n) + \sum_{r=0}^{n} k(n, r)x(r), \qquad (4.3.1)$$

and its perturbation

$$y(n+1) = A(n)y(n) + \sum_{r=0}^{n} k(n, r)y(r) + g(n), \qquad (4.3.2)$$

where $A(n)$, $B(n, r)$ are $d \times d$ matrix functions on Z^+ and $Z^+ \times Z^+$ respectively and $g: Z^+ \to R^d$. Here we denote Z, Z^+, Z^- by sets of integers, nonnegative integers and nonpositive integers respectively.

We develop the theory to find a periodic solution of the difference system with infinite delay

$$z(n+1) = A(n)z(n) + \sum_{r=-\infty}^{n} k(n, r)z(r) + g(n), \qquad (4.3.3)$$

where

$$A(n+N) = A(n), k(n+N, r+N) = k(n, r), g(n+N) = g(n), \quad (4.3.4)$$

for $n, m \in Z$ and $N \in Z^+$, $N \neq 0$.

We further assume that

$$\sum_{r=0}^{\infty} |k(n, n-r)| < \infty. \tag{4.3.5}$$

Consider $\phi: Z^{-} \to R^{d}$. A solution $z(n, 0, \phi)$ of (4.3.3) is a sequence that verifies (4.3.3) for $n \in z^{+}$ such that initially $z(r) = \phi(r)$ for $r \in Z^{-}$. Assume that ϕ is bounded on Z^{-}. Under the hypothesis (4.3.5), it is known that solutions of (4.3.3) exist and are unique. In case $\sum_{r=-\infty}^{1} k(n, r)\phi(r) = 0$ or if $\phi(r) = 0$ for all negative integers, then equation (4.3.3) reduces to (4.3.2).

The following result is crucial in our discussion.

Lemma 4.3.1: *If $y(n)$ is a solution of (4.3.2) bounded on Z^{+}, then there is a corresponding solution $z(n)$ of (4.3.3) such that for every $n \in Z$, $z(n)$ is the limit of some subsequence of $y(n)$.*

Proof: Let $y(n)$ be a bounded solution of (4.3.2). Then $\{y(r, N)\}_{r=1}^{\infty}$, N being given in (4.3.4) is bounded and hence has a convergent subsequence $\{y(r_{i0}N)\}$ which converges to a point in R^{d}, say $z(0)$. There exists another subsequence $\{r_{i1}N\}$ of $\{r_{i0}N\}$ such that both $\{y(1 + r_{i1}N)\}$ and $\{Y(-1, r_{i1}N)\}$ converge to $z(1)$ and $z(-1)$ respectively. By induction, one shows that for each $n \geq 0$, $\{y[\pm(n-1) + r_{i(n-1)}N\}$ converge to $z(n-1)$ and $z[-(n-1)]$ respectively and $\{(\pm n + r_{in}N)\}$ converge to $z(n)$ and $z(-n)$ respectively where $\{r_{in}\}$ is a subsequence of $\{r_{i(n-1)}\}$.

We claim that $\{z(n)\}_{-\infty}^{\infty}$ is actually a solution of (4.3.3). From (4.3.2), we have

$$y[n + 1 + r_{i(n+1)}N] = A(n)y[n + r_{i(n+1)}N]$$

$$+ \sum_{u=0}^{n+r_{i(+1)}N} k(n + r_{i(+1)}N, j)y(j) + g(n, r_{i(n+1)}N)$$

$$= A(n)y(n + r_{i(+1}N) + \sum_{j=-r_{i(n1)}N}^{n} k(n, j)y(j + r_{i(n+1)}N) + g(n). \tag{4.3.6}$$

Let $r_{i(n+1)} \to \infty$. Then the LHS of (4.3.6) converges to $z(n+1)$ and the RHS converges to

$$A(n)z(n) + \sum_{j=-\infty}^{n} k(n, j)z(j) + g(n).$$

Hence $z(n)$ is a solution of (4.3.3).

The proof of the following lemma is a simple observation.

Lemma 4.3.2: *Let $L(n, s)$ be the matrix defined in (4.1.9) and the condition (4.3.4) holds. Then*

$$L(n + N, s + N) = L(n, s). \tag{4.3.7}$$

We are now in a position to prove the main result of this section.

Theorem 4.3.1:   *Suppose that the trivial solution of* (4.3.1) *is uniformly asymptotically stable. Then equation* (4.3.3) *has the unique N-periodic solution*

$$z(n) = \sum_{m=-\infty}^{n-1} L(n, m+1)g(m) \tag{4.3.8}$$

where the resolvent matrix L is defined in (4.1.9).

Proof: The uniform asymptotic stability of the trivial solution of (4.3.1) yields

$$|L(n,m)| \le R\gamma^{n-m}, \ R > 0, \gamma \in (0,1)$$

which implies $\lim_{n\to\infty} L(n,0) = 0$. The VPF proved in Theorem 4.1.3, shows that solutions of (4.3.2) are bounded. Following Lemma 4.3.1, one constructs a sequence $\{y(n + r_{in}N\}$ from a bounded solution $y(n)$ of (4.3.2) such that $\{y(n + r_{in}N)\}$ converges to a solution $z(n)$ of (4.3.3). from the VPF (4.1.10) we obtain

$$y(n + r_{in}N) = L(n + r_{in}N, 0)y(0) + \sum_{m=-r_{in}N}^{n-1} L(n, m+1)g(m).$$

Hence,

$$z(n) = \lim_{r_{in}\to\infty} y(n + r_{in}N) = \sum_{m=-\infty}^{\infty} L(n, m+1)g(m).$$

By Lemma 4.3.2, $z(n)$ is N-periodic.

It remains to show that $z(n)$ given by (4.3.8) is the only N-periodic solution of (4.3.3). For this purpose, assume that $\bar{z}(n)$ is another N-periodic solution of (4.3.3). Set $\psi(n) = z(n) - \bar{z}(n)$. Clearly, $\psi(n)$ is an N-periodic solution of the equation

$$\psi(n + 1) = A(n)\psi(n) + \sum_{r=0}^{n} k(n, r)\psi(r) + \sum_{r=-\infty}^{-1} k(n, r)\psi(r).$$

By the VPF (4.1.10), we have

$$\psi(n) = L(n, 0)\psi(0) + \sum_{j=0}^{n-1} L(n, j+1)\left[\sum_{r=-\infty}^{-1} k(n, r)\psi(r)\right]$$

$$= L(n, 0)\psi(0) + \sum_{j=0}^{n-1} L(n, j+1)\left[\sum_{r=j+1}^{\infty} k(j, j-r)\psi(j-r)\right].$$

Hence

$$|\psi(n)| \le |L(n,0)|\,|\psi(0)| + M\sum_{j=0}^{n-1}|L(n, j+1)|\sum_{r=+1}^{\infty}|k(j, j-r)|,$$

where $M = \sup\{\psi(n), n \in Z\}$.

Hence $\lim_{n \to \infty} \psi(n) = 0$. Since $\psi(n)$ is periodic, it follows that $\psi(n)$ is identically zero. Hence $z(n) = \overline{z}(n)$. The proof is complete.

Remark 4.3.1: Suppose that system (4.3.2) has an N-periodic solution $y(n)$ and $z(n)$ is the unique N-periodic solution (4.3.8) of the equation (4.3.3). Then $w(n) = z(n) - y(n)$ is a solution of the equation

$$w(n+1) = A(n)w(n) + \sum_{r=0}^{n} k(n,r)w(r) + \sum_{r=-\infty}^{-1} k(n,r)\phi(r) \qquad (4.3.9)$$

where ϕ is the initial function of $z(n)$.

By the VPF (4.1.10), we have

$$w(n) = L(n,0)w(0) + \sum_{j=0}^{n-1} L(n,j+1) \left[\sum_{r=-\infty}^{-1} k(j,r)\phi(r) \right].$$

Hence $\lim_{n \to \infty} w(n) = \lim_{n \to \infty} (z(n) - y(n)) = 0$ implying that $z(n) \equiv y(n)$. It then follows from (4.3.9) that

$$\sum_{r=-\infty}^{-1} L(n,r)\phi(r) = 0.$$

4.4 BOUNDARY VALUE PROBLEMS

In the present section, we give a simple application for VPF for linear difference equations obtained in Theorem 4.1.1. We consider Sturm-Liouville problem and obtain Green's matrix representation.

The discrete analog of the Sturm-Liouville problem is the following:

$$\Delta(p_{k-1}\Delta y_{k-1}) + (q_k + \lambda r_k)y_k = 0, \qquad (4.4.1)$$

$$\alpha_0 y_0 + \alpha_1 y_1 = 0, \ \alpha_M y_M + \alpha_{M+1} y_{M+1} = 0, \qquad (4.4.2)$$

where all the sequences are of real numbers, $r_k > 0$, $\alpha_0 \neq 0$, $\alpha_M \neq 0$ and $0 \leq k \leq M$. The problem can be treated by using arguments very similar to the continuous case. We shall transform the problem into a vector form, and reduce it to a problem of linear algebra. Note that the equation (4.4.1) can be rewritten as

$$p_k y_{k+1} - (p_k p_{k-1})y_k + p_{k-1}y_k + (q_k + \lambda r_k)y_k = 0.$$

Let

$$a_k = p_k + p_{k-1} - q_k, \quad k = 2, \ldots, M-1,$$

$$a_1 = p_1 + p_0 - q_1 + \frac{\alpha_1}{\alpha_0}p_0, \ a_M = p_M + p_{M-1} - q_M + \frac{\alpha_M}{\alpha_{M+1}}p_M,$$

$$y = (y_1, y_2, \ldots, y_M)^T, \ R = \text{diag}(r_1, r_2, \ldots, r_M),$$

and

$$
A = \begin{pmatrix}
a_1 & -p_1 & 0 & \cdots & & 0 \\
-p_1 & a_2 & -p_2 & & & \vdots \\
0 & & & & & 0 \\
\vdots & & & & & -p_{M-1} \\
0 & \cdots & 0 & -p_{M-1} & & a_M
\end{pmatrix}, \qquad (4.4.3)
$$

then the problem (4.4.1), (4.4.2) is equivalent to

$$
Ay = \lambda Ry. \qquad (4.4.4)
$$

This is a generalized eigenvalue problem for the matrix A. The condition for existence of solutions to this problem is

$$
\det(A - \lambda R) = 0, \qquad (4.4.5)
$$

which is a polynomial equation in λ.

Theorem 4.4.1: *The generalized eigenvalues of (4.4.4) are real and distinct.*

Proof: Let $S = R^{-1/2}$. It then follows that the roots of (4.4.5) are roots of

$$
\det(SAS - \lambda I) = 0. \qquad (4.4.6)
$$

Since the matrix SAS is symmetric, it will have real and distinct eigenvalues.

For each eigenvalue λ_i, there is an eigenvector y^i, which is the solution of (4.4.4). By using standard arguments, it can be proved that if y^i and y^j are two eigenvectors associated with two distinct eigenvalues, then

$$
(y^i, Ry^i) \equiv \sum_{s=1}^{M} r_s y_s^i y_s^j = 0. \qquad (4.4.7)
$$

This completes the proof.

Definition 4.4.1: Two vectors u and v such that $(u, Rv) = 0$ are called **R**-orthogonal.

Since the Sturm-Liouville problem (4.4.1), (4.4.2) is equivalent to (4.4.4), we have the following result.

Theorem 4.4.2: *Two solutions of the discrete Sturm-Liouville problem corresponding to two distinct eigenvalues are **R**-orthogonal.*

Consider now the more general problem

$$
y_{n+1} = A(n)y_n + b_n, \qquad (4.4.8)
$$

where $y_n, b_n \in R^s$ and $A(n)$ is an $s \times s$ matrix. Assume the boundary conditions are given by

$$\sum_{i=0}^{N} L_i y_{n_i} = w, \tag{4.4.9}$$

where $n_i \in N_0^+$, $n_i < n_{i+1}$, $n_0 = 0$, w is a given vector in R^s and L_i are given $s \times s$ matrices. Let $\Phi(n, j)$ be the fundamental matrix for the homogeneous problem

$$y_{n+1} = A(n) y_n, \tag{4.4.10}$$

such that $\Phi(0, 0) = I$. The solutions of (4.4.8) are given

$$y_n = \Phi(n, 0) y_0 + \sum_{j=0}^{n-1} \Phi(n, j + 10 b_j, \tag{4.4.11}$$

where y_0 is the unknown initial condition. The conditions (4.4.9) will be satisfied if

$$\sum_{i=0}^{N} L_i y_{n_i} = \sum_{i=0}^{N} L_i \Phi(n_i, 0) y_0 + \sum_{i=0}^{N} L_i \sum_{j=0}^{n_i-1} \Phi(n_i, j+1) b_j = w,$$

which can be written as

$$\sum_{i=0}^{N} L_i \Phi(n_0, 0) y_0 + \sum_{i=0}^{N} L_i \sum_{j=0}^{n_N-1} \phi(n_i, j+1) T(j+1, n_i) b_j = w, \tag{4.4.12}$$

where the step matrix $T(j, n)$ is defined by

$$T(j, n) = \begin{cases} I & \text{for } j \leq n, \\ 0 & \text{for } j > n. \end{cases}$$

By introducing the matrix $Q = \sum_{i=0}^{N} L_i \Phi(n_i, 0)$, the previous formula becomes

$$Q y_0 = w - \sum_{i=0}^{N} \sum_{j=0}^{n_N-1} L_i \Phi(n_i, j+1) T(j+1, n_i) b_j. \tag{4.4.13}$$

Theorem 4.4.3: *If the matrix Q is nonsingular, then the problem (4.4.8) with boundary conditions (4.4.9) has only one solution given by*

$$y_n = \Phi(n, 0) Q^{-1} w + \sum_{s=0}^{n_N-1} G(n, s) b_s, \tag{4.4.14}$$

where the matrices $G(n, j)$ are defined by:

$$G(n, j) = \Phi(n, j+1) T(j+1, n)$$

$$- \Phi(n, 0) Q^{-1} \sum_{i=0}^{N} L_i \Phi(n_i, j+1) T(j+1, n_i). \tag{4.4.15}$$

Proof: Since Q is nonsingular, from (4.4.13), we see that (4.4.11) solves the problem if the initial condition is given by

$$y_0 = Q^{-1}w - Q^{-1}\sum_{i=0}^{N}\sum_{j=0}^{n_N-1}L_i\Phi(n_i, j+1)T(j+1, n_i)b_j. \qquad (4.4.16)$$

By substituting (4.4.11), one has

$$y_n = \Phi(n,0)Q^{-1}w - \Phi(n,0)Q^{-1}\sum_{i=0}^{N}\sum_{j=0}^{n_N-1}L_i\Phi(n_i, j+1)T(j+1, n_i)b_j$$

$$+ \sum_{j=0}^{n_N-1}\Phi(n, j+1)T(j+1, n)b_j$$

$$= \Phi(n,0)Q^{-1}w$$

$$+ \sum_{j=0}^{n_N-1}\left[\Phi(n,j+1)T(j+1,n) - \Phi(n,0)Q^{-1}\sum_{i=0}^{N}L_i\Phi(n_i, j+1)T(j+1, n_i)\right]b_j,$$

from which by using the definition (4.4.14) of $G(n, j)$, the conclusion follows. Thus the proof is complete.

The matrix $G(n, i)$ is called Green's matrix and it has some interesting properties. For example,

(i)   for fixed j, the function $G(n,j)$ satisfies the boundary conditions $\sum_{i=0}^{N}L_iG(n_i, j) = 0$,

(ii)   for fixed j and $n \neq j$, the function $G(n, j)$ satisfies the autonomous equation $G(n+1, j) = A(n)G(n, j)$, and

(iii)   for $n = j$, one has $G(j+1, j) = A(j)G(j, j) + I$.

If the matrix Q is singular, then the equation (4.4.15) can have either an infinite number of solutions or no solutions Suppose, for simplicity, we indicate by b the right-hand side of (4.4.13), then the problem is reduced to establishing the existence of solutions for the equation

$$Qy_0 = b. \qquad (4.4.17)$$

Let $R(Q)$ and $N(Q)$ be respectively the range and the null space of Q. Then (4.4.17) will have solutions if $b \in R(Q)$. In this case, if c is any vector in $N(Q)$ and $\overline{y}_0$ any solution of (4.4.17), the vector $c + \overline{y}_0$ will also be a solution. Otherwise if $b \notin R(Q)$, the problem will not have solutions.

In the first case ($b \in R(Q)$), a solution can be obtained by introducing the generalized inverse of Q, defined as follows. Let $r = \text{rank } Q$. The generalized inverse Q^I of Q is the only matrix satisfying the relations

$$QQ^I = P, \quad Q^IQ = P_1, \qquad (4.4.18)$$

$$QQ^IQ = Q, \quad Q^IQQ^I = Q^I, \qquad (4.4.19)$$

where P and P_1 are the projections on $R(Q)$ and $R(Q^*)$ (Q^* is the conjugate transpose of Q) respectively. It is well known that if F is an $s \times s$ matrix whose columns span $R(Q)$, then P is given by

$$F(F^*F)^{-1}F^*. \tag{4.4.20}$$

By using Q^I the solution $\overline{y}_0$ of (4.4.17), when $b \in R(Q)$ is given by

$$\overline{y}_0 = Q^I b. \tag{4.4.21}$$

In fact, we have $Q\overline{y}_0 = QQ^I b = Pb = b$. A solution $\overline{y}_n$ of the boundary value problem can now be given in a form similar to (4.4.14) and (4.4.15) with Q^{-1} replaced by Q^I. This solution, as we have already seen, is not unique.

In fact if $c \in N(Q)$, $y_k = \Phi(n,0)c + \overline{y}_n$ will also be a solution satisfying the boundary conditions since

$$\sum_{i=0}^{N} L_i y_{n_i} = \sum_{i=0}^{N} \Phi(n_i,0)c + \sum_{i=0}^{n} L_i \overline{y}_{n^i} = Qc + w = w.$$

When $b \notin R(Q)$, the relation (4.4.21) has the meaning of least square solution because $\overline{y}_0$ minimizes the quantity $\| Qy_0 - b \|_2$ and the sequence $\overline{y}_n$ defined consequently may serve as an approximate solution.

4.5 STABILITY PROPERTIES

We present below some applications of linear and nonlinear VPF obtained in Sections 4.1, 4.2 and investigate stability properties of linear and nonlinear difference equations of Volterra type. To begin with we give an application of Theorem 4.1.4 and Corollary 4.1.2.

Theorem 4.5.1: *Consider the equation* (4.1.7) *and assume the hypothesis of Theorem* 4.1.4. *Let for* $n \geq n_0$, $0 < \alpha < 1$, *the following estimates hold:*
(i) $| L(n,s) | \leq k_0 e^{n-s}$,
(ii) $| F(n) | \leq k_0 \alpha^n$,
(iii) $| \Phi(n,s) | \leq k_0 \alpha^{n-2}$.
Then every solution of (4.1.7) *tends to zero as* $n \to \infty$.

Proof: Let $y(n, n_0, x_0)$ be any solution of (4.1.7). Then by Theorem 4.1.4, it is also a solution of (4.1.12). Hence, in view of Corollary 4.1.2, we get

$$| y(n, n_0, \eta) | \leq | \Phi(n, n_0) | \, | \eta | + \sum_{s=n_0}^{n-1} | \Phi(n, s+1) | \, | p(s) ds \tag{4.5.1}$$

where $p(s) = H(s) + L(s, n_0)\eta$. We now use the estimates given above and obtain

$$| p(s) | \leq | F(s) | + | L(s, n_0) | \, | \eta | + \sum_{\sigma=n_0}^{n-1} | \Phi(n, s+1) | \, | p(s) | \qquad (4.5.2)$$

$$\leq k_0 \alpha^s + k_0 \alpha^{s-n_0} | x_0 | + \sum_{\sigma=n_0}^{s-1} k_0 \alpha^{s-\sigma-1} k_0 \alpha^\sigma$$

$$\leq k_0 \alpha^s \left[1 + \alpha^{-n_0} | \eta | + \alpha^{-1} (k_0 \sum_{\sigma=n_0}^{s-1} (1)) \right]$$

$$\leq k_0 \alpha^s [1 + \alpha^{-n_0} | \eta | + \alpha^{-1} k_0 (s - n_0)].$$

Hence, (4.5.1) gives us

$$| y(n, n_0, \eta) |$$

$$\leq k_0 \alpha^{n-n_0} | \eta | + \left(\sum_{s=n_0}^{n-1} k_0 \alpha^{n-s-1} [k_0 \alpha^s (1 + \alpha^{-n_0} | \eta | + \alpha^{-1} k_0 (s - n_0))] \right),$$

$$\leq | \eta | (k_0 + k_1 n) \alpha^{n-n_0} + k_2 (n + n^2) \alpha^{n-n_0}, \ n \geq n_0;$$

where k_1 and k_2 are positive constants. Clearly the right-hand side tends to zero. Now we consider the nonlinear difference equation of Volterra type given by

$$\Delta x(n) = f(n, x(n)) + \sum_{s=n_0}^{n-1} g(n, s, x(s)), \ x(n_0) = \eta, \qquad (4.5.3)$$

where $f: N_{n_0}^+ \times S(\rho) \to R^d$, $g: N_{n_0}^+ \times N_{n_0}^+ \times S(\rho) \to R^d$, $f(n, x)$, $g(n, s, x)$ are continuous in x, and $S(\rho) = \{x \in R^d : \| x \| < \rho\}$. Suppose that
(i) $f(n, 0) \equiv 0, g(n, s, 0) \equiv 0$, and
(ii) f_x, g_x exist and are continuous in x.
Setting $f_x(n, 0) \equiv A(n), g_x(n, s, 0) \equiv K(n, s)$, and using the mean value theorem, Equation (4.5.3) can be rewritten as

$$\Delta x(n) = A(n)x(n) + F(n, x(n)) + \sum_{s=n_0}^{n-1} [K(n, s)x(s) + G(n, s, x(s))],$$

$$x(n_0) = n, \qquad (4.5.4)$$

where

$$F(n, x) = \int_0^1 [f_x(n, x\theta) - f_x(n, 0)] d\theta \cdot x,$$

$$G(n, s, x) = \int_0^1 [g_x(n, s, x\theta) - g_x(n, s, 0)] d\theta \cdot x.$$

We shall prove the following stability result.

Theorem 4.5.2: *Assume that the conditions of Theorem 4.1.4 hold. Suppose further that,*

(H_1) *the trivial solution of the linear difference equation of Volterra type*

$$\Delta x(n) = A(n)x(n) + \sum_{s=n_0}^{n-1} K(n, s)x(s), \ x(n_0) = n, \qquad (4.5.5)$$

is exponentially asymptotically stable.

(H_2) *for $(n, x) \in N_{n_0}^+ \times s(\rho)$ we have $| F(n, x) | \leq w_1(n, | x |)$ and $| G(n, s, x) | \leq w_2(n, s, | x |)$, where $w_1 \colon N_{n_0}^+ \times R^+ \to R^+$, $w_2 \colon N_{n_0}^+ \times N_{n_0}^+ \times R^+ \to R^+$, and $w_1(n, u)$, $w_2(n, s, u)$ are continuous and nondecreasing in u for each n, s.*

Then the stability properties of the trivial solution of the difference equation

$$\Delta u(n) = -u(n) + M\left(w_1(n, u(n)) + \sum_{s=n_0}^{n-1} w(n, s, u(n)) \right),$$

$$u(n_0) = u_0 \geq 0, \qquad (4.5.6)$$

where

$$w(n, s, u(s)) = w_2(n, s, u(s)) + | L(n, s) | w_1(s, u(s))$$

$$+ \sum_{\sigma=n_0}^{s-1} | L(s, \sigma) | w_2(s, \sigma, u(\sigma)), \qquad (4.5.7)$$

$L(n, s)$ being any solution of (4.1.11) imply the corresponding stability properties of the trivial solution of (4.5.3).

Proof: By Theorem 4.1.4, Equation (4.5.3) is equivalent to

$$\Delta x(n) = B(n)x(n) + L(n, n_0)\eta + F(n, x(n))$$

$$+ \sum_{s=n_0}^{n-1} \left(L(n, s)F(s, x(s)) + G(n, s, x(s)) + \sum_{\sigma=n_0}^{s-1} L(s, \sigma)G(s, \sigma, x(\sigma)) \right) (4.5.8)$$

with $x(n_0) = \eta$, where $B(n) = A(n) - L(n, n)$ and $L(n, s)$ is a solution of (4.1.11). Also, it is clear that (4.5.5) is equivalent to

$$\Delta x(n) = B(n)x(n) + L(n, n_0)\eta, \ x(n_0) = \eta, \qquad (4.5.9)$$

and hence, by (H_1), it implies that the trivial solution of (4.5.9) is also exponentially asymptotically stable. Consequently, there exists a Lyapunov function V such that

(a) $V: N_{n_0}^+ \times S(\rho) \to R^+$, $V(n, x)$ is Lipschitzian in x for a constant $M > 0$, and
$|x| \le V(n, x) \le M |x|$, $(n, x) \in N_{n_0}^+ \times S(\rho)$;

(b) $V(n + 1, \widetilde{x}(n + 1)) - V(n, \widetilde{x}(n)) \le - V(n, \widetilde{x}(n))$, $(n, x) \in N_{n_0}^+ \times S(\rho)$,
where $\widetilde{x}(n)$ is the solution of (4.5.9).

Let $x(n) = x(n, n_0, x_0)$ be any solution of (4.5.8). Then, using (a) and (b) and setting $p(n) = V(n, x(n))$, we get the difference inequality of Volterra type

$$\Delta p(n) \le - p(n) + M \left(w_1(n, p(n)) + \sum_{s=n_0}^{n-1} w(n, s, p(s)) \right),$$

where $w(n, s, u)$ is given by (4.5.7). Setting $Z(n) = p(n)\alpha^{-(n-n_0+1)}$, $0 < \alpha < 1$, we obtain

$$\Delta Z(n) \le M\alpha^{-(n-n_0)} \left(w_1(n, z(n)\alpha^{(n-n_0)-1}) + \sum_{s=n_0}^{n-1} w(n, s, z(s))\alpha^{s-n_0-1}) \right),$$

which implies

$$Z(n) \le r(n, n_0, z(n_0)), \ n \ge n_0,$$

where $r(n, n_0, u_0)$ is the solution of

$$\Delta u(n) = M\alpha^{-(n-n_0)} \left(w_1(n, u(n)\alpha^{(n-n_0)-1}) + \sum_{s=n_0}^{n-1} w(n, s, u(s))\alpha^{s-n_0-1}) \right),$$

with $u(n_0) = u_0$. As a result, we have

$$V(n, x(n)) \le r(n, n_0, V(n_0, x_0))\alpha^{n-n_0-1}, \ n \ge n_0,$$

and it is easy to see that $R(n, n_0, u_0) = r(n, n_0, u_0)\alpha^{n-n_0-1}$ is the solution of the difference equation (4.5.6). Hence, the desired stability properties of the trivial solution of (4.5.3) follow from the corresponding stability properties of the trivial solution of (4.5.6). The proof is complete.

Finally, we study stability properties of nonlinear difference equation of Volterra type by using nonlinear VPF proved in Theorem 4.2.3.

Consider the difference equations

$$\Delta u(n) = f \left(n, u(n), \sum_{i=n_0}^{n-1} k(n, i, u(i)) \right), u(n_0) = u_0 \qquad (4.5.10)$$

$$\Delta v(n) = f \left(n, v(n), \sum_{i=n_0}^{n-1} k(n, i, v(i)) \right),$$

$$+ g\left(n, v(n), \sum_{i=n_0}^{n-1} h(n, i, v(i))\right), v(n_0) = v_0 \qquad (4.5.11)$$

where the functions f, g, k, h are as given in Theorem 4.2.3.

We investigate stability properties of solutions of (4.5.11) by assuming certain estimates on f, g, k, h. Let $|\cdot|$ denote a suitable norm in R^d. We need the following stability definitions and the difference inequality.

Definitions 4.5.1: (i) The solution $w(n, n_0, u_0)$ of (4.5.10) or (4.5.11) is said to be uniformly stable if there exists a constant $M > 0$ such that $|w(n, n_0, u_0)| \leq M |u_0|$ for all $n \in N_{n_0}^+$ and $|u_0| < \infty$.

(ii) The solution $w(n, n_0, u_0)$ of (4.5.10) or (4.5.11) is said to be exponentially asymptotically stable if there exist constants $M > 0$, $\alpha > 0$ such that $|w(n, n_0, u_0| \leq M |u_0| \exp\{-\alpha(n - n_0)\}$ for all $n \in N_{n_0}^+$ and $|u_0| < \infty$.

(iii) The solution $w(n, n_0, u_0)$ of (4.5.10) or (4.5.11) is said to be uniformly exponentially growing if for every $\alpha > 0$, there exists a constant $M > 0$ such that $|w(n, n_0, u_0| \leq M |u_0| \exp\{\alpha(n - n_0)\}$ for all $n \in N_{n_0}^+$ and $|u_0| < \infty$.

Lemma 4.5.1: *Let $w(n)$, $a(n)$ and $b(n)$ be real-valued nonnegative functions defined on $N_{n_0}^+$ and $w(n_0) = w_0$. If*

$$w(n) \leq w_0 + \sum_{j=n_0}^{n=1} a(j)\left(w(j) + \sum_{i=n_0}^{j-1} b(i)w(i)\right)$$

holds for all $n \in N_{n_0}^+$, then

$$w(n) \leq w_0 \prod_{j=n_0}^{n-1}\left(1 + a(j)\left(1 + \sum_{i=n_0}^{j-1} b(i)\right)\right), \ n \in N_{n_0}^+.$$

Theorem 4.5.3: *Suppose the solution $u(n, n_0, u_0)$ of (4.5.10) is uniformly stable and the hypotheses of Theorem 4.2.3 hold. Further, suppose that*

$$|\Phi(n, n_0, p(j))\Phi^{-1}(j+1, n_0, p(j))| \leq M_0, \ n_0 \leq j \leq n - 1, \qquad (4.5.12)$$

and

$$|f(n, x, y)| \leq s_1(n)(|x| + |y|), n \in N_{n_0}^+ \qquad (4.5.13)$$

$$|g(n, x, y)| \leq t_1(n)(|x| + |y|), n \in N_{n_0}^+ \qquad (4.5.14)$$

$$|k(n, i, w)| \leq s_2(i)|w|, \ |h(n, i, w)| \leq t_2(i)|w|, ni \in N_{n_0}^+ \quad (4.5.15)$$

where $s_1, s_2, t_1, t_2: N_{n_0}^+ \to R_+$. If there exist positive constants M_1, M_2 such that for all $N_{n_0}^+$

$$\sum_{j=n_0}^{n-1} s_1(j) \mid p(j) \mid \sum_{i=n_0}^{j-1} s_2(i) \le M_1 \mid u_0 \mid \qquad (4.5.16)$$

and

$$M(1 + M_0 M_1) \prod_{j=n_0}^{n-1} \left(1 + M_0(2s_1(j) + t_1(j)) \left(1 + \sum_{i=n_0}^{j-1} \max\{s_2(i), t_2(i)\} \right) \right)$$

$$\le M_2, \qquad (4.5.17)$$

then all solutions of (4.5.11) are uniformly stable on $N_{n_0}^+$.

Proof: From Theorem 4.3.6 and the hypotheses we have

$$\mid v(n, n_0, u_0) \mid \; \le \; \mid u(n, n_0, u_0) \mid \; + \sum_{j=n_0}^{n-1} \mid \Phi(n, n_0, p(j)) \Phi^{-1}(j+1, n_0, p(j))$$

$$\times \; \Omega(j, n_0, u_0, p(j)) \mid$$

$$\le M \mid u_0 \mid \; + M_0 \sum_{j=n_0}^{n-1} \mid \Omega(j, n_0, u_0, p(j)) \mid, \quad n \in N_{n_0}^+ \qquad (4.5.18)$$

Further, from the definition of Ω, (4.2.16) and conditions (4.5.13) to (4.5.15), we find that

$$\mid \Omega(j, n_0, u_0, p(j)) \mid \; \le s_1(j) \left(\mid v(j) \mid \; + \sum_{i=n_0}^{j-1} \mid k(j, i, v(i)) \mid \right)$$

$$+ t_1(j) \left(\mid v(j) \mid \; + \sum_{i=n_0}^{j-1} \mid h(j, i, v(i)) \mid \right)$$

$$+ s_1(j) \left(\mid v(j) \mid \; + \sum_{i=n_0}^{j-1} \mid k(j, i, u(i, n_0, p(j))) \mid \right)$$

$$\le s_1(j) \left(\mid v(j) \mid \; + \sum_{i=n_0}^{j-1} s_2(i) \mid v(i) \mid \right) + t_1(j) \left(\mid v(j) \mid \; + \sum_{i=n_0}^{j-1} t_2(i) \mid v(i) \mid \right)$$

$$+ s_1(j) \left(\mid v(j) \mid \; + M \sum_{i=n_0}^{j-1} s_2(i) \mid p(j) \mid \right), \; n_0 \le j < n.$$

Substituting the above estimate into (4.5.18), arranging the terms and applying Lemma 4.5.1 successively lead to

$$| v(n, n_0, u_0) | \leq M | u_0 | + M_0 \left\{ \sum_{j=n_0}^{n-1} (s_1(j) + t_1(j)) | v(j) | \right.$$

$$+ \sum_{j=n_0}^{n-1} (s_1(j) + t_1(j)) \sum_{i=n_0}^{j-1} \max\{s_2(i), t_2(i)\} | v(i) | \left. \right\}$$

$$+ M_0 \sum_{j=n_0}^{n-1} s_1(j) | v(j) | + M_0 M \sum_{j=n_0}^{n-1} s_1(j) | p(j) | \sum_{i=n_0}^{j-1} s_2(i)$$

$$\leq M(1 + M_0 M_2) | u_0 | + M_0 \left\{ \sum_{j=n_0}^{n-1} (2s_1(j) + t_1(j)) | v(j) | \right.$$

$$+ \sum_{j=n_0}^{n-1} (2s_1(j) + t_1(j)) \sum_{i=n_0}^{j-1} \max\{s_2(i), t_2(i)\} | v(i) | \left. \right\}$$

$$\leq M(1 + m_0 M_1) \prod_{j=n_0}^{n-1} \left(1 + m_0 (2s_1(j) + t_1(j)) \left(1 + \sum_{i=n_0}^{j-1} \max\{s_2(i), t_2(i)\} \right) \right) | u_0 |$$

$$\leq M_2 | u_0 |, \quad n \in N_0^+.$$

The proof is complete.

Theorem 4.5.4: *Suppose the solution* $u(n, n_0, u_0)$ *of* (4.5.10) *is exponentially asymptotically stable and the hypotheses of Theorem 4.2.3 hold. Further, suppose that for* $n_0 \leq j < n$ *conditions*

$$| \Phi(n, n_0, p(j)) \Phi^{-1}(j+1, n_0, p(j)) | \leq M_0 \exp\{ - \alpha(n - j)\}, \quad (4.5.19)$$

$$| k(n, i, w) | \leq s_2(i) | w | \exp\{ - \alpha(n - i)\}, \, | h(n, i, w) |$$

$$\leq t_2(i) | w | \exp\{ - \alpha(n - i)\}, \quad (4.5.20)$$

and (4.5.13), (4.5.14) *are satisfied, where* s_1, s_2, t_1, t_2 *are the same as in Theorem 4.5.3. If there exist positive constants* M_1, M_2 *such that for all* $n \in N_{n_0}^+$ (4.5.16), (4.5.17) *hold, then all solutions of* (4.5.11) *are exponentially asymptotically stable on* $N_{n_0}^+$.

Proof: Using Theorem 4.2.3 and the hypotheses we find that

$$| v(n, n_0, u_0) | \leq | M | u_0 | \exp\{ - \alpha(n - n_0)\}$$

$$+ M_0 \sum_{j=n_0}^{n=1} | \Omega(j, n_0, u_0, p(j)) | \exp\{ - \alpha(n - j)\}, n \in I(n_0). \quad (4.5.21)$$

Further, as in Theorem 4.5.3, we have

$$| \Omega(j, n_0, u_0, p(j)) | \leq s(j) \left(| v(j) | + \sum_{i=n_0}^{j-1} s_2(i) | v(i) | e^{-\alpha(j-i)} \right)$$

$$+ t_1(j) \left(| v(j) | + \sum_{i=n_0}^{j-1} t_2(i) | v(i) | e^{-\alpha(j-i)} \right)$$

$$+ s_1(j) \left(| v(j) | + M \sum_{i=n_0}^{j-1} s_2(i) | p(j) | e^{-\alpha(j-n_0)} \right), \quad n_0 \leq j < n.$$

Substituting the above estimate into (4.5.21) and multiplying the both sides of the resulting inequality by $\exp\{\alpha n\}$, we obtain

$$| v(n, n_0, u_0 | e^{\alpha n}$$

$$\leq M | u_0 | e^{\alpha n_0} + M_0 \sum_{j=n_0}^{n-1} \left\{ s_1(j) \left(| v(j) | e^{\alpha j} + \sum_{i=n_0}^{j-1} s_2(i) | v(i) | e^{\alpha i} \right) \right.$$

$$+ t_1(j) \left(| v(j) | e^{\alpha j} + \sum_{i=n_0}^{j-1} t_2(i) | v(i) | e^{\alpha i} \right)$$

$$\left. + s_1(j) \left(| v(j) | e^{\alpha j} + M e^{\alpha n_0} \sum_{i=n_0}^{j-1} s_2(i) | p(j) | \right) \right\}, \quad n \in N_{n_0}^+.$$

Let $v(j, n_0, u_0) \exp\{\alpha j\} = w(j, n_0, u_0) = w(j)$, $n_0 \leq j \leq n$, and denote $u_0 \exp\{\alpha n_0\}$ as w_0. Then, in view of (4.5.16), we get

$$| w(n, n_0, u_0 | \leq M | w_0 | + M_0 \sum_{j=n_0}^{n-1} \left\{ s_1(j) \left(| w(j) | + \sum_{i=n_0}^{j-1} s_2(i) | w(i) | \right) \right.$$

$$+ t_1(j) \left(| w(j) | + \sum_{i=n_0}^{j-1} t_2(i) | w(i) | \right)$$

$$\left. + s_1(j) \left(| w(j) | + M e^{\alpha n_0} \sum_{i=n_0}^{j-1} s_2(i) | p(j) | \right) \right\},$$

$$\leq M(1 + M_0 M_1 e^{\alpha n_0}) | w_0 | + M_0 \sum_{j=n_0}^{n-1} \left\{ s_1(j) \left(2 | w(j) | + \sum_{i=n_0}^{j-1} s_2(i) | w(i) | \right) \right.$$

$$\left. + t_1(j) \left(| w(j) | + \sum_{i=n_0}^{j-1} t_2(i) | w(i) | \right) \right\}$$

$$\leq M(1 + m_0 M_1 e^{\alpha n_0}) \mid w_0 \mid + M_0 \sum_{j=n_0}^{n-1} (2s_1(j) + t_1(j))$$

$$\times \left\{ \mid w(j) \mid + \sum_{i=n_0}^{j-1} \max\{s_2(i), t_2(i)\} \mid w(i) \mid \right\}, \; n \in N_{n_0}^+.$$

Now, as in Theorem 4.5.3, an application of Lemma 4.5.1 to the above inequality leads to

$$\mid w(n, n_0, u_0) \mid$$

$$\leq M(1 + M_0 M_1 e^{\alpha n_0}) \prod_{j=n_0}^{n-1} \left(1 + M_0(2s_1(j) + t_1(j)) \left(1 + \sum_{i=n_0}^{j-1} \max\{s_2(i), t_2(i)\} \right) \right)$$

$$\times \mid w_0 \mid, \; n \in N_{n_0}^+,$$

that is,

$$\mid v(n, n_0, u_0) \mid \; \leq \widetilde{M}_2 \mid u_0 \mid \exp\{-\alpha(n - n_0)\}, \; n \in N_{n_0}^+,$$

where $\widetilde{M}_2$ is a suitable constant. The proof is complete.

The proof of the following result is similar to that of Theorem 4.5.4.

Theorem 4.5.5: *Suppose the solution $u(n, n_0, u_0)$ of (4.5.10) is uniformly exponentially growing and the hypotheses of Theorem 4.2.3 hold. Further, suppose that (4.5.13), (4.5.14), (4.5.19) and (4.5.20) with α replaced by $-\alpha$ are satisfied. If there exist positive constants M_1, M_2 such that for all $n \in N_{n_0}^+$ (4.5.16), (4.5.17) hold, then all solutions of (4.5.11) are uniformly exponentially growing on $N_{n_0}^+$.*

4.6 NUMERICAL METHODS

It is well known that differential equations can be discretized and then the methods of linear difference equations and then the methods of linear difference equations are employed to discuss the nature of solutions. Since recursive methods are natural for difference equations, differential equations, when approximated by difference equations, can be studied using such methods.

Let us consider nonlinear differential equation

$$y' = f(t, y), y(t_0) = y_0, t \in [t_0, T] \tag{4.6.1}$$

Let f be smooth enough to have a unique solution $y(t)$ existing on $[t_0, T]$. Let for $h > 0$, $t_i = t_0 + ih$, $i = 1, \ldots, N = \frac{T}{h}$. Let the discrete problem such approximates (4.6.1) can be denoted by

$$F_n(y_n, y_{n+1}, \ldots, y_{n+k}, f_n, f_{n+1}, \ldots, f_{n+k}) = 0, \tag{4.6.2}$$

where $y_i = y(t_i) + O(h^q)$, $q > 1$, $i = 0, 1, \ldots, k-1, n+k \leq N$. This is a difference equation of order k which needs k initial conditions which can be obtained by noting that the initial condition in (4.6.1) is $y(t_0) = y_0$. Assume that (4.6.2) has unique solution y_n.

Definition 4.6.1: The problem (4.6.2) is said to be consistent with problem (4.6.1) if

$$F_n(y(t_n), y(t_{n+1}), \ldots, y(t_{n+k}), f(t_n, y(t_n)), \ldots$$

$$f(t_{n+k}, y(t_{n+k}))) \equiv \tau_n = (h^{p+1}). \tag{4.6.3}$$

with $p \geq 1$. Here τ_n is the truncation error.

Equation (4.6.3) is a perturbed equation in relation to (4.6.2).

Definition 4.6.2: The discrete problem (4.6.2) is said to be convergent to the problem (4.6.1) if the solution y_n of (4.6.2) tends to the solution $y(t)$ of (4.6.1) for $n \to \infty$ and $t_n - t_0 = nh \leq T$.

Since the solution of the continuous problem satisfies (4.6.3) which is a perturbation of (4.6.2), the convergence is evident when (4.6.2) is stable under perturbation.

When F_n is linear in its arguments, one can apply multi-step methods. Let

$$F_n = \sum_{i=0}^{k} \alpha_i y_{n+i} - h \sum_{i=0}^{k} \beta_i f_{n+i} = 0, \tag{4.6.4}$$

where $\alpha_k = 1$ and α_i and β_i are real numbers. Define $Ey_k = y_{k+1}$ where E is a shift-operator, and two polynomials ρ and σ given by

$$\rho(z) = \sum_{i=0}^{k} \alpha_i z^i, \sigma(z) = \sum_{i=0}^{k} \beta_i z^i.$$

Then the equation (4.6.4) is transferred to

$$\rho(E)y_n - h\sigma(E)f_n = 0. \tag{4.6.5}$$

The polynomials ρ and σ characterize the method (4.6.2) uniquely. Here the relation (4.6.3) takes the form

$$\rho(E)y(t_n) - h\sigma(E)f(t_n, y(t_n)) = \tau_n. \tag{4.6.6}$$

At this stage, we recall Corollary 4.2.1, which provides nonlinear VPF for difference equations, to study the global error. Let $x(n, n_0, y_0)$ and $y(n, n_0, y_0)$ be the solutions of (4.6.5) and (4.6.1) respectively. Then the difference

$$E_n = y(n, n_0, y_0) - x(n, n_0, y_0) \tag{4.6.7}$$

is defined as the global error.

We prove the following result using nonlinear VPF for (4.6.1) given in Theorem 4.1.2.

Theorem 4.6.1: *Let*
(i) $y(t)$ be the solution of (4.6.1);
(ii) $y_0 = z_0, z_1, \ldots, z_n$ be some approximation of $y(t)$ at the points $t_1, t_2, \ldots, t_n$;
(iii) $p(t)$ be a sufficiently smooth function that interpolates the points z_i so that
$$p(t_i) = z_i, \ i = 0, 1, \ldots, n.$$
Then $E_n = y(t_n) - z_n$ satisfies the relation

$$E_n = -\int_{t_0}^{t_n} \Phi(t_n, s, p(s))[p'(s) - f(s, p(s))]ds$$

where $\Phi(t, t_0, y_0) = \frac{\partial y(t, t_0, y_0)}{\partial y_0}$.

Proof: Let $s \in [t_0, t_n]$ and $y(t_n, s, p(s))$ be the solution of (4.6.1) with initial values $(s, p(s))$. Then

$$\frac{\partial y(t_n, s, p(s))}{\partial s} = -\Phi(t_n, s, p(s))f(s, p(s)),$$

and

$$\frac{\partial y(t_n, s, p(s))}{\partial p} = \Phi(t_n, s, p(s)).$$

Consider the integral

$$\int_{t_0}^{t_n} \frac{d}{ds} y(t_n, s, p(s))ds = y(t_n, t_n, p(t_n)) - y(t_n, t_0, y_0)$$

$$= -E_n.$$

This integral is also equal to

$$\int_{t_0}^{t_n} \left[\frac{\partial y(t_n, s, p(s))}{\partial s} + \frac{\partial y(t_n, s, p(s))}{\partial p} \cdot p'(s)\right]ds$$

$$= \int_{t_0}^{t_n} \Phi(t_n, s, p(s))[p'(s) - f(s, p(s)))]ds,$$

which completes the proof.

4.7 NOTES

Extensive literature is available on the theory of difference equations. See the recent monographs by Lakshmikantham and Trigiante [1], Kelly and Peterson [1], Agarwal [2] and Agarwal and Lalli [1].

Theorems 4.1.1 and 4.1.2 have both been taken from Lakshmikantham and Trigiante [1], while Theorems 4.1.3 and 4.1.4 are due to Leela and Zouyousefain [1]. The result in Theorem 4.1.5 is new.

The nonlinear VPFs appearing in Theorems 4.2.1, 4.2.2 are adapted from Lakshmikantham and Trigiante [1]. The results in Theorems 4.2.3 and 4.2.4 are due to Agarwal [2]. These results have been extended to higher order difference equations by Sheng and Agarwal [3]. See also the work by Pachpattee [1, 2], Luca and Talpalaru [1].

The VPF has found applications for several properties of solutions of difference equations. Among these, the results in Section 4.3 are due to Elaydi [2]. See also Burton [1]. The results occurring in Section 4.4 on boundary value problems have been taken from Lakshmikantham and Trigiante [1]. Theorems 4.5.1 and 4.5.2 are taken from Leela and Zouyousefain [1] and for Theorems 4.5.3 and 4.5.4, see Sheng and Agarwal [2]. Also refer to Elaydi [1].

The results on numerical methods are taken from Lakshmikantham and Trigiante [1] and Piazza and Trigiante [1]. Also refer to Wanner and Reitberger [1] and Gear and Tu [1] for related results.

5. Abstract Differential Equations

5.0 INTRODUCTION

The theory of differential equations in abstract spaces has been developed considerably during the last three decades because of its applications to several areas of analysis and particularly to ordinary and partial differential equations. The powerful techniques of ODE and the elegant theories of nonlinear functional analysis renders simplicity in studying evolution equations representing many physical models.

One of the essential tools in several applications of abstract differential equations is the method of VP. The representation of solutions of perturbed equations together with other analytical methods such as differential and integral inequalities, fixed point results, iterative processes helps one to analyze the qualitative behavior of solutions of such equations. The aim of the present chapter is to organize the VPF in abstract differential equations and then to present few significant application.

In order to make reading of this chapter smooth, we include few essential preliminaries in Section 5.1. This section is then followed by linear variation of parameters. We study evolution equation by using simple properties of semi-group of linear operators. This is the content of Section 5.2. In order to develop the VPF for nonlinear equations, one needs to develop the theory of differentiability of solutions with respect to initial conditions. We establish these details in Section 5.3 and obtain the VPF which naturally includes as a particular case the VPF for linear case.

Section 5.4 analyzes the way in which the solution of nonlinear differential equation can grow to infinity. The essential tool is obviously VPF. The final Section 5.5 includes results on stability of solutions. Here we study stability and asymptotic stability. The applications given use linear as well as nonlinear VPF. The notes include the literature useful for current areas of interests.

5.1 NOTATION AND PRELIMINARIES

We devote this section to bring together the essential definitions used in the calculus of abstract functions required to follow the contents of following sections.

Let X denote a Banach space over real numbers R and for any $x \in X$, let $\| x \|$ denote the norm of X. The strong derivative of $x(t)$, $x: J \to X$, J an interval of R, is defined by

$$x'(t) = \lim_{\Delta t \to 0} \frac{[x(t+\Delta t) - x(t)]}{\Delta t}$$

where the limit is taken in the strong sense.

Let $f: S \to Y$ be a function from an open set S of the Banach space X into the Banach space Y. If at a point $x \in S$

$$f(x+h) - f(x) = L(x,h) + w(x,h), h \in X$$

where $L(x, \cdot): X \to Y$ is a linear operator and

$$\lim_{\| h \| \to 0} \frac{\| w(x,h) \|}{\| h \|} = 0,$$

then $L(x,h)$ is called the Fréchet differential of the function f at the point x with increment h and $w(x,h)$ is called the remainder of the differential. The operator $L(x, \cdot): X \to Y$ is called Fréchet derivative of f at x and is denoted by $f'(x)$.

Definition 5.1.1: A family $\{T(t)\}, 0 \le t < \infty$ of bounded linear operators mapping the Banach space X into X is called a strongly continuous semigroup of operators if the following conditions hold:

(*i*) $T(t + s) = T(t)T(s), t, s \ge 0$;
(*ii*) $T(0) = I$, I being the identity operator in X;
(*iii*) for each $x \in X$, $T(t)x$ is strongly continuous in t on $[0, \infty)$.

Let $\{T(t)\}, t > 0$ be a strongly continuous semigroup of operators in the Banach space X. For $h > 0$, define the linear operator A_h by

$$A_h x = \frac{T(h)x - x}{h}, x \in X.$$

Let $D(A)$ be the set of all $x \in X$ for which the $\lim_{h \to 0+} A_h x$ exists. Define the operator A on $D(A)$ as

$$Ax = \lim_{h \to 0+} A_h x, \quad x \in D(A).$$

Definition 5.1.2: The operator A with domain $D(A)$ is called the infinitesimal generator of the semigroup $\{T(t)\}, t \ge 0$. Given an operator A on $D(A)$, we say that it generates a strongly continuous semigroup $\{T(t)\} \ t \ge 0$ if A coincides with the infinitesimal generator of $\{T(t)\}, t \ge 0$.

Definition 5.1.3: A strongly continuous semigroup $\{\exp(-tA)\}$ is called an analytic semigroup if:

(*i*) $\exp(-tA)$ can be continued analytically as a strongly continuous semigroup into a section s_ω, $s_\omega = \{t \in C: \mid \arg t \mid \ < \omega, t \ne 0\}$ for some $\omega \in (0, \frac{\pi}{2})$;

(*ii*) for each $t \in s\omega$ the operators $A\exp(-tA)$ and $\frac{d}{dt}\exp(-tA)$ are in $B(X)$ and

$$\frac{d}{dt}\exp(-tA) = -A\exp(-tA)x, \ x \in X;$$

(*iii*) for any $0 < \epsilon < \omega$ the operators $\exp(-tA)$ and $tA\exp(-tA)$ are uniformly bounded in the sector $s_{\omega-\epsilon}$.

5.2 LINEAR VARIATION OF PARAMETERS

This section is concerned with the study of time-dependent Cauchy problem

$$\frac{du}{dt} + A(t)u = f(t), 0 < t \leq T, u(0) = u_0 \in X \tag{5.2.1}$$

and the associated homogeneous equation

$$\frac{du}{dt} + A(t)u = 0, 0 < t \leq T, u(0) = u_0 \in X \tag{5.2.2}$$

where the unknown function $u(t)$ maps $[0, T]$ into a Banach space X, $f \colon [0, T] \to X$ is a given function and for each t in $[0, T]$, $A(t)$ is a given closed linear operator in X with domain $D[A(t)] = D$ independent of t and dense in X. The operator $- A(t)$ generates an analytic semigroup. Since parabolic partial differential equations are realized in this form, (5.2.1) is said to be parabolic.

Definition 5.2.1: An operator-valued function $U(t, \tau)$ with values in $B(X)$, defined and strongly continuous jointly in t, τ for $0 \leq \tau \leq t \leq T$ is called a fundamental solution of the homogeneous equation (5.2.2), if
(i)   the partial derivative $\frac{\partial U(t,\tau)}{\partial t}$ exists in the strong topology of X, belongs to
      $B(X)$ for $0 \leq \tau \leq t \leq T$ and is strongly continuous in t for $0 \leq \tau \leq t \leq T$;
(ii)   the range of $U(t, \tau)$ is in D;
(iii) $\frac{\partial U(t,\tau)}{\partial t} + A(t)U(t, \tau) = 0, 0 \leq \tau < t \leq T$ and $U(\tau, \tau) = I$.
$U(t, \tau)$ is called evolution operator.

Definition 5.2.2: A function $u \colon [0, T] \to x$ is called a mild solution of (5.2.1) if it admits the integral representation

$$u(t) = U(t, 0)u_0 + \int_0^t U(t, s)f(s)ds. \tag{5.2.3}$$

It is to be remarked that (5.2.3) may not represent solution of (5.2.1) for arbitrary choice of u_0 and $f(t)$. The existence of $\frac{du}{dt}$ and $A(t)u$ for (5.2.3) can be proved only under certain assumptions on u_0 and $f(t)$. The known sufficient conditions for the existence of $U(t, \tau)$ are $- A(t)$ is assumed to be the infinitesimal generator of a strongly continuous semigroup of bounded linear operators on X. Further, $A(t)$ is assumed to depend on t smoothly in some sense. We assume the following:

(H_1) For each $\sigma \in [0, T]$, $A(\sigma)$ is closed operator in X with domain $D[A(\sigma)] = D$
        independent of σ and dense in X;
(H_2) For each $\sigma \in [0, t]$, the resolvent set $\rho[- A(\sigma)]$ contains
        $s = \{\lambda \in c, - \theta \leq \arg \lambda \leq \theta, \frac{\pi}{2} < \theta < \pi, \theta \text{ fixed}\}$ and

$$\| [\lambda I + A(\sigma)]^{-1} \| \leq \frac{\widetilde{c}}{(1+|\lambda|)}, \lambda \in S$$

where $\widetilde{c}$ is a positive constant independent of λ and σ;

(H_3) $\| [A(t) - A(\tau)]A^{-1}(s) \| \le c \,|\, t - \tau \,|^{\,\alpha}, 0 < \alpha \le 1$ where c, and α are positive constants independent of t, τ, s for $0 \le t, \tau, s \le T$;

(H_4) $\| f(t) - f(s) \| \le c \,|\, t - s \,|^{\,\beta}, 0 < \beta \le 1, 0 \le t, s \le T$, where c and β are positive constants independent of t and s.

We have the following result.

Theorem 5.2.1: *Let the hypothesis (H_1)-(H_4) hold. Then the evolution equation (5.2.1) has the unique solution*

$$u(t) = U(t,0)u_0 + \int\limits_0^t U(t,s)f(s)ds.$$

The proof of this theorem needs several properties of semigroup $\{\exp[-tA(\tau)]\}$ which are not given here. For complete mathematical rigor, reference may be made to the literature given in the notes. We also note that the VPF for linear abstract differential equations can be obtained as a special case of nonlinear abstract differential equation which we obtain in the following section.

5.3 NONLINEAR VARIATION OF PARAMETERS

In the current section, we study the nonlinear abstract Cauchy problem

$$\frac{du}{dt} = f(t,u) \text{ and } u(t_0) = u_0 \tag{5.3.1}$$

where $f: R_+ \times X \to X$ is a given function and X a Banach space. Our aim is to develop the VPF with respect to (5.3.1) and its perturbation

$$\frac{dv}{dt} = f(t,v) + F(t,v) \text{ and } v(t_0) = v_0 \tag{5.3.2}$$

where $F: R_+ \times X \to X$. As we shall see later, this result is a convenient tool in discussing the properties of solutions of the perturbed system (5.3.2). First, it is necessary to study the uniqueness of solutions, their continuity and differentiability with respect to initial conditions (t_0, u_0), and to show that the Fréchet derivatives of the solutions $u(t, t_0, u_0)$ of (5.3.1) with respect to initial values exist and satisfy the equation of variation of (5.3.1) along the solution $u(t, t_0, u_0)$. The treatment rests on the existence of an admissible functional in X, a mild one-sided estimate of f, and the theory of scalar differential inequalities. By $S(x_0, r)$ we shall denote the sphere $\{x \in X: \| x - x_0 \| \le r\}$. For a function $f: R_+ \times X \to X$ the Fréchet derivative with respect to x, if it exists, is denoted by $f_x(t, \cdot)$ and it belongs to $B(X)$. The notation $w(h) = O(\| h \|)$ for $h \in X$ stands for a vector in X satisfying the condition

$$\lim_{\| h \| \to 0} w(h)/\| h \| = 0.$$

Definition 5.3.1: A (nonlinear) continuous functional $\Phi: X \to R_+$ is said to be admissible in X if the following conditions hold:

(i) $\Phi(x) > 0$, $x \in X$, $x \ne 0$ and $\Phi(0) = 0$;

(ii) if $\lim_{n \to \infty} \Phi(x_n) = 0$ for $x_n \in X$, then $\lim_{n \to \infty} x_n = 0$;

(*iii*) there exists a mapping $M: X \times X \to R$ such that $M[x, h]$ is continuous in h, uniformly with respect to x in any sphere $S(x_0, r)$, and satisfying the properties

(*a*) $\Phi(x + h) - \Phi(x) \leq M[x, h] + 0(\| h \|)$, $x, h \in X$;

(*b*) $M[x, \lambda h] = \lambda M[x, h]$, $\lambda \geq 0$, $x, h \in X$;

(*c*) $M[x, h_1 + h_2] \leq M[x, h_1] + M[x, h_2]$, $x, h_1, h_2 \in X$.

The following lemma proves that the functional $M[x, \cdot]$ is bounded.

Lemma 5.3.1: *Let Φ be an admissible functional in X. Then for any sphere $S = S(x_0, r)$ there is a constant $K(r)$ such that*

$$| M[x, h] | < K(r) \| h \|, x \in S, h \in X.$$

Proof: In view of the continuity of $M[x, h]$ in h, uniformly with respect to $x \in S$, and the fact that $M[x, 0] = 0$, it follows that, given $\epsilon > 0$ there exists a $\delta = \delta(\epsilon, r)$ such that $\| \widetilde{h} \| \leq \delta$ implies $| M[x, \widetilde{h}] | < \epsilon$ for all $x \in S$. For an arbitrary h set $\widetilde{h}$ $(\delta / \| h \|)h$. Then $\| \widetilde{h} \| = \delta$ and

$$| M[x, (\delta / \| h \|)h]\} < \epsilon, x \in S, h \in X. \tag{5.3.3}$$

Since

$$M[x, (\delta / \| h \|)h] = (\delta / \| h \|)M[x, h], \tag{5.3.4}$$

it follows from (5.3.3) and (5.3.4) that

$$| M[x, h] | < (\epsilon/\delta) \| h \|, x \in S, h \in X.$$

The proof is completed with $K(r) = \epsilon/\delta$.

Consider the abstract Cauchy problem

$$u' = f(t, u), \tag{5.3.5}$$

$$u(t_0) = u_0, \tag{5.3.6}$$

and the related scalar initial value problem

$$r' = g(t, r), \tag{5.3.7}$$

$$r(t_0) = r_0, \tag{5.3.8}$$

where $f: R_+ \times X \to X$ and $g: R_+ \times R_+ \to R$.

Here, we assume that f and g are smooth enough to guarantee the existence of solutions of (5.3.5) and (5.3.7) for all $t \in R_+$. Actually, local existence would suffice and our results can be easily restated to hold locally. Of course the mere continuity of f will not suffice even for local existence of solutions of (5.3.5).

A solution of (5.3.5) and (5.3.6) will be denoted by $u(t, t_0, u_0)$. The maximal solution of (5.3.7) and (5.3.8) will be denoted by $r(t, t_0, r_0)$.

We shall also assume, the existence of an admissible functional Φ in X satisfying the properties (i)-(iii) of Definition 5.3.1.

For easy reference we state the following hypotheses.

(H_1) $$M[x - y, f(t, x) - f(t, y)] \leq g[t, \Phi(x - y)] \qquad (5.3.9)$$

for $t \in R_+$ and all $x, y \in X$.

(H_2) The function $f(t, x)$ has a continuous Fréchet derivative $f_x(t, x)$ with respect to x and

$$M[h, f_x(t, z)h] \leq g[t, \Phi(h)], \ t \geq 0, h \in X \qquad (5.3.10)$$

and all z in any sphere $S(x_0, r)$.

We shall prove that (H_2) implies (H_1). First we shall prove that under (H_1), the system (5.3.5) and (5.3.6) has a unique solution $u(t, t_0, u_0)$ which depends continuously on the initial conditions (t_0, u_0) provided that the scalar initial value problems (5.3.7) and (5.3.8) has these properties.

We need the following lemma. The symbol $D_+r(t)$ denotes the lower right-Dini derivative of the function $r(t)$.

Lemma 5.3.2: *For any differentiable function $x: R_+ \rightarrow X$ the following inequality holds:*

$$D_+\Phi[x(t)] \leq M[x(t), x'(t)], \ t \in R_+. \qquad (5.3.11)$$

Proof: From the definition of $D_+\Phi(t)$, the admissibility of Φ and Lemma 5.3.1 we obtain

$$D_+\Phi[x(t)] = \liminf_{h \to 0_+} h^{-1}[\Phi(x(t + h)) - \Phi(x(t))]$$

$$= \liminf_{h \to 0_+} h^{-1}(\Phi[x(t) + hx'(t) + O(h^2)] - \Phi[x(t)])$$

$$\leq \liminf_{h \to 0_+}[h^{-1}(M[x(t), hx'(t)] + O(h^2)) + O(\| hx'(t) + O(h^2) \|)]$$

$$\leq M[x(t), x'(t)] + \liminf_{h \to 0^+} M[x(t), O(h)]$$

$$= M[x(t), x'(t)].$$

The proof is complete.

The following lemma is employed to prove uniqueness.

Lemma 5.3.3: *Let (H_1) hold. Assume that $\Phi(u_0 - v_0) \leq r_0$. Then*

$$\Phi[u(t, t_0, u_0) - v(t, t_0, v_0)] \leq r(t, t_0, r_0), t \geq t_0. \qquad (5.3.12)$$

Proof: Let $u(t) = u(t, t_0, u_0)$ and $v(t) = v(t, t_0, v_0)$ be solutions of (5.3.5) through (t_0, u_0) and (t_0, v_0) respectively and $r(t, t_0, r_0)$ be the maximal solution of (5.3.7) and (5.3.8). Define $z(t) = u(t) - v(t)$. From Lemma 5.3.2 and (H_1) we obtain

$$D_+\Phi(z(t)) \le M[z(t), z'(t)]$$

$$= M[u(t) - v(t), f[t, u(t)] - f[t, v(t)]]$$

$$\le g(t, \Phi(z(t))), \quad t \ge t_0.$$

Also

$$\Phi[z(t_0)] = \Phi(u_0 - v_0) \le r_0.$$

From these inequalities the estimate (5.3.12) follows.

Theorem 5.3.1: *Let (H_1) be satisfied. Assume that $r(t, t_0, 0) \equiv 0$. Then the system* (5.3.5) *and* (5.3.6) *has a unique solution.*

Proof: Let $u_1(t) = u_1(t, t_0, u_0)$ and $u_2(t) = u_2(t, t_0, u_0)$ be two solutions of (5.3.5) and (5.3.6). It follows from (5.3.12) that

$$\Phi[u_1(t) - u_2(t)] \le r(t, t_0, 0) \equiv 0.$$

Hence $u_1 \equiv u_2(t)$, for $t \ge t_0$ and the proof is complete.

Next, we prove the continuous dependence of solutions of (5.3.5) and (5.3.6) with respect to the initial conditions.

Theorem 5.3.2: *Let (H_1) hold. Assume that the maximal solution of* (5.3.7) *and* (5.3.8) *depends continuously on* (t_0, r_0) *for each* $(t_0, r_0) \in R_+ \times R_+$ *and* $r(t, t_0, 0) \equiv 0$, *for* $t \ge t_0$. *Then* $u(t, t_0, u_0)$ *depends continuously on* (t_0, u_0).

Proof: Let $u_1(t) = u(t, t_1, u_1)$ and $u_2(t) = u(t, t_2, u_2)$ be solutions of (5.3.5) through (t_1, u_1) and (t_2, u_2), respectively. Let $t_2 \ge t_1$. Define

$$z(t): = u_1(t) - u_2(t).$$

From Lemma 5.3.2 and (H_1), we obtain

$$D_+\Phi(z(t)) \le M[z(t), z'(t)]$$

$$= M[z(t), f[t, u_1(t)] - f[t, u_2(t)]]$$

$$\le g(t, \Phi[z(t)]), \quad t \ge t_1. \tag{5.3.13}$$

Also

$$\Phi[z(t_1)] = \Phi[u_1 - u(t_1, t_2, u_2)]. \tag{5.3.14}$$

From (5.3.13) and (5.3.14), it follows that

$$\Phi[z(t)] \le r(t, t_1, \Phi[u_1 - u(t_1, t_2, u_2)]). \qquad (5.3.15)$$

Since $\Phi(x)$ is continuous in x, $u(t, t_2, u_2)$ is continuous in t and $r(t, t_1, r_1)$ is by hypothesis continuous in (t_1, r_1), it follows from (5.3.15) that

$$\lim_{\substack{t_1 \to t_2 \\ u_1 \to u_2}} \Phi[z(t)] \le r[t, t_2, \Phi(u_2 - u_2)]$$

$$= r(t, t_2, 0) \equiv 0.$$

Hence, from the definition of Φ

$$\lim_{\substack{t_1 \to t_2 \\ u_1 \to u_2}} z(t) = 0$$

and the proof is complete.

Now we shall prove that under (H_2) and $r(t, t_0, 0) \equiv 0$ the solutions $u(t, t_0, u_0)$ of (5.3.5) and (5.3.6) are continuously differentiable with respect to initial conditions (t_0, u_0) and the Fréchet derivatives $(\partial/\partial u_0)u(t, t_0, u_0)$ and $(\partial/\partial t_0)u(t, t_0, u_0)$ exist and satisfy the equation of variation of (5.3.5) along the solution $u(t, t_0, u_0)$. From the existence and continuity of $f_x(t, x)$ and from the mean value theorem for Fréchet differential functions, it follows that $f(t, x)$ is locally Lipschitzian in x, and consequently the local existence of solutions of (5.3.5) and (5.3.6) is secured. The following lemmas are needed.

Lemma 5.3.4:   *Let $f \in C[R_+ \times S(x_0, r), X]$ and let $f_x(t, x)$ exist and be continuous for $x \in S(x_0, r)$. Then for $x_1, x_2 \in S(x_0, r)$ and $t \ge 0$*

$$f(t, x_1) - f(t, x_2) - \int_0^1 f_x(t, sx_1 + (1 - s)x_2)(x_1 - x_2)ds. \qquad (5.3.16)$$

Proof: Define

$$F(s) = f[t, sx_1 + (1 - s)x_2], \ 0 \le s \le 1.$$

The convexity of $S(x_0, r)$ implies that $F(s)$ is well defined. Using the chain rule for Fréchet derivatives, we obtain

$$F'(s) = f_x[t, sx_1 + (1 - s)x_2](x_1 - x_2). \qquad (5.3.17)$$

Since $F(1) = f(t, x_1)$ and $F(0) = f(t, x_2)$ the result follows by integrating (5.3.17) with respect to s from 0 to 1.

Lemma 5.3.5: *Let (H_2) holds. Then*

$$M\left[h, \int_0^1 f_x[t, sx_1 + (1-s)x_2]h\,ds\right]$$

$$\leq g[t, \Phi(h)], \quad t \geq 0, \quad h, x_1, x_2 \in s(x_0, r). \qquad (5.3.18)$$

Proof: Let $\pi: 0 = s_0 < s_1 < \ldots < s_n = 1$ be any partition of $[0, 1]$. From the definition of Riemann integral for continuous abstract functions it follows that

$$\int_0^1 f_x[t, sx_1 + (1-s)x_2]h\,ds = \lim_{n \to \infty}\left[\sum_{i=0}^{n-1} f_x[t, \tau_i x_1 + (1 - \tau_i)x_2]h\Delta s_i\right] \quad (5.3.19)$$

where $\Delta s_i = s_{i+1} - s_i$ and $s_i \leq \tau_i \leq s_{i+1}$, for $i = 0, 1, \ldots, n-1$.

From the continuity of $M[x, h]$ in h (uniformly with respect to $x \in S(x_0, r)$), (5.3.19), and (H_2) it follows from

$$M\left[h, \int_0^1 f_x[t, sx_1 + (1-s)x_2]h\,ds\right]$$

$$= \lim_{n \to \infty} M\left[h, \sum_{i=0}^{n-1} f_x(t, \tau_i x_1 + (1 - \tau_i)x_2)h\Delta s_i\right]$$

$$\leq \lim_{n \to \infty} \sum_{i=0}^{n-1} \Delta s_i M[h, f_x(t, \tau_i x_1 + (1 - \tau_i)x_2)h]$$

$$\leq g(t, \Phi(h)) \lim_{n \to \infty} \sum_{i=0}^{n-1} \Delta s_i = g(t, \Phi(h)).$$

The proof is complete.

Lemma 5.3.6: (H_2) *implies* (H_1).

Proof: In view of Lemmas (5.3.4) and (5.3.5) we have

$$M[x - y, f(t, x) - f(t, y)] = M\left[x - y, \int_0^1 f_x[t, sx + (1-s)y](x - y)ds\right]$$

$$\leq g[t, \Phi(x - y)].$$

The proof is complete.

Corollary 5.3.1: *Let (H_2) hold. Assume that the maximal solution of (5.3.7) and (5.3.8) depends continuously on (t_0, r_0) for each $(t_0, r_0) \in R_+ \times R_+$ and $r(t, t_0, 0) \equiv 0$ for $t \geq t_0$. Then the solutions of (5.3.5) and (5.3.6) exist locally, are unique, and depend continuously on initial conditions.*

Theorem 5.3.2: *Let* (H_2) *hold. Assume that the maximal solution of* (5.3.7) *through any point* $(t_0, 0)$ *is identically zero for* $t \geq t_0$.
   Then
(i) *the Fréchet derivative* $(\partial/\partial u_0)u(t, t_0, u_0) \equiv U(t, t_0, u_0)$ *exists and satisfies the operator equation*

$$U' = f_u[t, u(t, t_0, u_0)]U, \ t \geq t_0, \tag{5.3.20}$$

$$U(t_0) = I; \tag{5.3.21}$$

(ii) *the Fréchet derivative* $(\partial/\partial t_0)u(t, t_0, u_0) \equiv V(t, t_0, u_0)$ *exists and satisfies*

$$V' = f_u(t, u(t, t_0, u_0))V, \ t \geq t_0, \tag{5.3.22}$$

$$V(t_0) - f(t_0, u_0). \tag{5.3.23}$$

Furthermore

$$V(t, t_0, u_0) = -U(t, t_0, u_0)f(t_0, u_0). \tag{5.3.24}$$

Proof: (i) Since $f_u(t, \cdot) \in C[J \times S(u_0, r), B(X)]$, (5.3.20) and (5.3.21) has a unique solution which we denote by $U(t)$.
   Define the function

$$z(t) = u(t, t_0, u_0 + h) - u(t, t_0, u_0) - U(t)h,$$

$$t \geq t_0, \ u_0, u_0 + h \in S(u_0, r).$$

Then

$$D_+\Phi(z(t)/\parallel h \parallel) \leq M[z(t)/\parallel h \parallel, z'(t)/\parallel h \parallel$$

$$= M[z(t)/\parallel h \parallel, f[t, u(t, t_0, u_0 + h)] - f[t, u(t, t_0, u_0)]/\parallel h \parallel$$

$$- f_u[t, u(t, t_0, u_0 + h)]U(t)h/\parallel h \parallel]. \tag{5.3.25}$$

From the Fréchet differentiability of f with respect to $u \in S(u_0, r)$, we have

$$f[t, u(t, t_0, u_0 + h)] - f[t, u(t, t_0, u_0)]$$

$$= f_u[t, u(t, t_0, u_0)][u(t, t_0, u_0 + h) - u(t, t_0, u_0)]$$

$$+ O(\parallel u(t, t_0, u_0 + h) - u(t, t_0, u_0) \parallel)$$

$$= f_u[t, u(t, t_0, u_0)][z(t) + U(t)h]$$

$$+ O(\parallel u(t, t_0, u_0 + h) - u(t, t_0, u_0 \parallel). \tag{5.3.26}$$

Define

$$\omega(h) = O(\parallel u(t, t_0, u_0 + h) - u(t, t_0, u_0) \parallel).$$

From (5.3.25) and (5.3.26) we get

$$D_+\Phi(z(t)/\parallel h \parallel) \le M[z(t)/\parallel h \parallel, f_u[t, u(t, t_0, u_0)]z(t)/\parallel h \parallel + \omega(h)/\parallel h \parallel]$$

$$\le M[z(t)/\parallel h \parallel, f_u[t, u(t, t_0, u_0)]z(t)/\parallel h \parallel]$$

$$+ M[z(t)/\parallel h \parallel, \omega(h)/\parallel h \parallel]$$

$$\le g[t, \Phi(z(t)/\parallel h \parallel)] + M[z(t)/\parallel h \parallel, \omega(h)/\parallel h \parallel]. \qquad (5.3.27)$$

Next we shall prove that in any compact interval of t

$$\lim_{\parallel h \parallel \to 0} M[z(t)/\parallel h \parallel, \omega(h)/\parallel h \parallel] = 0. \qquad (5.3.28)$$

Set

$$m(t) = \parallel u(t, t_0, u_0 + h) - u(t, t_0, u_0) \parallel. \qquad (5.3.29)$$

It follows that

$$D_+m(t) \le \parallel u'(t, t_0, u_0 + h) - u'(t, t_0, u_0) \parallel$$

$$= \parallel f[t, u(t, t_0, u_0 + h)] - f[t, u(t, t_0, u_0)] \parallel.$$

In view of Lemma 5.3.4 and the continuity of f we obtain from (5.3.29)

$$D_+m(t) \le K_1 m(t) \qquad (5.3.30)$$

where K_1 is a constant such that

$$\parallel f_u(t, z) \parallel \le K_1$$

for t in a compact interval I around t_0 and z being in the line segment joining the solutions $x(t, t_0, x_0)$ and $x(t, t_0, x_0 + h)$. Also from (5.3.29)

$$m(t_0) = \parallel h \parallel. \qquad (5.3.31)$$

By (5.3.30) and (5.3.31) we get

$$m(t) \le \parallel h \parallel \exp K_1(t - t_0)$$

$$\le K_2 \parallel h \parallel, \quad t \in I \qquad (5.3.32)$$

where K_2 is a constant.
    Let K_3 be a constant such that

$$\| U(t) \| \leq K_3, \ t \in I. \tag{5.3.33}$$

From the definition of $z(t)$, (5.3.32), and (5.3.33) we obtain

$$\| z(t) \| / \| h \| \leq K_2 + K_3 \equiv K_4, \ t \in I, \ \| h \| \ \text{sufficiently small.} \tag{5.3.34}$$

In view of Lemma 5.3.1 we get

$$M[z(t)/ \| h \|, \omega(h)/ \| h \|] \leq K \| \omega(h) \| / \| h \| . \tag{5.3.35}$$

Finally, from the definition of $\omega(h)$ and (5.3.32) we obtain

$$\| \omega(h) \| / \| h \| \leq K_2 \| \omega(h) \| / m(t)$$

$$= K_2 \| \omega(h) \| / \| u(t, t_0, u_0 + h) - u(t, t_0, u_0) \| \ \to 0 \ \text{as} \ h \to 0$$

and (5.3.28) has been established.

From (5.3.27) and (5.3.28) we obtain

$$D_+\Phi(z(t)/ \| h \|) \leq g[t, \Phi(z(t)/ \| h \|)] + O(1). \tag{5.3.36}$$

Also

$$\Phi(z(t_0)/ \| h \|) = 0. \tag{5.3.37}$$

In view of (5.3.36) and (5.3.37) it follows that

$$\lim_{\| h \| \to 0} \Phi(z(t)/ \| h \|) = r(t, t_0, 0) \equiv 0.$$

Hence, from Definition 5.3.1

$$\lim_{\| h \| \to 0} z(t)/ \| h \| = 0,$$

which proves that the Fréchet derivative $(\partial/\partial u_0)u(t, t_0, u_0)$ exists and it is equal to $U(t)$. The proof of (i) is complete.

(ii) Let $U(t)$ be as in (i) the solution of (5.3.20) and (5.3.21). Define the function

$$z(t) = u(t, t_0 + h, u_0) - u(t, t_0, u_0) + U(t)f(t_0, u_0)h.$$

Then as in (i), we have

$$D_+\Phi(z(t)/h) \leq M[z(t)/h]$$

$$= M[z(t)/h, (f[t, u(t, t_0 + h, u_0)] - f[t, u(t, t_0, u_0)])/h$$

$$+ f_u[t, u(t, t_0, u_0)]U(t)f(t_0, u_0)]$$

$$= M[z(t)/h, f_u[t, u(t, t_0, u_0)]z(t)/h + \omega(h)/h]$$

$$\leq M[z(t)/h, f_u[t, u(t, t_0, u_0)](z(t)/h)] + M[z(t)/h, \omega(h)/h]$$

$$\leq g[t, \Phi(z(t)/h)] + O(1) \text{ as } h \to 0. \tag{5.3.38}$$

Also

$$\Phi(z(t_0)/h) = \Phi([u(t_0, t_0 + h, u_0) - u_0]/h + f(t_0, u_0))$$

$$= \Phi([u(t_0, t_0 + h, u_0) - u(t_0 + h, t_0 + h, u_0)]/h + f(t_0, u_0))$$

$$= O(1) \quad \text{as } h \to 0. \tag{5.3.39}$$

From (5.3.38), (5.3.39), it follows as in (i) that

$$\lim_{h \to 0} \Phi(z(t)/h) = r(t, t_0, 0) \equiv 0.$$

Hence

$$\lim_{h \to 0_+} z(t)/h = 0.$$

which proves (ii). In addition

$$(\partial/\partial t_0)u(t, t_0, u_0) = -U(t)f(t_0, u_0)$$

$$= -(\partial/\partial u_0)u(t, t_0, u_0)f(t_0, u_0).$$

The proof is complete.

Theorem 5.3.4: *Under the hypotheses of Theorem 5.3.3 the following formula holds:*

$$u(t, t_0, v_0) - u(t, t_0, u_0) = \int_0^1 U[t, t_0, u_0 + s(v_0 - u_0)](v_0 - u_0)ds. \tag{5.3.40}$$

Proof: From Theorem 5.3.3 and the chain rule for abstract functions, we have

$$(d/ds)u[t, t_0, u_0 + s(v_0 - u_0)] = U[t, t_0, u_0 + s(v_0 - u_0)](v_0 - u_0). \tag{5.3.41}$$

Integrating (5.3.41) from 0 to 1 with respect to s the desired result follows.

Now we shall establish the variation of constants formula with respect to (5.3.5) and (5.3.6) and its nonlinear perturbation

$$v' = f(t, v) + F(t, v), \tag{5.3.42}$$

$$v(t_0) = v_0 \tag{5.3.43}$$

where $f, F: R_+ \times X \to Y$ are smooth enough to guarantee the existence of solutions of (5.3.42) and (5.3.43) for $t \geq t_0$. A solution of (5.3.42) and (5.3.43) is denoted by $v(t, t_0, v_0)$.

Theorem 5.3.5: *Let $f, F \in C[R_+ \times X; X]$ and let f satisfy (H_1). Let $u(t, t_0, u_0)$ and $v(t, t_0, u_0)$ be solutions of (5.3.5) and (5.3.42) through (t_0, u_0), respectively. Then, for $t \geq t_0$*

$$v(t, t_0, u_0) = u(t, t_0, u_0) + \int_{t_0}^{t} U[t, s, v(s, t_0, u_0)] F[s, v(s, t_0, u_0)] ds \qquad (5.3.44)$$

where

$$U(t, t_0, u_0) \doteq (\partial/\partial u_0) u(t, t_0, u_0).$$

Proof: Denote $v(t) = v(t, t_0, u_0)$. Then, in view of Theorem 5.3.6

$$(d/ds) u[t, s, v(s)] = (\partial/\partial s) u[t, s, v(s)] + (\partial/\partial v) u[t, s, v(s)] v'(s)$$

$$= -U[t, s, v(s)] f(s, v(s))$$

$$+ U[t, s, v(s)] (f[s, v(s)] + F[s, v(s)])$$

$$= U[t, s, v(s)] F[s, v(s)]. \qquad (5.3.45)$$

Since the right-hand side of (5.3.45) is continuous, we can integrate from t_0 to t, obtaining the VPF (5.3.44).

5.4 GROWTH PROPERTIES OF SOLUTIONS

The objective of this section is to analyze the way in which the solutions of nonlinear differential equations can grow to infinity. Such equations apply to early stage growth processes during which nonlinearities account for transition or interaction. Late stage growth models may involve different nonlinearities which stabilize or limit growth. The method we use supposes that the nonlinear equation is a perturbation of a linear equation for which growth estimates are known. We employ a variation of constants formula to establish conditions which yield similar growth estimates for the nonlinear problem.

The problem we consider has the form

$$z'(t) = Az(t) + F(z(t)), \quad t \geq 0, z(0) \in X, \qquad (5.4.1)$$

where X is a Banach space, A is the infinitesimal generator of a strongly continuous semigroup $T(t)$, $t \geq 0$ of bounded linear operators in X, and F is a nonlinear operator in X. Assume that

(i) there exist a nonnegative integer n, a nonnegative constant ω, and a positive constant M such that

$$\limsup_{t \to \infty} e^{-\omega t} \mid T(t) \mid /t^n \leq M, \qquad (5.4.2)$$

(ii) F is bounded on bounded sets, continuous, and there exists $r_0 > 1$ and a nonincreasing function

$$c: [r_0, \infty) \to [0, \infty) \text{ such that } \| F(z) \| < c(\| z \|) \| z \| \qquad (5.4.3)$$

for all $z \in X$ such that $\| z \| \geq r_0$.

We suppose that equation (5.4.1) has a mild solution $z(t)$ for $t \geq 0$; that is, there is a continuous function $z(t)$ from $[0, \infty)$ to X satisfying

$$z(t) = T(t)z(0) + \int_0^t T(t-s)F(z(s))ds, \; t \geq 0. \qquad (5.4.4)$$

Theorem 5.4.1: *Let (5.4.2) and (5.4.3) hold. If $\gamma > \omega$ and $\lim_{r \to \infty} c(r) = 0$, then there exists a constant M_γ such that $\| z(t) \| \leq M_\gamma e^{\gamma t}$, $t \geq 0$.*

Proof: Let $\delta \in (\omega, \gamma)$. There exists a constant $\widehat{M}$ such that $e^{-\delta t} | T(t) | \leq \widehat{M}$ for $t \geq 0$. There exists $\eta > r_0$ such that if $\| z \| > \eta$, then $c(\| z \|) < (\gamma - \delta)/\widehat{M}$. There exists a constant C such that $\| F(z) \| \leq C$ for $\| z \| \leq \eta$. From (5.4.4) we obtain

$$\| z(t) \| \leq | T(t) | \| z(0) \| + \int_0^t | T(t-s) | \| F(z(s)) \| ds, \; t \geq 0. \quad (5.4.5)$$

If $\| z(s) \| \leq \eta$, then $\| F(z(s)) \| \leq C$, and if $\| z(s) \| > \eta$, then

$$\| F(z(s)) \| \leq c(\| z(s) \|) \| z(s) \| \leq ((\gamma - \delta)/\widehat{M}) \| z(s) \| .$$

Thus,

$$e^{-\delta t} \| z(t) \| \leq \widehat{M} \| z(0) \| + \int_0^t \widehat{M}[C + \tfrac{\gamma - \delta}{\widehat{M}} \| (s) \|]e^{-\delta s}ds$$

$$\leq \widehat{M} \| z(0) \| + \widehat{M}C/\delta + (\gamma - \delta)\int_0^t e^{-\delta s} \| z(s) \| ds, \; t \geq 0.$$

By Gronwall's Lemma,

$$e^{-\delta t} \| z(t) \| \leq \widehat{M}(\| z(0) \| + C/\delta)e^{(\gamma - \delta)t}, \; t \geq 0,$$

which implies the conclusion.

Theorem 5.4.2: *Let (5.4.2) and (5.4.3) hold. If $\omega > 0$ and*

$$\int_{r_0}^\infty \frac{c(r)}{r}(\log r)^n dr < \infty, \qquad (5.4.6)$$

then there exists a positive constant M' such that

$$\limsup_{t \to \infty} e^{-\omega t} \| z(t) \| /t^n \leq M'. \qquad (5.4.7)$$

Proof: Let $\widehat{M} > 0$ such that $|T(t)| \leq \widehat{M}e^{\omega t}(1 + t^n)$, $t \geq 0$, $M(t) = \widehat{M}(1 + t^n)$, $t \geq 0$, $\delta \in (0, \omega)$, $\Omega = \{t \geq 0 : \| z(t) \| \leq e^{\delta t}\}$, and for $t > 0$,

$$K_t = \sup_{0 \leq s \leq t} \{e^{-\omega s} \| z(s) \| / M(s)\},$$

and

$$C_t = (\widehat{M}/\delta) \int_{e^{\delta t}}^{\infty} \frac{c(r)}{r}(1 + (\log r)^n / \delta^n)dr.$$

Use (5.4.6) to choose $t_1 > (\log r_0)/\delta$ such that $C_{t_1} < 1$. From (5.4.4) we obtain

$$e^{-\omega t} \| z(t) \| / M(t) \leq \| z(0) \| + \int_0^t \frac{M(t-s)}{M(t)} e^{-\omega s} \| F(z(s)) \| ds, t \geq 0. \quad (5.4.8)$$

Let $t \geq t_1$ and write the integral in (5.4.8) as the sum of the three integrals

$$\int_{[0,t_1]} + \int_{[t_1,t] \cap \Omega} + \int_{[t_1,t] \cap \Omega^c}. \quad (5.4.9)$$

Since $z(s)$ is bounded on $[0, t_1]$ and F is bounded on bounded sets, there exists a constant C_1 independent of t such that the first integral in (5.4.9) is bounded by C_1. Let C_2 be a constant such that $\| F(z) \| \leq C_2$ if $\| z \| \leq r_0$. Then the second integral in (5.4.9) is bounded by

$$\int_{t_1}^t e^{-\omega s}(C_2 + c(r_0)e^{\delta s})ds \leq \frac{C_2}{\omega} + \frac{c(r_0)}{\omega - \delta}.$$

The third integral in (5.4.9) is bounded by

$$\int_{t_1}^t c(e^{\delta s})M(s)K_t ds = K_t \int_{e^{\delta t_1}}^{e^{\delta t}} c(r)M(\frac{\log r}{\delta})\frac{1}{\delta r} dr.$$

From (5.4.8) we then obtain for $t > t_1$

$$K_t \leq \| z(0) \| + C_1 + \frac{C_2}{\omega} + \frac{c(r_0)}{\omega - \delta} + C_{t_1}K_t.$$

Since $C_{t_1} < 1$, we have that for $t > t_1$

$$K_t \leq (\| z(0) \| + C_1 + \frac{C_2}{\omega} + \frac{c(r_0)}{\omega - \delta})/(1 - C_{t_1}),$$

which implies (5.4.7). The proof is complete.

Theorem 5.4.3: *Let (5.4.2) hold, let $p \in [0, 1]$, and let there exist a constant C such that $\| F(z) \| \leq C \| z \|^p$, $z \in X$. If $p = 1$, then $\| z(t) \| \leq \| w(t) \|$, $t \geq 0$, where $e^{-\omega t} |T(t)| \leq M(t) =: \sum_{k=0}^n M_k t^k$, $t \geq 0$, and*

$$w(t) = M(t)e^{\omega t} \| z(0) \| + C \int_0^t M(t - s)e^{\omega(t-s)}w(s)ds, \ t \geq 0. \quad (5.4.10)$$

If $p \in [0, 1)$ and $\omega > 0$, then (5.4.7) holds.

Proof: To prove the claim for $p = 1$ use (5.4.4) to obtain

$$\| z(t) \| \leq M(t)e^{\omega t} \| z(0) \| + C \int_0^t M(t - s)e^{\omega(t-s)} \| z(s) \| ds, \ t \geq 0.$$

Then, (5.4.10) follows immediately. To prove the claim for $p \in [0, 1)$ set $c(r) = C/r^{1-p}, r \geq 1$. Then $\| F(z) \| \leq C \| z \|^p = c(\| z \|) \| z \|$ for $\| z \| \geq 1$. Then (5.4.6) holds, since

$$\int_1^\infty \frac{c(r)}{r}(\log r)^n dr = \int_1^\infty \frac{C(\log r)^n}{r^{2-p}} dr < \infty,$$

and Theorem 5.4.2 applies.

Remark 5.4.1: In the results above we have hypothesized that the solution $z(t)$ of (5.4.1) exists for all $t \geq 0$. If we hypothesize instead that the solution exists on a maximal interval of existence $[0, t_0)$ with $t_0 < \infty$, then the results show that $z(t)$ is bounded on $[0, t_0)$. In certain cases, the boundedness of the solution on its maximal interval of existence $[0, t_0)$ implies $t_0 = \infty$. If $p > 1$ in Theorem 5.4.3 then the maximal interval of existence for the solution of (5.4.1) may be finite.

We will use the following hypothesis to show that $\lim_{t\to\infty}e^{-\omega t}z(t)/t^n$ exists:

$$\text{for } x \in X, \ \lim_{t \to \infty} e^{-\omega t}T(t)x/t^n = : Px \text{ exists}$$

$$(P \text{ is necessarily a bounded linear operator} \quad (5.4.11)$$

$$\in X \text{ which commutes with } T(t), t \geq 0).$$

Theorem 5.4.4: Let (5.4.2), (5.4.3), (5.4.6), (5.4.11) hold and let $\omega > 0$. Then

$$\lim_{t \to \infty} e^{-\omega t}z(t)/t^n = P\left[z(0) + \int_0^\infty e^{-\omega s}F(z(s))ds\right]. \quad (5.4.12)$$

Proof: We first claim that $\int_0^\infty e^{-\omega s}F(z(s))ds$ exists in X. By the continuity of F and $z(s)$ it suffices to show that $\int_0^\infty e^{-\omega s} \| F(z(s)) \| ds < \infty$. Let δ, Ω, and C_2 be as in the proof of Theorem 5.4.2. For $t \geq t_0 = : (\log r_0)/\delta$,

$$\int_{t_0}^t e^{-\omega s} \| F(z(s)) \| ds \leq \int_{t_0}^t (C_2 + c(r_0)e^{\delta s})e^{-\omega s}ds$$

$$+ \delta^{-(n+1)} \left(\sup_{s \geq t_0} \frac{e^{-\omega s}\|z(s)\|}{s^n} \right) \int_{r_0}^{e^{\delta t}} \frac{c(r)}{r} (\log r)^n dr,$$

which is bounded independently of t by (5.4.6) and (5.4.7).

The claim (5.4.12) will follow if we establish

$$\lim_{t \to \infty} \int_0^t \frac{e^{-\omega t}T(t-s)}{t^n} F(z(s))ds = \int_0^\infty e^{-\omega s}PF(z(s))ds. \qquad (5.4.13)$$

By (5.4.11) and the Principle of Uniform Boundedness there exists $M_1 > 0$ such that $e^{-\omega t} \mid T(t) \mid /t^n \leq M_1$ for $t \geq t_0$. By (5.4.7) there exists $M_2 > 0$ such that $e^{-\omega t} \| z(t) \| /t^n \leq M_2$ for $t \geq t_0$. Let $M_3: = \sup_{0 \leq s \leq t_0}\{e^{-\omega s} \mid T(s) \mid\}$. For $t_1 > t_0$ and $t \geq t_1 + t_0$,

$$\left\| \int_0^t \left(\frac{e^{-\omega t}T(t-s)}{t^n} - e^{-\omega s}P \right)F(z(s))ds \right\| \qquad (5.4.14)$$

$$\leq \int_0^{t_1} \frac{e^{-\omega(t-s)}\mid T(t-s) \mid}{(t-s)^n} \left(1 - \left(\frac{t-s}{t}\right)^n\right)e^{-\omega s} \| F(z(s)) \| ds$$

$$+ \int_0^{t_1} \left\| \left(\frac{e^{-\omega(t-s)}T(t-s)}{(t-s)^n} - P \right)e^{-\omega s}F(z(s)) \right\| ds$$

$$+ \int_{t_1}^t (M_1 + M_3 t_0^{-n} + \mid P \mid)[(C_2 + c(r_0)e^{\delta s})e^{-\omega s} + c(e^{\delta s})M_2 s^n]ds.$$

The last integral in (5.4.14) is bounded by

$$(M_1 + M_3 t_0^{-n} + \mid P \mid)\left[\frac{C_2 e^{-\omega t_1}}{\omega} + \frac{c(r_0)e^{-(\omega-\delta)t_1}}{\omega-\delta} + (M_2/\delta^{n+1})\int_{e^{\delta t_1}}^\infty \frac{c(r)(\log r)^n}{r} dr \right].$$

For fixed t_1 the first two integrals on the right side of (5.4.14) converge to 0 as $t \to \infty$. Claim (5.4.13) now follows from (5.4.6).

We illustrate the above with the following examples.

Example 5.4.1: Let $X = \mathbb{R}^2$ with norm $\left\| \begin{bmatrix} x \\ y \end{bmatrix} \right\| = \mid x \mid + \mid y \mid$ and consider the system

$$x' = y + \frac{x}{x+y}$$

$$y' = \frac{x}{x+y}$$

$$x(0) = 1, \; y(0) = 1.$$

For this example

$$A = \begin{bmatrix} 0 & 1 \\ 0 & 0 \end{bmatrix}, T(t) = \begin{bmatrix} 1 & t \\ 0 & 1 \end{bmatrix}, F\left(\begin{bmatrix} x \\ y \end{bmatrix}\right) = \frac{1}{x+y}\begin{bmatrix} x \\ x \end{bmatrix}, x \geq 1, y \geq 1,$$

$$F\left(\begin{bmatrix} x \\ y \end{bmatrix}\right) = \frac{1}{\max\{x,1\}+\max\{y,1\}}\begin{bmatrix} \max\{x,1\} \\ \max\{x,1\} \end{bmatrix}, \text{ otherwise,}$$

$n = 1$, $\omega = 0$, $|T(t)| = 1+t$, and $c(r) = \frac{1}{r}$, $r \geq 1$. Theorem 5.4.1 applies, but Theorem 5.4.2 does not. In fact the conclusion of Theorem 5.4.2 does not hold, since $\lim_{t\to\infty} x(t)/t^2 = \frac{1}{2}$, $\lim_{t\to\infty} y(t)/t = 1$.

Example 5.4.2: Consider the system in $\mathbb{R}^2$

$$x' = y + \frac{x^2}{x+y}$$

$$y' = \frac{xy}{x+y}$$

$$x(0) = 1, \ y(0) = 1.$$

$A, T(t)$, $t \geq 0$, n, ω and $|T(t)|$ are the same as in Example 5.4.1 and $F(\begin{bmatrix} x \\ y \end{bmatrix}) = \frac{x}{x+y}\begin{bmatrix} x \\ y \end{bmatrix}$, $x \geq 1$, $y \geq 1$. There does not exist a function $c(r)$ satisfying (5.4.3), so neither Theorem 5.4.1 nor Theorem 5.4.2 applies. Theorem 5.4.3 does apply with $M(t) = 1+t$, $C = 1$, and $w(t) = (.9\sqrt{5}+.t)\exp(.5(1+\sqrt{5})t) + (.5 - .9\sqrt{5})\exp(.5(1-\sqrt{5})t)$. The solution of this system is $x(t) = 2(1+t)e^t/(2+t)$, $y(t) = 2e^t/(2+t)$.

Example 5.4.3: Consider the system

$$x' = x + y + \sqrt{x+y}$$

$$y' = y + \sqrt{x+y}$$

$$x(0) = 1, \ y(0) = 1.$$

In this example $n = 1$, $\omega = 1$, $|T(t)| = (1+t)e^t$,

$$A = \begin{bmatrix} 1 & 1 \\ 0 & 1 \end{bmatrix}, T(t) = e^t\begin{bmatrix} 1 & t \\ 0 & 1 \end{bmatrix},$$

$$P = \begin{bmatrix} 0 & 1 \\ 0 & 0 \end{bmatrix}, F\left(\begin{bmatrix} x \\ y \end{bmatrix}\right) = \begin{bmatrix} \sqrt{x+y} \\ \sqrt{x+y} \end{bmatrix}, x \geq 1, y \geq 1.$$

Theorem 5.4.4 and the second part of Theorem 5.4.3 apply with $C = 2$, $p = 1/2$. Numerical solution of this system yields $\lim_{t\to\infty} x(t)/(te^t) \simeq 7.7$, $\lim_{t\to\infty} y(t)/e^t \simeq 7.7$.

Example 5.4.4: Consider the system

$$x' = x + y + \frac{x+y}{(\log(x+y))^p}$$

$$y' = y + \frac{x+y}{(\log(x+y))^p}$$

$$x(0) = 1,\ y(0) = 1.$$

$A, T(t),\ t \geq 0,\ P,\ n,\ \omega$ and $|T(t)|$ are the same as in Example 5.4.3 and $F([^x_y]) = \frac{1}{(\log(x+y))^p} [^{x+y}_{x+y}],\ x \geq 1,\ y \geq 1,\ p \geq 2$. Take $c(r) = 1/(\log r)^p,\ r \geq e$. If $p > 2$, then

$$\int\limits_e^\infty \frac{c(r)}{r} \log r\, dr = \int\limits_e^\infty \frac{1}{r(\log r)^{p-1}} dr < \infty,$$

and Theorems 5.42 and 5.4.4 apply. If $p = 3$, then numerical solution of this system yields $\lim_{t\to\infty} x(t)/(te^t) \simeq 2.53$, $\lim_{t\to\infty} y(t)/e^t \simeq 2.4$. If $p = 2$, then

$$\int\limits_e^\infty \frac{c(r)}{r} \log r\, dr = \int\limits_e^\infty \frac{1}{r \log r} dr = \infty,$$

and Theorem 5.4.2 does not apply. Numerical solution for the case $p = 2$ yields $\lim_{t\to\infty} (x(t) + y(t))/(te^t) = \infty$.

Example 5.4.5: Let A be the infinitesimal generator of a C_0-semigroup in the Banach space X, let B and C be bounded linear operators in X such that $(A + B)Cx = C(A + B)x$ for $x \in D(A)$, let $T(t),\ t \geq 0$ be the C_0-semigroup in X generated by $A + B$, let $T(t),\ t \geq 0$ satisfy (5.4.2) with $\omega > 0$, and let $c_i(r)$, $i = 1, 2, 3$ be scalar functions on $[0, \infty)$ such that $|c_i(r) - 1|$ is nonincreasing and

$$\int\limits_0^\infty \frac{|c_i(r)-1|}{r} (\log r)^{n+1} er < \infty.$$

Let $\chi = X \times X$ and consider the abstract nonlinear system in X

$$u' = Au + c_1(\|u\| + \|v\|)Bu + c_2(\|u\| + \|v\|)Cv,\ u(0) \in X,$$

$$v' = Av + c_3(\|u\| + \|v\|)Bv,\ v(0) \in X.$$

Define

$$\mathcal{A} = \begin{bmatrix} A + B & C \\ 0 & A + B \end{bmatrix},\ \mathcal{T}(t) = \begin{bmatrix} T(t) & tT(t)C \\ I & T(t) \end{bmatrix},$$

$$\mathcal{F}\left(\begin{bmatrix} x \\ y \end{bmatrix}\right) = \begin{bmatrix} (c_1(\|x\| + \|y\|) = 1)Bx & (c_2(\|x\| + \|y\|) - 1)Cy \\ 0 & (c_3(\|x\| + \|y\|) - 1)By \end{bmatrix}$$

Then, $T(t)$, $t \geq 0$ is a C_0-semigroup in χ, $\mathcal{A}$ is the infinitesimal generator of $T(t)$, $t \geq 0$, and the nonlinear system can be written as

$$\begin{bmatrix} u \\ v \end{bmatrix}' = \mathcal{A} \begin{bmatrix} u \\ v \end{bmatrix} + \mathcal{F}\left(\begin{bmatrix} u \\ v \end{bmatrix} \right).$$

By Theorem 5.4.2 there exist a positive constant M' such that

$$\limsup_{t \to \infty} e^{-\omega t} (\parallel u(t) \parallel + \parallel v(t) \parallel)/t^{n+} \leq M'.$$

If (5.4.11) holds, then Theorem 5.4.4 yields

$$\lim_{t \to \infty} e^{-\omega t} \begin{bmatrix} u(t) \\ v(t) \end{bmatrix} /t^{n+1} = \mathcal{P}\left(\begin{bmatrix} u(0) \\ v(0) \end{bmatrix} + \int_0^\infty e^{\omega s} \mathcal{F}\left(\begin{bmatrix} u(s) \\ v(s) \end{bmatrix} \right) ds \right),$$

where $\mathcal{P} = \begin{bmatrix} 0 & PC \\ 0 & 0 \end{bmatrix}$.

5.5 STABILITY BEHAVIOR

In the present section, we consider the stability behavior of some parabolic differential equations. The result is analogous known for ordinary differential equations.

Let X be a Banach space and a c_0-semigroup $T(t)$, $t \geq 0 \to L(X)$ satisfying the estimate

$$| T(t) | \leq M e^{\beta t}, \forall t \geq 0 \text{ and } \beta \leq 0. \tag{5.5.1}$$

Let $f(t, x)$ be given function such that $f \colon [0, \infty) \times X \to X$, continuous in t and X and satisfies the Lipschitz estimate

$$| f(t, x_1) = f(t, x_2) | \leq g(t) | x_1 - x_2 |, \ t \geq 0$$

$$x_1, x_2 \in X, g \in C[R_+, R_+]. \tag{5.5.2}$$

If A is the infinitesimal generator of the semigroup $T(t)$, strong solutions on $[0, \infty)$ of the equation

$$u'(t) = Au(t) + f(t, u(t)), t \geq 0. \tag{5.5.3}$$

belong to the space $C^1 R_+, X)$ and $u(t) \in D(A)$, $t \geq 0$ while the mild solutions of (5.5.3) are functions $u \colon [0, \infty) \to X$ such that

$$u(t) = T(t)u(0) + \int_0^t T(t - \sigma)f(\sigma, u(\sigma))d\sigma \tag{5.4.5}$$

and these mild solutions belong to the space $C[[0, \infty), X]$.
We prove the following result.

Theorem 5.5.1: *The mild solutions of (5.5.3) or (5.5.4) are stable if*
(i) $\beta = 0$ *and* $\int_0^\infty g(t)dt < \infty$ *or* $g(t) = L$, $\beta = -LM$; *they are asymptotically stable if*
(ii) $\beta < 0$ *and* $\int_0^\infty g(t)dt < \infty$ *or* $g(t) = L$ *and* $\beta < = LM$.

Proof: Let $u_i(t)$, $i = 1, 2$ be the solutions such that

$$u_i(t) = T(t)u_i(0) + \int_0^t T(t - \sigma)f(\sigma, u_i(\sigma))d\sigma, \quad i = 1, 2, t \geq 0.$$

It follows that

$$| u_1(t) - u_2(t) |$$

$$\leq Me^{\beta t} | u_1(0) - u_2(0)) | + \int_0^t Me^{\beta(t-\sigma)} | f(\sigma, u_1(\sigma)) - f(\sigma, u_2(\sigma)) | d\sigma$$

$$\leq Me^{\beta t} | u_1(0) - u_2(0) | + Me^{\beta t} \int_0^t e^{-\beta\sigma} g(\sigma) | u_1(\sigma) - u_2(\sigma) | d\sigma,$$

and hence

$$| u_1(t) - u_2(t) | e^{-\beta t} \leq M | u_1(0) - u_2(0) |$$

$$+ \int_0^t Mg(\sigma)e^{-\beta\sigma} | u_1(\sigma) - u_2(\sigma) | d\sigma, t \geq 0$$

$$\leq M(| u_1(0) - u_2(0) |)\exp\left[\int_0^t Mg(s)ds \right],$$

by using Gronwall's integral inequality. We now obtain

$$| u_1(t) - u_2(t) | \leq Me^{\beta t} | u_1(0) - u_2(0) | \exp\left[\int_0^t Mg(s)ds \right].$$

Now, let $\beta = 0$ and $\int_0^\infty g(\sigma)d\sigma < \infty$. Then

$$| u_1(t) - u_2(t) | \leq M | u_1(0) - u_2(0) | \exp\left[\int_0^\infty Mg(s)ds \right]$$

which implies stability. let $g(t) = L$, then we have

$$| u_1(t) - u_2(t) | \le M | u_1(0) - u_2(0) | e^{\beta t} \cdot e^{MLt} = M | u_1(0) - u_2(0) |$$

if $\beta + ML = 0$,, which again implies stability. If $\beta < 0$ and $\int_0^\infty g(\sigma)d\sigma < \infty$,, we obtain

$$| u_1(t) - u_2(t) | \le M e^{\beta t} | u_1(0) - u_2(0) | \exp\left[\int_0^\infty Mg(s)ds\right] \to 0$$

as $t \to \infty$. Finally, let $g(t) = L$, then we have

$$| u_i(t) - u_2(t) | \le M | u_1(0) - u_2(0) | e^{\beta t + MLt}$$

which tends to zero as $t \to \infty$ when $\beta + ML < 0$. The proof is complete.

Let us consider the abstract differential equation

$$u' = f(t, u) \text{ and } u(t_0) = u_0 \tag{5.5.5}$$

and its perturbation

$$v' = f(t, v) + F(t, v) \text{ and } v(t_0) = v_0 \tag{5.5.6}$$

where $f, F \in C[R_+ \times S(p), X]$, $S(p)$ being the sphere $[u \in X : \| u \| < p]$ in the Banach space X. We assume that the functions f and F are smooth enough to ensure of solutions $u(t, t_0, u_0)$ and $v(t, t_0, v_0)$ of (5.5.5) and (5.5.6), respectively, on $[t_0, \infty)$. When $f(t, 0) \equiv 0$, (5.5.5) has the trivial solution. In this case we have the following.

Definition 5.5.1: The trivial solution of (5.5.5) is said to be
(i) stable if for every $\epsilon > 0$ and $t_0 \in R_+$, there exists a $\delta > 0$ such that
$\| u_0 \| < \delta$ implies $| u(t, t_0, u_0) \| < \epsilon$ for all $t \ge t_0$;
(ii) asymptotically stable if it is stable and there exists a $\delta_0 > 0$ such that
$| u_0 | < \delta_0$ implies $\lim_{t\to\infty} u(t, t_0, u_0) = 0$;
(iii) uniformly stable in variation if for every $\epsilon > 0$ and $t_0 \in R_+$, there exists an $M(\epsilon) > 0$ such that $\| u_0 \| < \epsilon$ implies $\| U(t, t_0, u_0) \| < M(\epsilon)$ for all $t \ge t_0$, where $U(t, t_0, u_0)$ is the solution of the variational equation (5.3.20) and (5.3.21).

Definition 5.5.2: Let $A \in B(X)$. The logarithmic norm of the operator A is defined by

$$\mu(A) \equiv \lim_{h \to 0_+} (\| I + hA \| - 1)/h. \tag{5.5.7}$$

We need the following result.

Lemma 5.5.1: *Let $A(t) \in B(X)$ for each $t \in R_+$ and suppose that $u(t)$ is the solution of*

$$u' = A(t)u \text{ and } u(t_0) = u_0.$$

Then

$$\| u(t) \| \leq \| u_0 \| \exp \left(\int_{t_0}^{t} \mu[A(s)]ds \right), t \geq t_0. \qquad (5.5.8)$$

Proof: Define $m(t) = \| u(t) \|$. Then, for small $h > 0$,

$$m(t+h) - m(t) \leq \| u(t) + hA(t)u(t) \| - \| u(t) \| + \epsilon(h)$$

$$\leq (\| I + hA(t) \| - 1)m(t) + \epsilon(h)$$

where $\epsilon(h)/h \to 0$ as $h \to 0_+$. Hence

$$D_+ m(t) \leq \mu[A(t)]m(t) \text{ and } m(t_0) = \| u_0 \|$$

from which the estimate (5.5.8) follows.

Lemma 5.5.2: *Let the hypotheses of Theorem 5.3.3 hold. Suppose further that there exists a function $\alpha \in C[R_+, R]$ such that*

$$\mu[f_u(t,u)] \leq \alpha(t), \ (t,u) \in R_+ \times S(p). \qquad (5.5.9)$$

Then for $u_0, v_0 \in S(p)$ we have the estimates

$$\| u(t,t_0,v_0) - u(t,t_0,u_0) \| \leq \| v_0 - u_0 \| \exp \left[\int_{t_0}^{t} \alpha(s)ds \right], \ t \geq t_0 \qquad (5.5.10)$$

and

$$\| v(t,t_0,v_0) - u(t,t_0,u_0) \|$$

$$\leq \| v_0 - u_0 \| \exp \left[\int_{t_0}^{t} \alpha(s)ds \right]$$

$$+ \int_{t_0}^{t} \exp \left[\int_{s}^{t} \alpha(\xi)d\xi \right] \| F[s, v(s,t_0,v_0)] \| ds, \ t \geq t_0. \qquad (5.5.11)$$

Proof: From Theorem 5.5.4 we have

$$u(t,t_0,v_0) - u(t,t_0,u_0) = \int_{0}^{1} U[t,t_0,u_0 + s(v_0 - u_0)](v_0 - u_0)ds. \qquad (5.5.12)$$

By virtue of (5.5.9) and Lemma 5.5.1, it follows that

$$\max_{0 \leq s \leq 1} \| U[t,t_0, u_0 + s(v_0 - u_0)] \| \leq \exp \int_{t_0}^{t} \alpha(s)ds.$$

This and (5.5.12) yield (5.5.10).

Next, from Theorem 5.3.5 we have

$$v(t, t_0, v_0) - u(t, t_0, u_0)$$

$$= u(t, t_0, v_0) - u(t, t_0, u_0)$$

$$+ \int_{t_0}^{t} U[t, s, v(s, t_0, v_0)] F[s, v(s, t_0, v_0)] ds. \tag{5.5.13}$$

Again, from Lemma 5.5.1

$$\max_{t_0 \leq s < t} \| U[t, s, v(s, t_0, v_0)] \| \leq \exp \int_{s}^{t} \alpha(\xi) d\xi.$$

This, together with (5.3.10) and (5.3.13) yields the estimate (5.5.11). The proof is complete.

Theorem 5.5.2: *Assume that*
(i) *the hypotheses of Theorem 5.3.3 hold;*
(ii) $f(t, 0) \equiv 0, t \in R_+;$
(iii) *the condition (5.5.9) holds with*

$$\sigma \equiv \limsup_{t \to \infty} (t - t_0)^{-1} \int_{t_0}^{t} \alpha(s) ds < 0. \tag{5.5.14}$$

Then the trivial solution of (5.5.5) is asymptotically stable.

Proof: The assumption (5.5.14) implies that

$$\int_{t_0}^{t} \alpha(s) ds \leq (\sigma/2)(t - t_0), \quad t \text{ sufficiently large.}$$

Therefore

$$\lim_{t \to \infty} \exp \left[\int_{t_0}^{t} \alpha(s) ds \right] = 0. \tag{5.5.15}$$

If we take $v_0 = 0$ in (5.5.10) we obtain

$$\| u(t, t_0, u_0) \| \leq \| u_0 \| \exp \left[\int_{t_0}^{t} \alpha(s) ds \right].$$

This and (5.5.11) yield the desired conclusion.

Theorem 5.5.3: *Assume that*
(i) *the hypotheses of Theorem 5.3.3 hold;*
(ii) $f(t, 0) \equiv F(t, 0) \equiv 0, t \in R_+;$

(*iii*) $\sigma < 0$;
(*iv*) $\| F(t,v) \| = O(\| v \|)$ *as* $v \to 0$ *uniformly in* t.
Then, the trivial solution of (5.5.6) *is asymptotically stable.*

Proof: Let $\epsilon > 0$ and sufficiently small. Then (*iii*) implies that

$$\lim_{t \to \infty} \exp \left[\epsilon(t - t_0) + \int_{t_0}^{t} \alpha(s)ds \right] = 0. \tag{5.5.16}$$

Hence, there exists a positive constant K such that

$$\exp \left[\epsilon(t - t_0) + \int_{t_0}^{t} \alpha(s)ds \right] \leq K, \ t \geq t_0. \tag{5.5.17}$$

With the foregoing ϵ and because of (*iv*) there exists a $\delta > 0$ such that $\| v \| < \delta$ implies $\| F(t,v) \| < \epsilon \| v \|$. Now for $\| v_0 \| < \delta/K$ and from (5.5.11) with $u_0 = 0$ we obtain

$$\| v(t, t_0, v_0) \| \leq \| v_0 \| \exp \left[\int_{t_0}^{t} \alpha(s)ds \right]$$

$$+ \epsilon \int_{t_0}^{t} \exp \left[\int_{s}^{t} \alpha(\xi)d\xi \right] \| v(s, t_0 v_0 \| \ ds \tag{5.5.18}$$

as long as $\| v(t, t_0, v_0) \| < \delta$. Multiplying both sides of (5.5.18) by $\exp[-\int_{t_0}^{t} \alpha(s)ds]$ and applying Gronwall's inequality we get

$$\| v(t, t_0, v_0) \| \leq \| v_0 \| \left[\epsilon(t - t_0) + \int_{t_0}^{t} \alpha(s)ds \right] \tag{5.5.19}$$

as long as $\| v(t, t_0, v_0) \| < \delta$. Now (5.5.19) shows that $\| v(t, t_0, v_0) \| < \delta$ for all $t \geq t_0$. Otherwise there exists a T such that $\| v(T, t_0, v_0) \| = \delta$ and $\| v(t, t_0, v_0) \| \leq \delta$ for $t_0 \leq t \leq T$. Then from (5.5.19) and (5.5.17), we get the contradiction

$$\delta < (\delta/K)K = \delta$$

which proves our claim. Thus (5.5.19) holds for all $t \geq t_0$ and this together with (5.5.16) yields the desired conclusion.

Theorem 5.5.4: *Assume that*
(*i*)   *the hypotheses of Theorem 5.3.3 are satisfied;*
(*ii*) $f(t,0) \equiv 0, t \in R_+$;
(*iii*) *the trivial solution of* (5.5.5) *is uniformly stable in variation;*
(*iv*) *given* $\alpha > 0$ *there exists a function* $\lambda_\alpha \in L_1[0, \infty)$ *such that* $\| F(t,u) \| \leq$

$\lambda_\alpha(t)$ *for* $\| u \| \leq \alpha$.
Then, for every $\epsilon > 0$ there exists positive numbers $\delta = \delta(\epsilon)$ and $T = T(\epsilon)$ such that $\| u_0 \| < \delta$ implies $\| v(t, t_0, v_0) \| < \epsilon$ for $t \geq t_0 \geq T$.

Proof: Let $\epsilon > 0$ be given. Choose δ and T such that $\delta < \epsilon$, $2M(\epsilon)\delta < \epsilon$ and $\int_T^\infty \lambda_\epsilon(s)ds < \epsilon/2M(\epsilon)$. Assume that $\| u_0 \| < \delta$ and $t_0 \geq T$. Using (5.3.40) with $v_0 = 0$ and *(iii)* we get

$$\| u(t, t_0, u_0) \| \leq \| u_0 \| M(\epsilon) \leq \epsilon/2, t \geq t_0.$$

We claim that

$$\| v(t, t_0, u_0) \| < \epsilon, \ t \geq t_0.$$

If this is not true, let $t_1 > t_0$ be such that $\| v(t_1, t_0, u_0) \| = \epsilon$ and $\| v(t, t_0, u_0) \| \leq \epsilon$ for $t_0 \leq t \leq t_1$. Then from (5.3.44)

$$\epsilon \leq \epsilon/2 + M(\epsilon) \int_{t_0}^t \lambda_\epsilon(s)ds < \epsilon.$$

This contradiction proves the theorem.

5.6 NOTES

Several classical books deal with the differential equations in Banach spaces. Monograph by Martin [1] contains preliminaries needed for differential equations in Banach spaces, and includes many examples. The other books which cover these areas are by Lakshmikantham and Ladas [1], Deimling [1], Lakshmikantham and Leela [2], Hille and Phillips [1].

The preliminaries of Section 5.1 are taken from Lakshmikantham and Ladas [1]. Also see Kato [1]. The contents of Section 5.2 appear in Kato [2] and Lakshmikantham and Ladas [1]. The nonlinear VPF in abstract spaces given in Section 5.3 is due to Ladas, Ladde, and Lakshmikantham [1]. See also Lakshmikantham and Ladas [1].

The material appearing in Section 5.4 is taken from Webb [2]. See also Webb [1], Gyellenberg and Webb [1]. The application of the VPF for the study of stability behavior is due to Zaidman [1]. For several other applications of the VPF refer to Mamedov [1], Zaidman [1], Hale and Perissinotto [1].

6. Impulsive Differential Equations

6.0 INTRODUCTION

Several evolution processes in real-life situations change the state abruptly. The perturbations causing the change of state are for a short duration and hence it becomes natural to assume that the perturbations act instantaneously which are in the form of impulses. The impulsive differential equations (ImDEs) therefore rightly describe such behaviors and hence the study of such equations has become essential. Extensive literature is now available analyzing several qualitative properties of solutions. In many situations, to study the behavior of solutions we need to employ VPFs.

In the present chapter, we bring together known results on VP method for linear and nonlinear impulsive differential and integrodifferential equations (ImIDE). There are several types of impulsive equations known in the literature. We shall, however, consider only some typical impulsive equations.

For convenience in Section 6.1, we bring together some essential details of ImDEs. In Section 6.2, we develop the VPF for ImDEs, whose unperturbed part is linear. This technique is then extended for ImIDEs. It is also shown that ImIDEs are equivalent to ImDEs under suitable conditions. In Section 6.3, we establish nonlinear variation of parameters. In order to obtain such a result, one needs to develop the theory of differentiability of solutions with respect to initial conditions. We also state the corresponding VPF for ImIDEs. These results are followed by generalized VP by using Lyapunov-like functions in Section 6.4.

Sections 6.5 to 6.7 include the applications of the method of VP to establish various properties of solutions of ImDEs and ImIDEs. In Section 6.5, periodic boundary value problems are studied while in Section 6.6, suitable conditions have been obtained to establish complete controllability property of solutions of ImDEs. Several stability properties of ImDEs are proved in Section 6.7.

6.1 PRELIMINARIES

We devote the present section to bring together some essential features of impulsive differential equations which we use to present the techniques of VP in the subsequent sections.

Let us consider an evolution process described by a system of differential equations

(i)
$$x' = f(t, x) \tag{6.1.1}$$

where $f: R_+ \times \Omega \to R^n$, $\Omega \subset R^n$ is an open set;
(ii) the sets $M(t)$ and $N(t) \subset \Omega$ for each $t \in R_+$ and
(iii) the operator $A(t): M(t) \to N(t)$ for each $t \in R_+$.

Let $x(t_0, x_0)$ be a solution of (6.1.1). The point $p_t = (t, x(t))$ begins its motion from the initial point $p_{t_0} = (t_0, x_0)$ and moves along the curve

$\{(t, x), t \geq t_0, x = x(t)\}$ until the time $t_1 > t_0$ at which point p_t meets the set $M(t)$. At $t = t_1$, the operator $A(t)$ transfers the point $p_{t_1} = (t_1, x(t_1))$ into $p_{t_1^+} = (t_1, x_1^+) \in N(t_1)$ where $x_1^+ = A(t_1)x(t_1)$. Then the point continuous its motion along the curve with $x(t, t_1, x_1^+)$ as the solution of (6.1.1) with initial data $p_{t_1} = (t_1, x_1^+)$ until it reaches $M(t)$ at the next moment $t_2 > t_1$. Now $p_{t_2} = (t_2, x(t_2))$ is transferred to $p_{t_{2c}^+} = (t_2, x_2^+) \in N(t_2)$ where $x_2^+ = A(t_2)x(t_2)$. Now p_t moves forward with $x(t) = x(t, t_2, x_2^+)$ as the solution of (6.1.1). The evolution process continues as long as the solution of (6.1.1) exists. The sets (i), (ii) and (iii) above define an impulsive differential system.

A mathematical model of a simple impulsive system may be described by

$$x' = f(t, x), t \neq t_k, \ k = 1, 2, \ldots,$$

$$\Delta x = I_k(x), t = t_k, \tag{6.1.2}$$

where for $t = t_k$, $\Delta x(t_k) = x(t_k^+) - x(t_k)$ and $x(t_k^+) = \lim_{h \to 0^+} x(t_k + h)$. Clearly any solution $x(t)$ of (6.1.2) satisfies

$$x'(t) = f(t, x(t)), \ t \in (t_k, t_{k+1}]$$

and

$$\Delta x(t_k) = I_k(x(t_k)), \ t = t_k, \ k = 1, 2, \ldots.$$

Note that here the impulsive effect is at fixed times.

A more general problem involving impulses at variable times is represented by the system

$$x' = f(t, x), \ t \neq \tau_k(x),$$
$$ k = 1, 2, \ldots$$
$$\Delta x = I_k(x), \ t = \tau_k(x), \tag{6.1.3}$$

where we define sequence of surfaces s_k given by $s_k \colon t \to \tau_k(x)$, $k = 1, 2, \ldots$ such that $\tau_k(x) < \tau_{k+1}(x)$ and $\lim_{k \to \infty} \tau_k(x) = \infty$, $f \colon D \to R^n$, $I_k \colon \Omega \to R^n$ and $D = R_+ \times \Omega$. Let in (6.1.3), the initial data be $x(t_0^+) = x_0$, $t_0 \geq 0$. We define the solution of (6.1.3) as follows.

Definition 6.1.1: A function $x \colon (t_0, t_0 + a) \to R^n$, $t_0 \geq 0$, $a > 0$ is said to be a solution of (6.1.3) if
(i) $\quad x(t_0^+) = x_0$ and $(t, x(t)) \in D$ for $t \in [t_0, t_0 + a)$,
(ii) $\quad x(t)$ is continuous differentiable and satisfies

$$x'(t) = f(t, x(t))$$

for $t \in [t_0, t_0 + a)$ and $t \neq \tau_k(x(t))$,
(iii) if $t \in [t_0, t_0 + a)$ and $t = \tau_k(x(t))$ then $x(t^+) = x(t) + I_k(x(t))$ and at such values of t we always assume that $x(t)$ is left continuous and $s \neq \tau_j(x(s))$ for any j, $t < s < \delta$ for some $\delta > 0$.

The literature on existence, uniqueness and several other analytical properties of solutions of impulsive differential equations is extensive and is referred to in the notes at the end of the chapter.

6.2 LINEAR VARIATION OF PARAMETERS

This section is devoted to the study of linear variation of parameters for impulsive differential equations (ImDE) and linear impulsive integro-differential equations (ImIDE).

Consider the linear system with fixed moments of impulsive effect

$$\left. \begin{array}{ll} x' = A(t)x, & t \neq t_k \\ \Delta x = B_k x, & t = t_k \end{array} \right\} \; x(t_0^+) = x_0$$ (6.2.1)

where $A(t)$ and B_k are $n \times n$ matrices and

(*i*) $\qquad\qquad 0 < t_1 < \ldots < t_k < \ldots$ and $\lim_{k \to \infty} t_k = \infty$. (6.2.2)

(*ii*) Suppose that $A(t)$ is continuous from R_+ to R having discontinuities of the first kind at $t = t_k$ and $A(t)$ is left continuous at $t = t_k$, $k = 1, 2, \ldots$.

Then in view of the hypothesis it is clear that, for any choice of (t_0, x_0), there exists a unique solution $x(t) = x(t, t_0, x_0)$ of (6.2.1) for $t \geq t_0$.

Theorem 6.2.1: *Let $v_k(t, s)$ denote the fundamental matrix solution of the linear differential equation*

$$x' = A(t)x, \; t_{k-1} < t \leq t_k.$$ (6.2.3)

Then the unique solution of (6.2.1) is given by

$$x(t, t_0, x_0) = w(t, t_0^+)x_0$$ (6.2.4)

where

$$w(t, s) = \begin{cases} u_k(t, s) \text{ for } (t, s) \in (t_{k-1}, t_k] \\[4pt] u_{k+1}(t, t_k)(I + B_k)u_k(t_k, s) \\ \qquad\qquad \text{ for } t_{k-1} < s \leq t_k < t \leq t_{k+1} \\[6pt] u_{k+}(t, t_k)\prod_{j=k}^{i+1}(I + B_j)u_j(t_j, t_{j-1})(I + B_i) \\ \qquad\qquad\qquad\qquad\qquad u_i(t_i, s) \\ \qquad\quad \text{ for } t_{i-1} < s \leq t_i < t_k < t \leq t_{k+1} \end{cases}$$ (6.2.5)

I being the identity matrix.

The proof of Theorem 6.2.1 can be obtained by the method of verification intervalwise. We present an outline below.

We study the nature of $w(t, s)$ intervalwise. In the interval $(t_0, t_1]$ we have

$$w(t, s) = u_1(t, s), \ (t, s) \in [t_0, t_1].$$

If $u_2(t, s)$ is the fundamental matrix of (6.2.1) in $(t_1, t_2]$, then clearly, $x(t) = u_1(t, t_0)x_0$ for t in $(t_0, t_1]$. Hence

$$x(t_1) = u_1(t_1, t_0)x_0.$$

From (6.2.1), we see that

$$\Delta x(t_1) = x(t_1^+) - x(t_1) = B_1 x(t_1).$$

Consequently,

$$x(t_1^+) = (I + B_1)x(t_1) = (I + B_1)u_1(t_1, t_0)x_0.$$

For $t \in (t_1, t_2]$, we obtain

$$x(t) = u_2(t, t_1)x(t_1^+)$$

$$= u_2(t, t_1)(I + B_1)u_1(t_1, t_0)x_0.$$

As a result, we get

$$w(t, s) = u_1(t, s), \ (t, s) \in (t_0, t_1]$$

$$w(t, s) = u_2(t, t_1)(I + B_1)u_1(t_1, s), \ t_0 < s \le t_1 < t \le t_2.$$

While considering the fundamental matrix $w(t, s)$ $(t_2, t_3]$, we need to consider the possibilities $(t, s) \in (t_1, t_2]$, $t_1 < s \le t_2 < t \le t_3$ and $t_0 < s \le t_1 < t_2 < t \le t_3$. Extending the ideas employed in the previous interval we arrive at

$w(t, s)$

$$\begin{cases} \quad\quad = u_2(t, s), & (t, s) \in (t_1, t_2], \\ \quad = u_3(t, t_2)(I + B_2)u_2(t_2, s), & t_1 < s \le t_2 < t \le t_3, \\ = u_3(t, t_2)(I + B_2)u_2(t_2, t_1)(I + B_1)u_1(t_1, s), & t_0 \le s \le t_1 < t_2 < t \le t_3. \end{cases}$$

The fundamental matrix $w(t, s)$ in the subsequent intervals can be written following similar argument.

An immediate verification yields

$$w(t, t) = I,$$

$$w(t_k^+, s) = (I + B_k)w(t_k, s),$$

$$\frac{\partial w}{\partial t}(t, s) = A(t)w(t, s), \ t \ne t_k.$$

Further

$$x(t, t_0, x_0 + y_0) = x(t, t_0, x_0) = y(t, t_0, y_0),$$

$$x(t, t_0, \lambda x_0) = \lambda x(t, t_0, x_0), \quad \lambda \subset R,$$

$$w(t, t_0^+) = x(t)x^{-1}(t_0^+),$$

where $x(t) = \{x_1(t), \ldots, x_n(t)\}$ be a matrix valued function whose columns are the solutions (6.2.1) and that these are linearly independent on R_+. We then have det $x(t_0^+) \neq 0$. If $\det(I + B_k) \neq 0$ for every $k \in N$ then these relations hold for all $t \in R_+$.

Now let us consider the perturbed equation for (6.2.1) given by

$$\begin{cases} y' = A(t)y + b(t), & t \neq t_k \\ \quad \Delta y = B_k y, & t = t_k \\ \quad y(t_0^+) = x_0 \end{cases} \tag{6.2.6}$$

where $A(t)$ and B_k are as given in (6.2.1) and $b: R_+ \to R^n$ is continuous on R_+.

We can now prove the following VPF for (6.2.6).

Theorem 6.2.2: *Assume the conditions given for the equation (6.2.6). Let $x(t)$ be any solution existing on $[t_0, \infty)$ of (6.2.1) and let $w(t, s)$ be the fundamental matrix solution of (6.2.1) given by (6.2.5). Then the solution $y(t, t_0, x_0)$ satisfies the relation*

$$y(t, t_0, x_0) = w(t, t_0^+)x_0 + \int_{t_0}^{t} w(t, s)b(s)ds. \tag{6.2.7}$$

Proof: The solution $x(t, t_0, x_0)$ of (6.2.1) is given by

$$x(t, t_0, x_0) = w(t, t_0^+)x_0,$$

where $w(t, s)$ is defined by the relation (6.2.5).

Let $x(t)$ be the fundamental matrix of the system (6.2.1) in R_+ and det $(I + B_k) \neq 0$ for each $k \in N$. Then by change of variables $x = X(t)y$, the system (6.2.6) is transformed into the system

$$\frac{dy}{dt} = X^{-1}(t)[A(t)X(t) = X'(t)]y + X^{-1}(t)b(t), \quad t \neq t_k,$$

$$\Delta y(t_k) = X^{-1}(t_k^+)[X(t_k) - X(t_k^+) + B_k X(t_k)]$$

which reduces to

$$\frac{dy}{dt} = X^{-1}(t)b(t), \quad t \neq t_k,$$

$$\Delta y(t_k) = 0. \tag{6.2.8}$$

As a result, we have the relation

$$y(t) = y(t_0^+) + \int_{t_0}^{t} X^{-1}(\tau)b(\tau)d\tau.$$

Now substituting for $y(t_0^+)$ and $X^{-1}(\tau)$ from the foregoing discussion (6.2.7) follows. The proof is complete.

A variant of the above VPF (6.2.7) is given in the next result which deals with nonlinear perturbation.

Theorem 6.2.3: *Consider the initial value problem for ImDE*

$$\left.\begin{array}{ll} y' = A(t)y + f(t,y), & t \neq t_k, \\ \Delta y = B_k y + I_k(y), & t = t_k, \\ y(t_0^+) = y_0, \end{array}\right\} \tag{6.2.9}$$

where A, B_k are as in (6.2.1) and in addition
(iii) $f: R_+ \times R^n \to R^n$ is continuous in $(t_{k-1}, t_k] \times R^n$ and for every $y \in R^n$,
* $k = 1, 2, \ldots,$*

$$\lim_{(t,y) \to (t_k, y)} f(t, y) \text{ exists for } t \geq t_k,$$

(iv)   for each k, $I_k: R^n \to R^n$ is continuous.
Let $y(t, t_0, x_0)$ be any solution of (6.2.9) existing on $[t_0, \infty)$ and let $w(t, s)$ be the fundamental matrix solution of (6.2.1) given by (6.2.5). Then $y(t, t_0, x_0)$ satisfies the interval equation for $t > t_0$

$$y(t, t_0, x_0 = w(t, t_0^+)x_0 + \int_{t_0}^{t} w(t, s)f(s, y(s))ds$$

$$+ \sum_{t_0 < t_k < t} w(t, t_k^+)I_k(y(t_k)).$$

The proof of this theorem needs minor changes in the proof of the previous theorem. Here (6.2.8) takes the form

$$\frac{dy}{dt} = X^{-1}(t)f(t, y)$$

$$\Delta y(t_k) = X^{-1}(t_k^+)I_k(y)$$

in view of (6.2.9). Integration then yields

$$y(t, t_0, x_0) = y(t, t_0^+, x_0) + \int_{t_0}^{t} X^{-1}(\tau)f(\tau, y(\tau)d\tau$$

$$+ \sum_{t_0 < t_k < t} X^{-1}(t_k^+)I_k(y(t_k)).$$

Recalling the substitution for X^{-1}, we get the desired conclusion stated above.

Let us now consider the linear ImIDE

$$y'(t) = A(t)y(t) + \int_{t_0}^{t} k(t,s)y(s)ds + F(t), \ t \neq t_k$$

$$\Delta y(t_k) = B_k y(t_k),$$

$$y(t_0^+) = x_0, \tag{6.2.10}$$

where $0 \leq t_0 < t_1 < \ldots < t_k < \ldots$ and $t_k \to \infty$ as $k \to \infty$, $A \in PC^+[R_+, R^{n^2}]$, $k \in PC^+[R_+^2, R^{n^2}]$, $F \in PC^+[R_+, R^n]$ and $B_k \geq 0$ is a $n \times n$ matrix for each k such that $(I + B_k)^{-1}$ exists, I being the identity matrix. We shall assume the existence and uniqueness of IVP (6.2.10).

To investigate the rich behavior of solutions of ImIDE, we first establish the related ImDE and then develop the corresponding VPF. With this aim in mind, we proceed to prove the following useful result.

Theorem 6.2.4:   *Assume that there exists an* $n \times n$ *matrix function* $L \in PC^+[R_+^2, R^{n^2}]$ *such that* $L_s(t,s)$ *exists and is continuous for* $t_{k-1} < s \leq t_k < t$ *and satisfies*

$$k(t,s) + L_s(t,s) + L(t,s)A(s) + \int_{s}^{t} L(t,\sigma)k(\sigma,s)d\sigma = 0, \quad s,t \neq t_k,$$

$$L(t, t_k^+) = (I + B_k)^{-1}L(t, t_k). \tag{6.2.11}$$

Then system (6.2.10) is equivalent to ImDE

$$\left. \begin{array}{c} y'(t) = c(t)y(t) + L(t,t_0)x_0 + H(t), \quad t \neq t_k \\ \Delta y(t_k) = B_k y(t_k) \\ y(t_0^+) = x_0 \end{array} \right\} \tag{6.2.12}$$

where $c(t) = A(t) - L(t,t)$ *and* $H(t) = F(t) + \int_{t_0}^{t} L(t,s)F(s)ds$.

Proof: Let $y(t)$ be any solution of (6.2.10) existing on $[t_0, \infty)$. Define $p(s) = L(t,s)y(s)$ for $t_{k-1} < s < t_k < t$ so that we have

$$p'(s) = L_s(t,s)y(s) + L(t,s)y'(s).$$

We now substitute y' from (6.2.10), integrate between t_0 to t, and obtain

$$p(t) - p(t_0) - \sum_{t_0 < t_k < t} \Delta p(t_k) = \int_{t_0}^{t} [L_s(t,s)y(s) + L(t,s)A(s)y(s)$$

$$+ L(t,s)\int_{t_0}^{s} k(s,\sigma)y(\sigma)d\sigma + L(t,s)F(s)]ds.$$

We use Fubini's Theorem, (6.2.11), and (6.2.10) to get

$$L(t,t)y(t) - L(t,t_0)x_0 - \sum_{t_0 < t_k < t} \Delta p(t_k)$$

$$= \int_{t_0}^{t} [L_s(t,s) + L(t,s)A(s) + \int_{s}^{t} L(t,\sigma)k(\sigma,s)d\sigma]y(s)ds + \int_{t_0}^{t} L(t,s)F(s)ds$$

$$= -\int_{s}^{t} k(t,s)y(s)ds + \int_{t_0}^{t} L(t,s)F(s)ds$$

$$= -y'(t) + A(t)y(t) + F(t) + \int_{t_0}^{t} L(t,s)F(s)ds.$$

The foregoing relation now yields

$$y'(t) = c(t)y(t) + L(t,t_0)x_0 + H(t) + \sum_{t_0 < t_k < t} \Delta p(t_k), \quad t \neq t_k. \qquad (6.2.13)$$

For $s = t_k$, we get using (6.2.11) and (6.2.10),

$$\Delta p(t_k) = L(t,t_k^+)x(t_k^+) - L(t,t_k)x(t_k)$$

$$= [L(t,t_k^+)(I + B_k) - L(t,t_k)]x(t_k), \qquad (6.2.14)$$

Finally, (6.2.13) and (6.2.14) together show that $y(t)$ is the solution of (6.2.12).

To prove the converse, let $y(t)$ be any solution of (6.2.12) existing on $[t_0, \infty)$. Set

$$z(t) = y'(t) = A(t)y(t) - \int_{t_0}^{t} k(t,s)y(s)ds - F(t).$$

Note that $\Delta z(t_k) = \Delta y(t_k) - B_k y(t_k) = 0$.

We prove that $z(t) \equiv 0$ which will prove that $y(t)$ satisfies (6.2.10). Substitute $y'(t)$ from (6.2.12), use (6.2.11) and Fubini's Theorem to get

$$z(t) = [A(t) - L(t,t)]y(t) + L(t,t_0)x_0 + H(t) - A(t)y(t)$$

$$+ \int_{t_0}^{t} [L_s(t,s) + L(t,s)A(s) + \int_{s}^{t} L(t,\sigma)k(\sigma,s)d\sigma]y(s)ds - F(t), \quad t \neq t_k$$

$$= -L(t,t)y(t) - L(t,t_0)x_0 - \int_{t_0}^{t} L_s(t,s)y(s)ds]$$

$$+ \int_{t_0}^{t} L(t,s)[A(s)y(s) + \int_{t_0}^{t} L(s,\sigma)d\sigma + F(s)]ds, \quad t \neq t_k.$$

Set again, $p(s) = L(t, s)y(s)$ to obtain for $t_{k-1} < s < t_k < t$

$$p'(s) = L_s(t, s)y(s) + L(t, s)y'(s)$$

and after integration between t_0 to t,

$$L(t, t)y(t) - L(t, t_0)x_0 - \sum_{t_0 < t_k < t} \Delta p(t_k)$$
$$= \int_{t_0}^{t} [L_s(t, s)y(s) + L(t, s)y'(s)]ds.$$

Further, we have for $s = t_k$, using (6.10.11)

$$\Delta p(t_k) = [L(t, t_k^+)(I + B_k) - L(t, t_k)]y(t_k) = 0,$$

Hence, it follows that

$$z(t) = \int_{t_0}^{t} L(t, s)[-y'(s) + A(s)y(s) + \int_{t_0}^{s} k(s, \sigma)y(\sigma) + F(s)]ds$$
$$= -\int_{t_0}^{t} L(t, s)z(s)ds,$$

and $\Delta z(t_k) = 0$.
   Hence $z(t) \equiv 0$, by uniqueness of solutions of linear ImDe. The proof is complete.

Remarks 6.2.1: (i) Let $L(t, s)$ be the solution of the IVP

$$L_s(t, s) + L(t, s)A(s) + \int_{s}^{t} L(t, \sigma)k(\sigma, s)d\sigma = 0, \quad s, t \neq t_k,$$

$$L(t, t_k^+) = (I + B_k)^{-1}L(t, t_k), \quad s = t_k,$$

$$L(t, t) = I.$$

From Theorem 6.2.4, it follows that the unique solution $y(t, t_0, x_0)$ of (6.2.10) is given by

$$y(t, t_0, x_0) = L(t, t_0)x_0 + \int_{t_0}^{t} L(t, s)F(s)ds, \quad t \geq t_0 \qquad (6.2.14)$$

where $L(t, s)$ is the corresponding resolvent kernel. The relation (6.2.14) is therefore the VPF for the equation (6.2.10).
(ii) Let $k(t, s) \equiv 0$ in (6.2.10), then it takes the form of ImDE

$$y'(t) = A(t)y(t) + F(t), \quad t \neq t_k$$

$$\Delta y(t_k) = B_k y(t_k), \quad x(t_0^+) = x_0. \qquad (6.2.15)$$

The equation (6.2.15) is similar to (6.2.6). Hence (6.2.14) is also the VPF for linear ImDE, $L(t, s)$ being the fundamental matrix solution such that $L(t_0, t_0) = I$.

6.3 NONLINEAR VARIATION OF PARAMETERS

We are concerned in this section with nonlinear impulsive differential equations and the related VPF. Before we obtain the main result, we need to show that the solutions of such equations are continuously differentiable with respect to initial conditions. Consider ImDE

$$x' = f(t, x), \quad t \neq t_k,$$

$$\Delta x = I_k(x), \, t = t_k, \quad k = 1, 2, \ldots$$

$$x(t_0^+) = x_0, \; t_0 \geq 0, \tag{6.3.1}$$

where $t \in J$, $x \in \Omega \subset R^n$, Ω-open, $f \in R_+ \times \Omega \to R^n$, $I_k: \Omega \to R^n$

$$0 < t_1 < \ldots < t_k < \ldots \text{ and } \lim_{k \to \infty} t_k = \infty,$$

$$\Delta x \mid_{t=t_k} = x(t_k^+) - x(t_k^-).$$

Let $x(t, t_0, x_0)$ denote the solution of (6.3.1) such that $x(t_0^+, t_0, x_0) = x_0$. We assume that solutions $x(t, t_0, x_0)$ continuously depend on initial conditions. This is the conclusion of the following theorem stated without proof.

Theorem 6.3.1: *Let the following conditions be fulfilled*
(i)   *the function $f: J \times \Omega \to R^n$ is continuous in $(t_{k-1}, t_k] \times \Omega$, $k = 1, 2, \ldots$ and for every k and $x_0 \in \Omega$, there exists a finite limit of $f(t, x)$ as $(t, x) \to (t_k, x_0)$, $t > t_k$;*
(ii)   *the function f is locally Lipschitzian in x on $J \times \Omega$;*
(iii) *for $k = 1, 2, \ldots$, the mapping $\psi_k: \Omega \to \Omega$, $x \to z$, $z \equiv \psi_k(x) \equiv x + I_k(x)$ is a homeomorphism;*
(iv) *the system (6.3.1) had a solution $\phi(t)$ defined in $[\alpha, \beta]$, $(\alpha, \beta \neq t_k$, $k = 1, 2, \ldots)$.*
Then there exists a number $\xi > 0$ and a set

$$V = \{(t, x) \in I \times \Omega, \alpha \leq t \leq \beta, \mid x - \phi(t^+) \mid \; < \xi\}$$

such that:
(i)   *for every $(t_0, x_0) \in V$, there exists a unique solution $x(t, t_0, x_0)$ of the system (6.3.1) which is defined on $[\alpha, \beta]$,*
(ii) *the function $x(t, 0, x_0)$ is continuous for*

$$t \in [\alpha, \beta], (t_0, x_0) \in V, t \neq t_i, t_0 \neq t_k, i, k = 1, 2, \ldots,$$

(iii)   *for $k = 1, 2, \ldots, x_0 \in \Omega$, t, t_k belonging to the interval of existence of* *solution* $x(t, t_0, x_0)$ *of* $(6.3.1)$, $t \neq t_j$,

$$\lim_{\substack{\xi \to t_0 \\ \eta \to x_0}} x(t, \xi, \eta) = x(t, t_0, x_0).$$

Theorem 6.3.2: *Let the following conditions be fulfilled:*
(i) *the function* $f: I \times \Omega \to R^n$ *is continuous in* $(t_{k-1}, t_k] \times \Omega$, $k = 1, 2, \dots$ *and*
   $f_x(t, x)$ *is continuous in* $(t_{k-1}, t_k) \times \Omega$, $k = 1, 2, \dots$;
(ii) *for every* $x_0 \in \Omega$, $k = 1, 2, \dots$, *there exists finite limits of functions* f *and* f_x *as*
   $(t, x) \to (t_k, x_0)$, $t > t_k$ *and as* $(t, x) \to (t_k, x_0)$, $t > t_k$;
(iii) *for* $k = 1, 2, \dots$ *the mapping* $\psi_k: \Omega \to \Omega$, $x \to z$, $z = \psi_k(x) \equiv x + I_k(x)$ *is a*
   diffeomorphism and for $x \in \Omega$

$$\det\left(I + \tfrac{\partial I_k}{\partial x}(x)\right) \neq 0, \quad k = 1, 2, \dots. \tag{6.3.2}$$

Then
(i) *there exists* $\delta > 0$ *such that the solution* $x(t, t_0, x_0)$ *of* $(6.3.1)$ *has a*
   continuous derivative $\frac{\partial x}{\partial t}$, $\frac{\partial x}{\partial t_0}$, $\frac{\partial x}{\partial x_0}$ *in the domain*

$$V: \alpha < t < \beta, \alpha < t_0 < \beta, t, t_0 \neq t_k, k = 1, 2, \dots \tag{6.3.3}$$

$$|x_0 - \phi(t_0^+)| < \delta;$$

(ii) *the derivative* $\Phi(t, t_0, x_0) = \frac{\partial x}{\partial x_0}(t, t_0, x_0)$ *is a solution of the initial value*
   problem

$$\left.\begin{array}{rl} u' = f_x(t, \phi(t))u, & t \neq t_k \\ \Delta u = \tfrac{\partial I_k}{\partial x}(\phi(t_k))u, & t = t_k \\ u(t_0^+) = I \end{array}\right\} \tag{6.3.4}$$

   where $\phi(t)$ *is the solution of* $(6.3.1)$ *in* $[\alpha, \beta]$, $\alpha, \beta \neq t_k$, $k = 1, 2, \dots$;
(iii) *the derivative* $\frac{\partial x}{\partial t_0}$ *satisfies the relation*

$$\frac{\partial x}{\partial t_0}(t, t_0, x_0) = -\frac{\partial x}{\partial x_0}(t, t_0, x_0)f(t_0, x_0) \tag{6.3.5}$$

$$= -\Phi(t, t_0, x_0)f(t_0, x_0).$$

Proof: We first establish the existence of the derivative of the solution $x(t, t_0, x_0)$ in ith coordinate of x_0. We assume that there exists a $\delta > 0$ such that the function $x(t, t_0, x_0)$ is defined and continuous in the set V defined by the relation $(6.3.5)$. Let $|h| < r$, $r > 0$ be sufficiently small and t_0, x_0 are fixed and satisfy the relations $(6.3.3)$. We assume that

$$x(t, h) = x(t, t_0, x_0 + he_i)$$

$$\Delta x(t, h)x(t, h) - x(t, 0),$$

$$z(t, h) = \tfrac{1}{h}\Delta x(t, h),$$

where $e_i = (0, \ldots, 1, \ldots, 0)$ is the ith basis vector. In view of our assumptions,

$$\lim_{h \to 0} x(t, h) = x(t, 0) = x(t, t_0, x_0)$$

uniformly in $t \in [\alpha, \beta]$, $t \neq t_k$, $k = 1, 2, \ldots$.

For $h \neq 0$, we have

$$\frac{dz}{dt}(t, h) = \frac{1}{h}[f(t, x(t, h)) - f(t, x(t, 0))]$$

$$= \int_0^1 \frac{\partial f}{\partial x}(t, x(t, 0) + s\Delta x(t, h))ds \cdot z(t, h);$$

$$z(t_k^+, h) = \frac{1}{h}(x(t_k, h) - x(t_k, 0))$$

$$+ \frac{1}{h}(I_k(x(t_k, h)) - I_k(x(t_k)))$$

$$= z(t_k, h) + \int_0^1 \frac{\partial I_k}{\partial x}(x(t_k, 0) + s\Delta x(t_k, h))ds \cdot z(t_k, h);$$

and

$$z(t_0^+, h) = \frac{1}{h}(x(t_0^+, h) - x(t_0^+, 0)) = e_i.$$

The foregoing relations imply that $z(t, h)$ is a solution of the initial value problem

$$\frac{dz}{dt} = A(t, h)z, \quad t \neq t_k$$

$$\Delta z = B_k(h)z, \quad t = t_k \qquad\qquad (6.3.6)$$

$$z(t_0^+) = e_i$$

where

$$A(t, h) = \int_0^1 \frac{\partial f}{\partial x}(t, x(t, 0) + s\Delta x(t, h))ds$$

$$B_k(h) = \int_0^1 \frac{\partial I_k}{\partial x}(x(t_k^+) + s\Delta x(t_k, h))ds.$$

Hence it follows that

$$\lim_{h \to 0} z(t, h) = y(t),$$

where $y(t)$ is the solution of the initial value problem

$$\frac{dy}{dt} = \frac{\partial f}{\partial x}(t, x(t, t_0, x_0))y, \quad t \neq t_k$$

$$\Delta y = \tfrac{\partial I_k}{\partial x}(x(t_k, t_0, x_0))y \qquad (6.3.7)$$

$$y(t_0^+) = e_i$$

which is obtained from (6.3.6) setting $h = 0$.

We have proved the existence of derivative of the solution $x(t, t_0, x_0)$ with respect to ith coordinate of x_0 and along with it the existence of $\tfrac{\partial x}{\partial x_0}(t, t_0, x_0)$. In view of (6.3.6) we conclude that the matrix $u = \tfrac{\partial x}{\partial x_0} = \Phi(t, t_0, x_0)$. In view of (6.3.6) we conclude that the matrix $u = \tfrac{\partial x}{\partial x_0} = \Phi(t, t_0, x_0)$ satisfies the initial value problem (6.3.4) and conclude that the solution $\tfrac{\partial x}{\partial x_0}(t, t_0, x_0)$ of the initial value problem (6.3.4) is a continuous function for (t, t_0, x_0) in v.

After providing the scheme for constructing the proof of continuous differentiability with respect to x_0, it is easy to prove continuous differentiability with respect to the following corresponding proof of Theorem 2.2.3. Hence we omit the details. We are now in a position to prove the VPF for nonlinear system of impulsive differential equation. In addition to the system (6.3.1) consider the perturbed system

$$\left.\begin{array}{ll} y'(t) = f(t, y(t)) + F(t, y(t)), & t \neq t_k \\ \Delta y \mid_{t=t_k} = I_k(y(t_k)) + h_k(y(t_k)), & k = 1, 2, \ldots \\ y(t_0^+) = x_0, & t_0 \geq 0 \end{array}\right\} \qquad (6.3.8)$$

where $F: R_+ \times \Omega \to R^n$, $h_k: \Omega \to R^n$.

Theorem 6.3.4: *Let the systems (6.3.1), (6.3.8) satisfy the conditions of Theorem 6.3.3 and let $x(t, t_0, x_0)$ be a solution of (6.3.1) which is defined on J. Then for any solution $y(t) = y(t, t_0, x_0)$ of the system (6.3.8) which is defined on an interval I, the following formula is valid*

$$y(t, t_0, x_0) = x(t, t_0, x_0 + \int_{t_0}^{t} \Phi(t, s, y(s))F(s, y(s))ds$$

$$+ \sum_{t_0 < t_k < t} \int_0^1 \Phi(t, t_k^-, y(t_k) + I(y(t_k))$$

$$+ sh_k(y(t_k)))ds \cdot h_k(y(t_k)) \qquad (6.3.9)$$

for $t > t_0$, $t \neq t_k$, $k = 1, 2, \ldots$.

Proof: Set $p(s) = x(t, s, y(s))$ where $y(s) = y(s, t_0, x_0)$. Then for $s \neq t_k$, $k = 1, 2, \ldots$, we have

$$p'(s) = \tfrac{\partial x}{\partial s}(t, s, y(s)) + \tfrac{\partial x}{\partial y}(t, s, y(s))y'(s)$$

$$= -\Phi(t, s, y(s))f(t_0, x_0) + \Phi(t, s, y(s))[f(s, y(s)) + F(s, y(s))]$$

$$= \Phi(t, s, y(s))F(s, y(s)) \qquad (6.3.10)$$

and

$$\Delta p(s)\mid_{s=t_k} = x(t, t_k^+, y(t_k^+)) - x(t, t_k^-, y(t_k))$$

$$= x(t, t_k, y(t_k) + I_k(y(t_k)) + h_k(y(t_k)))$$

$$- x(t, t_k, y(t_k) + I_k(y(t_k)))$$

$$= \int_0^1 \Phi(t, t_k^-, y(t_k) + I_k(y(t_k)) + sh_k(y(t_k)))ds \cdot h_k(y(t_k)). \qquad (6.3.11)$$

Integrate (6.3.10) from t_0 to t and use (6.3.11) and the fact $p(t) = x(t, t, y(t)) = y(t)$, $t \neq t_k$, $k = 1, 2 \ldots$. We then obtain (6.3.9) completing the proof of the theorem.

6.4 GENERALIZED COMPARISON PRINCIPLE

In the present section, we combine widely used techniques of variation of parameters and the Lyapunov's second method to obtain a new comparison theorem, which connects the solutions of a known differential system through the solutions of the comparison equation. Such a result is useful in perturbation theory since it helps to preserve the nature of perturbations.

We need the following definitions for further discussions.

Definition 6.4.1: Let $0 \leq t_0 < t_1 < t_2 < \ldots < t_k <$ and $t_k \to \infty$ as $k \to \infty$. Then we say that $h \in PC[R_+ \times R^n, R^m]$ if $h(t_{k-1}, t_k] \times R^n \to R^n$ is continuous in $(t_{k-1}, t_k] \times R^n$ and for every $x \in R^n$

$$\lim_{(t,y) \to (t_k^+, x)} h(t, y) = h(t_k^+, x)$$

exists, $k = 1, 2, \ldots$.

Definition 6.4.2: We say that $V \in V_0$, if $V \in PC[R_+ \times S(\rho), R_+]$, $V(t, x)$ is locally Lipschitzian in x for $(t, x) \in (t_{k-1}, t_k] \times S(\rho)$, where $S(\rho) = \{x \in \mathbb{R}^n: \| x \| < \rho\}$.

Definition 6.4.3: $a \in \mathcal{K}$ implies $a \in [(0, \rho), \mathbb{R}_+]$, $a(0) = 0$ and $a(u)$ is strictly increasing in u.

$$y' = f(t, y), \ t \neq t_k,$$

$$\Delta y(t_k) = J_k(y(t_k)),$$

$$y(t_0^+) = x_0, \qquad (6.4.1)$$

and

$$x' = F(t, x), \ t \neq t_k,$$

$$\Delta x(t_k) = I_k(x(t_k)),$$

$$x(t_0^+) = x_0. \tag{6.4.2}$$

Suppose that the following assumptions hold true:

(A) (i) $0 \le t_0 < t_1 < t_2 < \ldots < t_k < \ldots$ and $t_k \to \infty$ as $k \to \infty$;
   (ii) $F, f \in PC[\mathbb{R}_+ \times \mathbb{R}^n, \mathbb{R}^n]$;
   (iii) $I_k, J_k \colon \mathbb{R}^n \to \mathbb{R}^n$.

The following hypothesis (H) is assumed relative to system $(6.4.1)$:

(H) The solutions $y(t, t_0, x_0)$ of $(6.4.1)$ exist for all $t \ge t_0$, unique, continuous
w.r.t. the initial data and $\mid y(t, t_0, x_0) \mid$ is locally Lipschitzian in x_0.

Let $\mid x_0 \mid \, < \rho$ and suppose that $\mid y(t, t_0, x_0) \mid \, < \rho$ for $t \ge t_0$. For any $V \in$
$PC[\mathbb{R}_+ \times S(\rho), \mathbb{R}_+]$ and for any fixed $t > t_0$, we define for $t_0 < s \le t$, $s \ne t_k$, and
$x \in S(\rho)$

$$D_-V(s, y(t, s, x)) \equiv \lim_{h \to h_-} \inf \tfrac{1}{h}[V(s + h, y(t, s + h, x + hF(s, x))$$

$$- V(s, y(t, s, x))]. \tag{6.4.3}$$

We now prove the comparison theorem, which is an important tool for relating
the solutions of $(6.4.2)$ to the solutions of $(6.4.1)$.

Theorem 6.4.1: *Suppose that the assumption (H) holds. Further that $V \in V_0$ and*
(i)   for $t_0 < s \le t, s \ne t_k, x \in S(\rho)$

$$D_-V(s, y(t, s, x)) \le g(s, V(s, y(t, s, x))), \tag{6.4.4}$$

(ii)   there exists a $\rho_0 = \rho_0(\rho) > 0$ such that $\mid x \mid \, < \rho_0$ implies $\mid x + I_k(x) \mid \, < \rho$
and

$$V(t_k^+, y(t, t_k^+, x(t_k) + I_k(x(t_k)))) \le \psi_k(V(t_k, y(t, t_k(x(t_k))))), \tag{6.4.5}$$

where $\psi_k \colon \mathbb{R}_+ \to \mathbb{R}_+$ are nondecreasing functions for all k;
(iii) $g \in PC[\mathbb{R}_+^2, \mathbb{R}]$ and the maximal solution, $r(t) = r(t, t_0, u_0)$ of the scalar
impulsive differential equation

$$u' = g(t, u), \ t \ne t_k,$$

$$u(t_k^+) = \psi_k(u(t_k)),$$

$$u(t_0^+) = u_0 \ge 0, \tag{6.4.6}$$

exists for $t \ge t_0$.
Then, if $x(t) = x(t, t_0, x_0)$ is any solution of $(6.4.2)$ and $V(t_0^+, y(t, t_0^+, x_0)) \le u_0$,
we get

$$V(t, x(t, x_0)) \le r(t, t_0, u_0), \ t \ge t_0. \tag{6.4.7}$$

Proof: Let $x(t) = x(t, t_0, x_0)$ be any solution of (6.4.2) such that $\mid x_0 \mid \ < \rho$. For $t_0 \le s \le t$, define

$$m(s) = V(s, y(t, s, x(s)))$$

so that $m(t_0) = V(t_0, y(t, t_0, x_0))$. Then using assumptions (H), (i) and (ii), we obtain

$$D_- m(s) \le g(s, m(s)), \ t_0 \le s \le t, \ s \ne t_k,$$

and

$$m(t_k^+) \le \psi_k(m(t_k)).$$

Consequently, it follows that (see Theorem 1.4.3 in Lakshmikantham, Bainov and Simeonov [1])

$$m(s) \le r(s, t_0, u_0), \ t_0 \le s \le t,$$

which implies, by taking $u_0 = V(t_0, y(t, t_0, x_0))$ and noting that $y(t, t, x(t)) = x(t)$, the desired result. Hence the proof is complete.

Corollary 6.4.1: *If in Theorem* 6.4.1, *we assume that*
(i) $g(t, u) = 0$, $\psi_k(u) = u$, then we get

$$V(t, x(t)) \le V(t_0, y(t)), \ t \ge t_0;$$

(ii) $g(t, u) = 0$ and $\psi_k(u) = d_k u$, $d_k \ge 0$, then one obtains

$$V(t, x(t)) \le \prod_{t_0 < t_k < t} d_k V(t_0, y(t)), \ t \ge t_0;$$

(iii) $g(t, u) = -\alpha u, \alpha > 0, \psi_k(u) = d_k u, d_k \ge 0$, then it follows that

$$V(t, x(t)) \le \left[\prod_{t_0 < t_k < t} d_k V(t_0, y(t)) \right] e^{-x(t-t_0)}, \ t \ge t_0;$$

(iv) $g(t, u) = \lambda'(t)u$, $\psi_k(u) = d_k u$, $d_k \ge 0$, $\lambda \in C^1[\mathbb{R}_+, \mathbb{R}_+]$, then

$$V(t, x(t)) \le \left[\prod_{t_0 < t_k < t < t} d_k V(t_0, y(t)) \right] \exp[\lambda(t) - \lambda(t_0)], \ t \ge t_0.$$

A number of remarks are now in order.
(i)   Assume that $f(t, y) \equiv 0$, $J_k(y) = 0$, for all k, in Theorems 6.4.1. Then since $y(t, t_0, x_0) = x_0$, (6.4.4) and (6.4.5) reduce to

$$D_-V(t,x) \equiv \lim_{h \to 0-} \inf \tfrac{1}{h}[V(t+h, x+hF(t,x)) - V(t,x)]$$

$$\leq g(t, V(t,x)), \ t \neq t_k$$

$$V(t_k^+, x(t_k) + I_k(x(t_k))) \leq \psi_k(V(t_k, x(t_k))),$$

which are the usual definition of the generalized derivative of the Lyapunov function relative to the system (6.4.2).

(ii)   If $f(t,y) \equiv 0$ and $J_k(y) = c_k y$, $0 \leq c_k \leq 1$, in Theorem 6.4.1, then $y(t, t_0, x_0) = \prod_{t_0 < t_k < t} c_k x_0$. The assumption (H) is clearly verified. Hence we get, from Theorem 6.4.1 the estimate

$$V(t, x(t)) \leq r\left(t, t_0, V\left(t_0, \prod_{t_0 < t_k < t} c_k x_0\right)\right), t \geq t_0.$$

(iii)   If $f(t,y) = A(t)y$ and $J_k(y) = c_k y$, $0 \leq c_k \leq 1$ in Theorem 6.4.1 then $y(t, t_0, x_0) = w(t, t_0)x_0$, where $w(t, t_0)$ is the fundamental matrix solution of the impulsive differential system (6.4.1) (see Theorem 6.2.1). Then we have

$$V(t, x(t)) \leq r(t, t_0, V(t_0, w(t, t_0)x_0)), \ t \geq t_0.$$

If, in addition, $g(t,u) \equiv 0$ and $\psi_k(u) = d_k u, d_k \geq 0$, we get by Corollary 6.4.1 (ii)

$$V(t, x(t)) \leq \prod_{t_0 < t_k < t} d_k V(t_0, w(t, t_0)x_0), \ t \geq t_0,$$

and if $d_k = 1$, by Corollary 6.4.1 (i)

$$V(t, x(t)) \leq V(t_0, w(t, t_0)x_0), \ t \geq t_0.$$

(iv) Suppose $F(t,y) \equiv F(t,y) + R(t,y)$, and system (6.4.1) satisfies the conditions of Theorem 6.3.2. Then,

$$\tfrac{\partial x}{\partial x_0}(t, t_0, x_0) = \Phi(t, t_0, x_0), \quad \text{and}$$

$$\tfrac{\partial x}{\partial t_0}(t, t_0, x_0) = -\Phi(t, t_0, x_0) \cdot f(t_0, x_0), \ t \geq t_0, \tag{6.4.8}$$

where $\Phi(t, t_0, x_0)$ is the matrix solution of the corresponding variational equation. If we assume that $V(t, x) \in V_0$ is differentiable for $s \neq t_k$, for a fixed t, we have

$$D_-V(s, y(t, s, x)) \equiv V_s(s, y(t, s, x)) + V_x(s, y(t, s, x))$$

$$\times \left[\tfrac{\partial y}{\partial t_0}(t, s, x(s)) + \tfrac{\partial y}{\partial x_0}(t, s, x(s)) \cdot x'(s)\right]. \tag{6.4.9}$$

Setting $V(s, x) = |x|^2$ and using (6.4.8) in (6.4.9), we get

$$D_- V(s, y(t, s, x)) = 2y(t, s, x)\Phi(t, s, x) \cdot R(s, x),$$

which shows how the perturbation term is involved.

6.5 PERIODIC BOUNDARY VALUE PROBLEM

This section is devoted to study the periodic boundary value problem involving impulsive effect. By using monotone iterative technique we formulate monotone sequences which converge to solutions of the PBVP. The method is constructive.

Consider the following PBVP for impulsive differential system

$$x' = f(t, x), t \neq t_k, t \in [0, T] = J \tag{6.5.1}$$

$$x(t_k^+ 0 = I_k(x(t_k)), \quad t_k \in J - 0, \quad k = 1, 2, \ldots \tag{6.5.2}$$

$$x(0) = x(T), \tag{6.5.3}$$

where

$$f \in J \times R^n \to R^n, \ I_n \colon R^n \to R^n.$$

We need the following result for further discussions.

Lemma 6.5.1: *Let* $m \in PC^1[J, R]$ *and*

$$m'(t) \leq -Mm(t), \ t \neq t_k$$

$$m(t_1^+) \leq L_k m(t_k), \ k = 1, 2, \ldots,$$

$$m(0) \leq M(T),$$

with $M > 0$, $L_k > 0$ *and* $\prod_{k=1}^{p} L_k e^{-MT} < 1$. *Then*

$$m(t) \leq 0, \ t \in J.$$

Proof: Under the given hypothesis, it is easy to show that

$$m(t) \leq m(0) \prod_{0 < t_k < t} L_k e^{-Mt} \quad t \in J.$$

To obtain the conclusion, it is enough to prove that $m(0) \leq 0$. If this is not true, let $m(0) > 0$. Then

$$m(0) \leq M(T) \leq m(0) \prod_{k=1}^{p} L_k e^{-MT} < m(0)$$

which is a contradiction. The proof is complete.

Definition 6.5.1:   Let $\Omega = [x: J \to R^n, x$ continuously differentiable for $t \neq t_k$, $x(t_k^+), x(t_k^-)$ exist and $x(t_k^+) = x(t_k^-)]$.

A function $v \in \Omega$ is called a lower solution of the PBVP (6.5.1)-(6.5.3) if

$$v'(t) \leq f(t, v(t)), \ t \neq t_k$$

$$v(t_k^+) \leq I_k(v(t_k))$$

$$v(0) \leq v(T).$$

Analogously, we can define the upper solution of (6.5.1)-(6.5.3) by reversing the above inequalities.

We are now in a position to prove the main result of this section.

Theorem 6.5.1: *Assume that*
(a)   *v, w are lower and upper solutions of (6.5.1)-(6.5.3) such that $v(t) \leq w(t)$ on J;*
(b)   *$f: J \times R^n \to R^n, f(t, x)$ is continuous for $t \neq t_k$,*
$\displaystyle \sup_{\substack{v(t) \leq x < w(t) \\ x \leq w(t),}}$ *$\mid f(t, x) \mid \leq \lambda(t)$, a.e. on J, where $\lambda \in L^1(J)$ and for $v(t) \leq y \leq$*

$$f(t, x) - f(t, y) \geq \ - M(x - y), \ M > 0; \tag{6.5.4}$$

(c)   *$I_k: R^n \to R^n$ is continuous and for $v(t_k) \leq y \leq x \leq w(t_k)$,*

$$I_k(x) - I_k(y) \geq L_k(x - y), \ k = 1, 2, \ldots, p, L_k > 0, \tag{6.5.5}$$

with $\displaystyle \prod_{k-1}^{p} L_k e^{-MT} < 1$.

Then there exists monotone sequences $\{v_n(t)\}, \{w_n(t)\}$, with $v_0 = v$, $w_0 = w$, such that $\displaystyle \lim_{n \to \infty} v_n(t) = \rho(t), \lim_{n \to \infty} w_n(t) = r(t)$ uniformly on J, and ρ, r are the minimal and maximal solutions of the PBVP (6.5.1) to (6.5.3) on J.

We would like to point out that condition (6.5.4) and (6.5.5) imply that f and I_k are quasimonotone nondecreasing functions and therefore we have not explicitly assumed in Theorem 6.5.1 such a condition.

Proof: For any η such that $\eta \in [v, w]$, where

$$[v, w] = [x \in \Omega: v(t) \leq x \leq w(t) \text{ on } J],$$

we consider the linear uncoupled PBVP

$$x' = \ - Mx + f(t, \eta(t)) + M\eta(t), \ t \neq t_k,$$

$$x(t_k^+) = L_k x(t_k) + I_k(\eta(t_k)) - L_k \eta(t_k), \ k = 1, 2, \ldots, p,$$

$$x(0) = x(T). \tag{6.5.6}$$

Clearly, the problem (6.5.6) possesses a unique solution

$$x(t) = x(0) \prod_{0 < t_k < t} L_k e^{-Mt} + \int_0^t \prod_{s < t_k < t} L_k e^{M(s-t)}[f(s, \eta(s)) + M\eta(s)]ds$$

$$+ \sum_{0 < t_k < t} \prod_{t_k < t_i < t} L_i e^{M(t_k - t)}[I_k(\eta(t_k)) - L_k \eta(t_k)], \qquad (6.5.7)$$

where

$$x(T) = x(0) = \left[1 - \prod_{k=1}^p L_k e^{-MT}\right]^{-1} \int_0^T \prod_{s < t_k < T} L_k e^{M(s-T)}[f(s, \eta(s)) + M\eta(s)]ds$$

$$+ \sum_{k=1}^p \prod_{s < t_k < T} L_k e^{M(t_k - T)}[I_k(\eta(t_k)) - L_k \eta(t_k)].$$

We define the mapping $A: [v, w] \to \Omega$ by $A\eta = x$, where $x(t)$ is the unique solution of (6.5.6) given by (6.5.7). This mapping will be used to define the desired sequences. Let us first prove that
(i) $v \leq Av$, $w \geq Aw$,
(ii) A is a monotone mapping on $[v, w]$.

To prove (i), we set $Av_0 = v_1$ and $p = v_0 - v_1$. Then we have

$$p' = v_0^1 - v_1' \leq f(t, v_0) - M(v_1 - v_0) - f(t, v_0) = -Mp, \ t \neq t_k,$$

$$p(t_k^+) \leq I_k(v_0(t_k)) + L_k(v_0(t_k)) - I_k(v_0(t_k)) = L_k p(t_k),$$

$$p(0) \leq p(T),$$

which implies by Lemma 6.5.1, $p(t) \leq 0$ on J, proving $v \leq Av$. Similarly, we can show that $Aw \leq w$.

To prove (ii), let $\eta_1 \leq \eta_2, \eta_1, \eta_2 \in [v, w]$ so that $u_1 = A\eta_1$, $u_2 = A\eta_2$. Then setting $p = u_1 - u_2$, we get using (b) and (c),

$$p' = u_1' - u_2' = f(t, \eta_1) - M(u_1 - \eta_1) - f(t, \eta_2) + M(u_2 - \eta_2)$$

$$\leq M(\eta_2 - \eta_1) - M(u_1 - \eta_1) + M(u_2 - \eta_2 0 = -Mp, t \neq t_k,$$

$$p(t_k^+) = u_1(t_k^+) - u_2(t_k^+) = I_k(t_k)) + L_k(u_1(t_k) - \eta_1(t_k)) = I_k(\eta_2(t_k))$$

$$- L_k(u_2(t_k) - \eta_2(t_k))$$

$$\leq -L_k(\eta_2(t_k) - \eta_1(t_k)) + L(u_1(t_k) - \eta_1(t_k)) - L(u_2(t_k) - \eta_2(t_k))$$

$$= L_k p(t_k), \ k = 1, 2, \ldots p,$$

$$p(0) = P(T),$$

which yields, by Lemma 6.5.1, $p(t) \leq 0$ on J proving (ii).

Define the sequences $\{v_n(t)\}, \{w_n(t)\}$ by $A_{v_n} = v_{n+1}$, $A_{w_n} = w_{n+1}$ and conclude from (i) and (ii) that

$$v_0 \leq v_1 \leq v_2 \leq \ldots \leq v_n \leq w_n \leq \ldots \leq w_2 \leq w_1 \leq w_0 \text{ on } J.$$

It then follows from standard arguments that $\lim_{n \to \infty} v_n(t) = \rho(t)$, $\lim_{n \to \infty} w_n(t) = r(t)$ uniformly on J and that ρ, r are solutions of (6.5.1) to (6.5.3) in view of the fact that v_n, w_n satisfy (6.5.6).

To prove that ρ, r are extremal solutions of (6.5.1) to (6.5.3), let $x(t)$ be any solution of (6.5.1) to (6.5.3) such that $x \in [v, w]$. Suppose that for some n, we have $v_n(t) \leq x(t) \leq w_n(t)$ on J. Then setting $p(t) = v_{n+1}(t) - x(t)$, we get

$$p' = f(t, v_n) - M(v_{n+1} - v_n) - f(t, x)$$

$$\leq M(x - v_n) - M(v_{n+1} - v_n) = -Mp, \quad t \neq t_k,$$

$$p(t_k^+) = I_k(v(t_k)) + L_k(v_{n+1}(t_k) - v_n(t_k)) - I_k(x(t_k))$$

$$\leq -L_k(x(t_k) - v(t_k)) + L_k(v_{n+1}(t_k) - v_n(t_k))$$

$$= Lp(t_k), \quad k = 1, 2, \ldots, p,$$

$$p(0) = p(T)$$

By Lemma 6.5.1, we have $p(t) \leq 0$ on J which implies that $v_{n+1}(t) \leq x(t)$ on J. Similarly, we get $x(t) \leq w_{n+1}(t)$ on J. Since $v_0 \leq x \leq w_0$ on J, by induction we conclude that $v_n \leq x \leq w_n$ on J for all n. Hence it follows that $\rho(t) \leq x(t) \leq r(t)$ on J by taking limit as $n \to \infty$. The proof is complete.

6.6 COMPLETE CONTROLLABILITY

Consider the impulsive control system

$$x' = Ax + Bu, \quad t \neq t_k, \quad t \in [0, T],$$

$$x(t_k^+) = [I + D^k u(t_k)] x(t_k), \quad t = t_k,$$

$$x(0) = v_0. \tag{6.6.1}$$

Assume that $x \in \mathbb{R}^n$, $u \in \mathbb{R}^r$, A, B are $n \times n$, $n \times r$ matrices,

$$0 < t_1 < t_2 < \ldots < t_\rho < T,$$

for each $k = 1, 2, \ldots, \rho$, $D^k u(t_k)$ is a diagonal matrix such that $D^k u(t_k) = \sum_{i=1}^{r} d_i^k u_i(t_k) I$ and I is the identity matrix.

The control $u = u(t)$ is said to be impulsive control if at $t = t_k$, the pulses are regulated and in the rest of the given domain of definition, $u(t)$ is chosen arbitrarily.

The solution of (6.6.1) is given by

$$x(t) = e^{At} \prod_{0 < t_k < t} [I + K^k u(t_k)] x_0 + \int_0^t \prod_{s < t_k < t} [I + D^k u(t_k)] e^{A(t-s)} Bu(s) ds, \quad (6.6.2)$$

for $0 \le t \le T$.

To test conditions of complete controllability of (6.6.1), we assume without loss of generality, that the final state is in the origin in $\mathbb{R}^n$, and the final time is $T > 0$. The complete controllability implies that

$$0 = x(T) = \prod_{0 < t_k < T} [I + D^k u(t_k)] x_0 + \int_0^T \prod_{s < t_k < T} [I + D^k u(t_k)] e^{-As} Bu(s) ds. \quad (6.6.3)$$

Note that e^{-As} can be written as

$$e^{-As} = \sum_{i=0}^{n-1} \alpha_i(s) A^i.$$

Define

$$\beta_i = \int_0^T \prod_{s < t_k < T} [I + D^k u(t_k)] \alpha_i(s) u(s) ds. \quad (6.6.4)$$

The relation (6.6.3) reduces to

$$\prod_{0 < t_k \le T} [I + D^k u(t_k)] x_0$$

$$= -\sum_{i=0}^{n-1} A^i B \beta_i = -[B \mid AB \mid \ldots \mid A^{n-1}B] \begin{bmatrix} \beta_0 \\ \beta_1 \\ \vdots \\ \beta_{n-1} \end{bmatrix}. \quad (6.6.5)$$

If the system (6.6.1) is controllable, then given any initial value $x(0) = x_0$, the relation (6.6.5) must be satisfied. The following cases need to be examined.

Case 1: Suppose that the $n \times nr$ matrix $[B \mid AB \mid \ldots \mid A^{n-1}B]$ is of rank n. Prescribe $u(t)$ at $t = t_k$ such that $E(j, \rho) \ne 0$, where

$$E(j, \rho) = \prod_{k=j}^{\rho} [I + D^k u(t_k)] \text{ for } 1 \le j \le \rho. \quad (6.6.6)$$

It is not difficult to see that β_i defined by (6.6.4) is of the form

$$\beta_i = \int_0^{t_i} \prod_{k=1}^{\rho}[I + D^k u(t_k)]\alpha_i(s)u(s)ds + \int_{t_1}^{t_2} \prod_{k=2}^{\rho}[I + D^k u(t_k)]\alpha_i(s)u(s)ds$$

$$+ \cdots + \int_{t_{\rho-1}}^{t_\rho} [I + D^\rho u(t_\rho)]\alpha_i(s)u(s)ds + \int_{t_\rho}^{T} \alpha_i(s)u(s)ds, \qquad (6.6.7)$$

and consequently, the definition of $E(j,\rho)$ implies that

$$\beta_i = \beta_i(1,\rho) + \beta_i(2,\rho) + \ldots + \beta_i(\rho,\rho) + \beta_i(0,0), \qquad (6.6.8)$$

where

$$\beta_i(j,\rho) = \int_{t_{j-1}}^{t_j} E(j,\rho)\alpha_i(s)u(s)ds \text{ and } \beta_i(0,0) = \int_{t_\rho}^{T} \alpha_i(s)u(s)ds.$$

Since $u(t)$ is already chose such that $E(1,\rho) \neq 0$ at $t = t_k$, $k = 1, 2, \ldots, \rho$, setting

$$\beta_i^0 = E^{-1}(1,\rho)[\beta_i(1,\rho) + \beta_i(2,\rho) + \ldots + \beta_i(\rho,\rho) + \beta_i(0,0)], \qquad (6.6.9)$$

we see that (6.6.5) reduces to

$$x_0 = -\sum_{i=0}^{n-1} A^i B \beta_i^0 = -[B \mid AB \mid \ldots \mid A^{n-1}B] \begin{bmatrix} \beta_0^0 \\ \beta_1^0 \\ \vdots \\ \beta_{n-1}^0 \end{bmatrix}. \qquad (6.6.10)$$

Now $u(t)$ can be chosen uniquely. Hence the impulsive system (6.6.1) is completely controllable in Case 1.

Case 2: Suppose that the rank of $[B \mid AB \mid \ldots \mid A^{n-1}B]$ is d, $0 < d < n$. Then there exists an invertible $n \times n$ matrix S such that the matrices $\widetilde{A} = S^{-1}AS$, $\widetilde{B} = S^{-1}B$ have the block structure given by

$$\widetilde{A} = \begin{bmatrix} A_1, & A_2 \\ 0, & A_3 \end{bmatrix}, \quad \widetilde{B} = \begin{bmatrix} B_1 \\ 0 \end{bmatrix}. \qquad (6.6.11)$$

Here A_1, A_2, A_3 and B_1 are, respectively, $d \times d$, $d \times (n-d)$, $(n-d) \times (n-d)$, and $d \times r$ matrices. Thus with the change of variable $x = Sy$, the system (6.6.1) is transformed into

$$y_1' = A_1 y_1 + A_2 y_2 + B_1 u, \quad t \neq t_k,$$

$$y_2' = A_3 y_2, \quad t \neq t_k,$$

$$y_1(t_k^+) = N_k^{11} y_1(t_k), \quad t = t_k,$$

$$y_2(t_k^+) = N_k^{22} y_2(t_k), \quad t = t_k,$$

$$y_1(0) = y_{10}, \quad y_2(0) = y_{20}, \tag{6.6.12}$$

where N_k^{11}, N_k^{22} are, respectively, $d \times d$ and $(n-d) \times (n-d)$ diagonal matrices such that

$$S^{-1} N_k S = \begin{bmatrix} N_k^{11} & 0 \\ 0 & N_k^{22} \end{bmatrix}$$

and $N_k = N_k u(t_k) = [I + D^k u(t_k)]$. Then we obtain from (6.6.12) the relations

$$y_2(t) = e^{A_3 t} \prod_{0 < t_k < t} N_k^{22} y_{20},$$

$$y_1(t) = e^{A_1 t} \prod_{0 < t_k < t} N_k^{11} y_{10} + \int_0^t \prod_{s < t_k < t} N_k^{11} e^{A_1(t-s)} [B_1 u(s) + A_2 y_2(s)] ds \tag{6.6.13}$$

for $0 \leq t \leq T$. Complete controllability implies

(i)
$$0 = \prod_{0 < t_k < T} N_k^{22} y_{20},$$

$$\tag{6.6.14}$$

(ii)
$$0 = \prod_{0 < t_k < T} N_k^{11} y_{10} + \int_0^T \prod_{s < t_k < T} N_k^{11} e^{-A_1 s} [B_1 u(s) + A_2 y_2(s)] ds.$$

We choose next $u(t_k)$ such that $\prod_{0 < t_k < T} N_k^{22} = 0$ so that (6.6.14) (i) is satisfied. We then get from (6.6.14)(ii)

$$0 = \int_{t_\rho}^T B_1 u(s) ds.$$

Now, $u(t)$ can be chose such that $u(t) = 0$, $t_\rho < t \leq T$ and hence the system (6.6.1) is completely controllable in Case 2 also.

Case 3: If the matrix B in (6.6.1) is a null matrix so that the rank of $[B \mid AB \mid \ldots \mid A^{n-1}B]$ is zero, then the relations (6.6.5) are satisfied by choosing impulsive control $u(T)$ such that $\prod [I + D^k u(t_k)] = 0$. For example, a simple choice is $I + D^k u(t_k) \neq 0$ for $k = 1, 2, \ldots, \rho - 1$, and $I + D^\rho u(t_\rho) = 0$. Hence the system (6.6.1) is controllable completely in this case as well.

We have thus proved the following result.

Theorem 6.6.1: *The impulsive system* (6.6.1) *is always completely controllable.*

Remark 6.6.1: The simple choice of $d^k u(t_k)$ has become necessary to obtain the expression (6.6.2) which has helped the investigation. Any generalization of $D^k u(t_k)$ leads to difficulties in obtaining (6.6.2) and analyzing further until restrictive assumptions are imposed.

6.7 STABILITY CRITERIA

In the present section, we employ the VPF established in Section 6.4 to study stability properties of system (6.4.2) when properties of (6.4.1) and (6.4.6) are known. We prove the following result.

Theorem 6.7.1: *Assume that* (H) *holds and the conditions of Theorem 6.4.1 are verified. Also, suppose that*
(i) $f(t,0) \equiv 0$, $F(t,0) \equiv 0$, $g(t,0) \equiv 0$, *and* $J_k(0) = I_k(0) = \psi_k(0) = 0$ *for all* k;
(ii) $b(\parallel x \parallel) \leq V(t,x) \leq a(\parallel x \parallel)$ *on* $\mathbb{R}_+ \times S(\rho)$, *where* $a, b \in \mathcal{K}$.

Furthermore, suppose that the trivial solution of (6.4.1) *is uniformly stable and* $u = 0$ *of* (6.4.6) *is uniformly asymptotically stable. Then, the null solution of* (6.4.2) *is uniformly asymptotically stable.*

Proof: Let $0 < \epsilon < \rho^* = \min(\rho_0, \rho)$, $t_0 \in \mathbb{R}_+$ be given. Since $u = 0$ of (6.4.6) is uniformly stable, given $b(\epsilon) > 0$, $t_0 \in \mathbb{R}_+$ there exists a $\delta_1 = \delta_1(\epsilon) > 0$ such that $0 \leq u_0 < \delta_1$ implies

$$u(t, t_0, u_0) < b(\epsilon) \text{ for all } t \geq t_0. \qquad (6.7.1)$$

Let $\delta_2 = a^{-1}(\delta_1)$. Since we know that $y = 0$ of (6.4.1) is uniformly stable, hence given $\delta_2 > 0$ there exists a $\delta_0 = \delta_0(\epsilon) > 0$ such that

$$\parallel y(t, t_0, x_0) \parallel \, < \delta_2, \ \ t \geq t_0, \text{ whenever } \parallel x_0 \parallel \, < \delta_0. \qquad (6.7.2)$$

We claim that if $\parallel x_0 \parallel \, < \delta_0$, then

$$\parallel x(t, t_0, x_0) \parallel \, < \epsilon \text{ for all } t \geq t_0, \qquad (6.7.3)$$

where $x(t, t_0, x_0)$ is any solution of (6.4.2).
   If this is not the case, there would exist a solution $x(t, t_0, x_0)$ of (6.4.2) with $\parallel x_0 \parallel \, < \delta_0$ and a $t^* > t_0$ such that $t_k < t^* \leq t_{k+1}$ for some k, satisfying

$$\epsilon \leq \, \parallel x(t^*) \parallel \text{ and } \parallel x(t) \parallel \, < \epsilon \text{ for } t_0 \leq t \leq t_k.$$

Thus $\parallel x(t_k) \parallel \, < \epsilon < \rho_0$ and hence by condition (ii) of Theorem 6.4.1, we get

$$\parallel x(t_k^+) \parallel \, = \, \parallel x(t_k) + I_k(t_k)) \parallel \, < \rho.$$

Hence, we can find a t^0 such that $t_k < t^0 \leq t^*$ satisfying

$$\epsilon \leq \, \parallel x(t^0) \parallel \, < \rho. \qquad (6.7.4)$$

This means that for $t_0 \leq t \leq t^0$, $\parallel x(t) \parallel \, < \rho$ and therefore by Theorem 6.4.1, we get

$$V(t, x(t, t_0, x_0)) \leq r(t, t_0, y(t, t_0, x_0)) \qquad (6.7.5)$$

Using (ii), (6.7.1), (6.7.2), (6.7.4) and (6.7.5) we then obtain successively

$$b(\epsilon) \le b(\parallel x(t^0, t_0, x_0) \parallel) \le V(t^0, x(t^0, t_0, x_0))$$

$$\le r(t^0, t_0, V(t^0, y(t^0, t_0, x_0)))$$

$$\le r(t^0, t_0, \alpha(\parallel y(t^0, t_0, x_0) \parallel))$$

$$\le r(t^0, t_0, a(\delta_2))$$

$$= r(t^0, t_0, \delta_1)$$

$$< b(\epsilon)$$

which is a contradiction, proving uniform stability, (6.7.3).

Suppose that $u = 0$ of (6.4.6) is uniformly asymptotically stable. Then given $b(\epsilon) > 0$, $t_0 \in \mathbb{R}_+$, there exists a $\delta_1 > 0$ and $T(\epsilon) > 0$ such that $u_0 < \delta_1$ implies

$$u(t, t_0, u_0) < b(\epsilon), t \ge t_0 + T. \tag{6.7.6}$$

Because of (6.7.3) taking $\epsilon = \rho^*$, it follows that $\parallel x_0 \parallel < \delta_0 = \delta_0(\rho^*)$ implies $\parallel x(t, t_0, x_0) \parallel < \rho^* < \rho$, $t \ge t_0$. Choose $\delta^* = \min\{\delta_1, \delta_0\}$ and let $\parallel x_0 \parallel < \delta^*$. Our previous arguments then yield

$$b(\parallel x(t, t_0, x_0) \parallel) \le V(t, x(t, t_0, x_0))$$

$$\le r(t, t_0, V(t_0, y(t, t_0, x_0)))$$

$$\le r(t, t_0, \delta_1) < b(\epsilon), t \ge t_0 + T.$$

Hence the trivial solution of (6.4.2) is uniformly asymptotically stable.

Example 6.7.1: We assume the following result to discuss the examples. Consider

$$u' = p(t)\phi(u), \quad t \ne t_k$$

$$u(t_k^+) = G_k(u(t_k)),$$

$$u(t_0) = u_0 \ge 0, \tag{6.7.7}$$

where $p \in C[\mathbb{R}_+, \mathbb{R}_+]$, $\phi, G_k \in \mathcal{K}$ and suppose that there exists a $\rho_0 > 0$ such that for each $\sigma \in (0, \rho_0]$

$$\int_{t_k}^{t_{k+1}} p(s)ds + \int_{\sigma}^{G_k(\sigma)} \frac{ds}{\phi(s)} \le 0, \quad k = 1, 2, \ldots. \tag{6.7.8}$$

Then $u = 0$ of (6.7.7) is stable.

We first consider the unperturbed system without impulses and the perturbed one with impulses

$$y' = e^{-t}y^2,$$

$$y(0) = x_0 \geq 0, \tag{6.7.9}$$

and

$$x' = e^{-t}x^2 + \tfrac{x^2}{2}, \ t \neq t_k,$$

$$x(t_k^+) = \beta_k x(t_k), \ \ 0 < \beta_k \leq 1,$$

$$x(t_0^+) = x_0 \geq 0. \tag{6.7.10}$$

Clearly (6.7.9) and (6.7.10) possess nonnegative solutions and we shall only consider these solutions.

The solutions of (6.7.9) are given by

$$y(t, t_0, x_0) = \frac{1}{[1+x_0(e^{-t}-e^{-t_0})]}, t \geq t_0,$$

and the fundamental matrix solution of the corresponding variational equation is

$$\Phi(t, t_0, x_0) = \frac{1}{[1+x_0(e^{-t}-e^{-t_0})]^2}, \ \ t \geq t_0.$$

Choosing, $V(s, x) = x^2$, we obtain

$$D_-V(s, y(t, s, x)) = 2y(t, s, x) \cdot \Phi(t, s, x) \cdot R(s, x)$$

and

$$V(t_k^+, y(t, t_k^+, x(t_k^+))) = V(t_k^+, y(t_k^+, \beta_k x(t_k)))$$

$$\leq \beta_k^2 V(t_k, y(t, t_k, x(t_k))).$$

Consequently, the corresponding comparison system is given by

$$u' = u^{3/2}, \ \ t \neq t_k,$$

$$u(t_k^+) = \beta_k^2 u(t_k),$$

$$u(t_0) = u_0 \geq 0. \tag{6.7.11}$$

Suppose that there exists a $\rho_0 > 0$ such that the impulses and the moments of impulses are related by

$$(t_{k+1} - t_k) \leq \frac{2[1-\beta_k]}{\beta_k \rho_0^{1/2}}. \tag{6.7.12}$$

Let $\epsilon > 0$ and $t_0 \in (t_k, t_{k+1}]$ be given. Choose $\delta^* = \min(\epsilon, \beta_k \epsilon)$ and $\delta_1 = (\delta^*)^{1/2}/2$. Now for any $0 \leq x_0 < \delta_0$, $\delta_0 = \min(e^{t_0}/2, \delta_1)$, we get

$$0 \leq u_0 = [y(t, t_0, x_0)]^2 = \frac{x_0^2}{[1+x_0(e^{-t}-e^{-t_0})]^2} < \frac{4x_0^2}{[1+e^{-t-t_0}]^2}$$

$$< 4x_0^2 < 4\delta_0^2 \leq \delta^*.$$

Also, because of (6.7.12), it follows that

$$\int\limits_{t_k}^{t_{k+1}} ds + \int\limits_{\sigma}^{\beta_k \sigma} \frac{ds}{s^{3/2}} \leq 0.$$

Thus we conclude that $u(t, t_0, [y(t, t_0, x_0)]^2) < \epsilon$, $t \geq t_0$ and therefore Theorem 6.7.1 yields that the trivial solution of (6.7.10) is stable.

We consider next the example involving impulses

$$y' = e^{-t} y^2, \ t \neq t_k,$$

$$y(t_k^+) = c_k y(t_k), \ \ 0 < c_k \leq 1,$$

$$y(t_0^+) = x_0 \geq 0, \tag{6.7.13}$$

where c_k's are such that $\prod_{k=1}^{\infty} c_k = c_0 \in (0, 1]$. Suppose that there exists a $\rho_0 > 0$ such that the moments of impulses, t_k, and the impulses c_k's are related by

$$e^{-t_k} - e^{-t_{k+1}} \leq \frac{1 - c_k}{c_k \rho_0}. \tag{6.7.14}$$

The solutions of (6.7.13) are given by, for $t \in (t_k, t_{k+1}]$

$$y(t, t_0^+, x_0) = \frac{\prod_{t_0 < t_j < t} c_j x_0}{\left(\begin{array}{c} [1 + x_0 \{(e^{-t_1} - e^{-t_0}) + c_1(e^{-t_2} - e^{-t_1}) \\ + \ldots + \prod_{t_0 < t_j < t} c_j (e^{-t} - e^{-t_k})\}] \end{array} \right)}, \ \ t \geq t_0$$

and the fundamental matrix solution of the corresponding variational equation is

$$\Phi(t, t_0^+, x_0) = \frac{\prod_{t_0 < t_j < t} c_j}{\left(\begin{array}{c} [1 + x_0 \{(e^{-t_1} - e^{-t_0}) + c_1(e^{-t_2} - e^{-t_1}) \\ + \ldots + \prod_{t_0 < t_j < t} c_j (e^{-t} - e^{-t_k})\}]^2 \end{array} \right)}, \ \ t \geq t_0.$$

Choosing $V(s, x) = 2c_0 x^2$, we get for $s \neq t_k$

$$D_- V(s, y(t, s, x)) = \frac{2c_0}{\prod_{t_0 < t_j < t} c_j} [y(t, s, x)]^3$$

$$\leq [y(t, s, x)]^3$$

$$\leq [V(s, y(t, s, x))]^{3/2},$$

and

$$V(t_k^+, y(t, t_k^+, x(t_k^+))) = V(t_k^+, y(t, t_k, \beta_k x(t_k)))$$

$$\leq 2c_0 \beta_k^2 V(t_k, y(t, t_k, x(t_k))).$$

Now the corresponding comparison equation is

$$u' = u^{3/2}, \quad t \neq t_k,$$

$$u(t_k^+) = d_k^2 u(t_k), \quad \text{where } d_k^2 = 2c_0 \beta_k^2,$$

$$u(t_0) = u_0 \geq 0. \tag{6.7.15}$$

Assume that there exists a $\rho_0 > 0$ such that the impulses d_k's and the moments of impulses t_k's are such that

$$t_{k+1} - t_k \leq \frac{2(1-d_k)}{d_k \rho_0^{1/2}}. \tag{6.7.16}$$

We claim that the trivial solutions of (6.7.13) and (6.7.15) are stable. This follows from (6.7.14) and (6.7.16) which imply

$$\int_{t_k}^{t_{k+1}} e^{-s} ds + \int_{\sigma}^{c_k \sigma} \frac{ds}{s^2} \leq 0,$$

and

$$\int_{t_k}^{t_{k+1}} ds + \int_{\sigma}^{d_k^2 \sigma} \frac{ds}{s^{3/2}} \leq 0.$$

Thus, from Theorem 6.7.1, we conclude the stability of the trivial solution of (6.7.10).

If the perturbed system corresponding to (6.7.9) is $x' = e^{-t}x^2 - x^2/2$, we know the asymptotic stability of the trivial solution. Hence, when we study the corresponding impulsive differential equations, we can relax some constraints.

6.8 NOTES

The theory of impulsive differential equations is relatively of recent origin. Among the earlier publications in this area are due to Milman and Myshkis [1] and Halany and Wexler [1]. For applications of such equations we refer to the work of Kruger-Thiemer [1] and Bellman [2].

For preliminaries and notation refer to Lakshmikantham, Bainov and Simeonov [1]. The contents of Section 6.2 are taken from Bainov and Simeonov [1]. Differentiability of solutions with respect to initial conditions and corresponding Alekseev type of VPF proved in Section 6.3 is due to Simeonov and Bainov [1]. Similar results for ImIDEs have been proved by Kulev and Bainov [1]. The generalized comparison principle in Section 6.4 has been established by Vasundhara Devi [1].

Among the applications, the study of periodic boundary value problems included in Section 6.5 has been adopted from Lakshmikantham, Bainov and Simeonov [1]. The results on complete controllability in Section 6.6 are due to Leela, McRae, and Sivasundaram [1]. The application of generalized comparison principle to stability behavior given in Section 6.7 is due to Vasundhara Devi [1].

For application of nonlinear VPF to two point BVP for impulsive equations see Murty, Howell, and Sivasundaram [2]. The VPF for integro-differential equations established by Kulev and Bainov [1] is employed in the study of stability properties of such equations by Lakshmikantham, Bainov and Simeonov [1]. See also the results proved by Rao and Sivasundaram [1].

7. Stochastic Differential Equations

7.0 INTRODUCTION

Mathematical problems in real-life situations quite often involve uncertainties. It is then natural to expect that the differential system representing these situations depend on random variables. When randomness is absent, we obtain deterministic models. The stochastic models may occur in linear as well as nonlinear form and hence study of both-types of stochastic systems is essential. Various analytical methods used in the study of deterministic systems need to be extended to probabilistic models. One of the useful tools used in the study of differential systems is the method of VPF. The aim of the present chapter is to develop various known results in this direction and employ them in meaningful applications.

In Section 7.1, we present essential preliminary results for smooth reading of this chapter. Section 7.2 includes the linear variation of parameters for stochastic systems of equations as well as stochastic integrodifferential systems. In Section 7.3, we first prove the differentiability of solutions of stochastic differential systems with respect to initial data and then prove the nonlinear VPF. Here again, the results åre extended to integro-differential equations.

Section 7.4 is devoted to the study of VPF in terms of Lyapunov-like functions for systems of differential as well as integro-differential equations. The content of Section 7.5 is the extension of VPFs for Ito-type of stochastic differential equations.

Section 7.6 provides an application of nonlinear VPF for boundedness of solutions of stochastic differential systems.

7.1 PRELIMINARIES

We give below necessary probabilistic analysis in order to develop VPF for stochastic differential equations.

Let us consider a random experiment E whose outcome, the elementary events ω are the elements of set Ω which is called sample space. Let $\mathcal{F}$ denote a σ-algebra of subsets of Ω whose elements are called events. Let P denote the probability measure. The triplet $(\Omega, \mathcal{F}, P)$ is called a probability space. It is said to be complete if every null event is an event.

Let $\Omega = (\Omega, \mathcal{F}, P)$ be a complete probability space. A function $x: \Omega \to R^n$ is said to be a R^n-valued measurable function if the preimages of measurable sets in R^n are measurable sets in Ω. Here x is called a random vector and collection of all such vectors is denoted by $R[\Omega, R^n]$.

Let x be a n-dimensional random vector. It is said to be P-integrable if the integral $\int_\Omega x dP$ is finite and it is called the expectation or mean of x, and is denoted by

$$E(x) = \int_\Omega x P(d\omega).$$

Jensen's Inequality: If ϕ is a real-valued continuous and concave function defined on a convex domain $D \subseteq R^n$, then $E[\phi(x)] \leq \phi(E(x))$.

Let I denote an arbitrary index set and let $\Omega \equiv (\Omega, \mathcal{F}, P)$ denote a probability space and $(R^n, \mathcal{F})$ denote the state-space.

Definition 7.1.1: A family $\{x(t), t \in I\}$ of R^n-valued random variables defined on a probability space $(\Omega, \mathcal{F}, P)$ is called a stochastic process or random process or random function with parameter set I and state space $(R^n, \mathcal{F})$. The class of random functions defined on I into $R[\Omega, R^n]$ is denoted by $R[I, R[\Omega, R^n]]$.

Definition 7.1.2: A random process $x \in R[[a, b], R[\Omega, R^n]]$ is said to be continuous at $t \in (a, b)$ if for every $\eta > 0$, $P\{\omega: \parallel x(t+h, \omega) - x(t, \omega) \parallel \geq \eta\} \to 0$ as $h \to 0$.

A random process $x \in R[[a, b], R[\Omega, R^n]]$ is said to be

(i)   almost-surely continuous at $t \in (a, b)$ if

$$P\{\omega; \lim_{h \to 0}[\parallel x(t+h, \omega) - x(t, \omega) \parallel] \neq 0\} = 0,$$

(ii)   sample continuous in $t \in (a, b)$ if

$$P \left\{ \bigcup_{t \in (a,b)} \left\{ \omega; \lim_{h \to 0}[\parallel x(t+h, \omega) - x(t, \omega) \parallel] \neq 0 \right\} \right\} = 0.$$

Definition 7.1.3: A random function $x \in C[[a, b], R[\Omega, R^n]]$ is said to possess

(i)   an almost-sure derivative $x'(t, \omega)$ at $t \in (a, b)$ if

$$P \left\{ \omega: \lim_{h \to 0} \left[\parallel \frac{x(t+h,\omega)-x(t,\omega)}{h} - x'(t, \omega) \parallel \right] \neq 0 \right\} = 0,$$

(ii)   a sample derivative $x'(t, \omega)$ in $t \in (a, b)$ if

$$P \left\{ \bigcup_{t \in (a,b)} \left\{ \omega: \lim_{h \to 0} \left[\parallel \frac{x(t+h,\omega)-x(t,\omega)}{h} - x'(t, \omega) \parallel \right] \neq 0 \right\} \right\} = 0.$$

Definition 7.1.4: A random process $x \in R[[a, b], R[\Omega, R^n]]$ is said to be measurable if and only if $x(t, \omega)$ is an $(\mathcal{F}' \times \mathcal{F})$-measurable function defined on $[a, b] \times \Omega$ with values in R^n and where $\mathcal{F}'$ denotes the σ-algebra of Lebesgue-measurable sets in $[a, b]$.

Let $M[[a, b], R[\Omega, R^n]]$ a collection of random functions in $R[[a, b], R[\Omega, R^n]]$ that are product measurable on $([a, b] \times \Omega, \mathcal{F}' \times \mathcal{F}, m \times P)$ where $\Omega = (\Omega, \mathcal{F}, P)$ and $([a, b], \mathcal{F}', m)$ are a complete probability space and a Lebesgue-measurable space respectively.

Definition 7.1.5: A random process $x \in M[[a, b], R[\Omega, R^n]]$ is said to be sample Lebesgue-integrable if $x(t, \omega)$ is Lebesgue-integrable on $[a, b]$ with probability a (w.p.1) and is denoted by

$$\int_a^b x(t,\omega)dt.$$

Let $B(z,\rho) = [x \in R^n \mid \| z - x \| < \rho]$, $z \in R^n$ is fixed. Let $\bar{B}(z,\rho)$ denotes the closure of $B(z,\rho)$. The following notation is in use:

(i) $C[R_+ \times B(z,\rho), R[\Omega, R^n]]$ is the class of sample continuous R^n-valued random function $f(t,x)$ whose realizations are denoted by $f(t,x,\omega)$;

(ii) $M[R_+ \times B(z,\rho), R[\Omega, R^n]]$ is the class of R^n-valued random functions $f(t,x)$ such that $f(t,x(t))$ is product measurable, whenever $x(t)$ is product measurable;

(iii) $IB[I, R[\Omega, R_+]]$ is the class $k \in M[I, R[\Omega, R_+]]$ whose sample Lebesgue integral is bounded with probability 1. Hence I denotes any interval on R_+.

Definition 7.1.6: A process $x \in R[I; R[\Omega, R^n]]$ is said to be Gaussian if it has a normal distribution

$$P\{x(t) < a\} = \frac{1}{\sqrt{(2\pi)^n \det (V(t))}} \int_{-\infty}^{a_1} \cdots \int_{-\infty}^{a_n} \exp\Big[-\tfrac{1}{2}(u - m(t))^T$$

$$(V(t))^{-1}(u - m(t))] \, du_1, \ldots, du_n.$$

A Gaussian process with independent increments is called a Wiener process. In particular, a Gaussian process with independent increments is called a process of Brownian motion or normalized Wiener process if

$$E[x(t) - x(s)] = 0,$$

$$E[(x(t) - x(s)^T (x(t) - x(s))] = \tilde{I}\,(t - s), \; t, s \in \text{interval } I$$

and $\tilde{I}$ is the identity matrix.

Definition 7.1.7: A Markov process $x(t)$ for $t \in I$ with values in R^n and a.s. sample continuous is called a diffusion process if its transition probability $P(s,x,t,B)$ satisfies the three conditions (due to Hincin) for every $s \in I$, $x \in R^n$ and $\epsilon > 0$:

(i)
$$\int_{\|y-x\|>\epsilon} P(s,x,t,dy) = o(t-s)$$

(ii)
$$\int_{\|y-x\|\le\epsilon} (y-x)P(s,x,t,dy) = f(s,x)(t-s) + o(t-s),$$

(iii)
$$\int_{\|y-x\|\le\epsilon} (y-x)(y-x)^T p(s,x,t,dy) = b(s,x)(t-s) + o(t-s),$$

where $f(s,x)$ is an R^n-valued function and $b(s,x)$ is an $n \times n$ matrix-valued function. Here f is called the drift vector and b is called the diffusion matrix.

Consider a system of first-order sample differential equation

$$x' = f(t, x, \omega), \; x(t_0, \omega) = x_0(\omega),$$

where the prime denotes the sample derivative of x and $f \in M[R_+ \times B(z, \rho), R[\Omega, R^n]]$.

Definition 7.1.8: A random process $x(t)$ is said to be a sample solution process of the above IVP on J if it satisfies the following conditions.
(i) $x(t_0) = x_0$,
(ii) $x(t)$ is sample continuous,
(iii) $x(t)$ is product measurable,
(iv) $x'(t, \omega) = f(t, x(t, \omega), \omega)$ w.p.1 for almost every $t \in J$.

7.2 LINEAR VARIATION OF PARAMETERS

We begin the linear random differential system

$$x'(t, \omega) = A(t, \omega)x(t, \omega), x(t_0, \omega) = y_0(\omega), t \in J = [t_0, t_0 + a], t_0 \in R^+, \quad (7.2.1)$$

where $A(t, \omega) = (a_{ij}(t, \omega))$ is a product measurable random matrix function defined on $J \times \Omega \to R^{n^2}$. Assume that $A(t, \omega)$ is a.s. sample L-integrable on J. Let $\Phi(t, \omega)$ be the $n \times n$ matrix whose columns are the n-vector solutions of (7.2.1) with the initial condition $y_0(\omega) = e_k, k = 1, \dots, n$. Clearly $\Phi(t_0, \omega) = I$, the unit matrix and $\Phi(t, \omega)$ satisfies the random matrix differential equation

$$\Phi'(t, \omega) = A(t, \omega)\Phi(t, \omega), \Phi(t_0, \omega) = I. \quad (7.2.2)$$

The following result is analogous to the corresponding deterministic result which we merely state here.

Lemma 7.2.1: *Let $A(t, \omega)$ be a product measurable random matrix function defined on $J \times (\Omega, \mathcal{F}, P)$ into R^n and sample Lebesgue-integrable on J. Then the fundamental matrix solution process $\Phi(t, \omega)$ of (7.2.2) is a.s. nonsingular on J. Moreover,*

$$\det \Phi(t, \omega) = \exp\left[\int_{t_0}^{t} tr A(s, \omega)ds\right], t \in J, \quad (7.2.3)$$

where $tr A(t, \omega) = \sum_{i=1}^{n} a_{ii}(t, \omega).$

Consider the random perturbed system of (7.2.1),

$$y'(t, \omega) = A(t, \omega)y(t, \omega) + F(t, y(t, \omega), \omega), y(t_0, \omega) = y_0(\omega) \quad (7.2.4)$$

where $A(t, \omega)$ is as defined above and $F \in M[J \times R^n, R[\Omega, R^n]]$.

The following result provides the integral representation for the solution process of (7.2.4) in terms of the solution process of (7.2.1).

Theorem 7.2.1: *Suppose that*
(i) $A(t, \omega)$ satisfies the hypothesis of Lemma 7.1.1,
(ii) $F \in M[J \times R^n, R[\Omega, R^n]]$ and $F(t, x, \omega)$ is sample continuous i x for fixed t,
(iii) $k \in IB[J; R[\Omega, R_+]]$ and satisfies

$$| F(t, x, \omega) | \leq k(t, \omega), \ (t, x) \in J \times \overline{B}(z, \rho).$$

Then any solution process $y(t, \omega) = y(t, t_0, y_0(\omega), \omega)$ of (7.2.4) satisfies the random integral equation

$$y(t, \omega) = x(t, \omega) + \int_{t_0}^{t} \Phi(t, \omega)\Phi^{-1}(s, \omega)F(s, y(s, \omega), \omega)ds, \qquad (7.2.5)$$

for $t \geq t_0$ where $\Phi(t, \omega)$ is the fundamental matrix solution of (7.2.1).

Proof: Let $x(t, \omega) = \Phi(t, \omega)y_0(\omega)$ be a solution process of (7.2.1) existing for $t \geq t_0$. The method of VP involves the determination of an a.s. sample differentiable function $z(t, \omega)$ so that

$$y(t, \omega) = \Phi(t, \omega)z(t, \omega), \ z(t_0, \omega) = y_0(\omega) \qquad (7.2.6)$$

is a solution of (7.2.4). The sample differentiation yields

$$A(t, \omega)y(t, \omega) + F(t, y(t, \omega), \omega) = \Phi'(t, \omega)z(t, \omega) + \Phi(t, \omega)z'(t, \omega).$$

This, because of (7.2.2), (7.2.6) gives

$$z'(t, \omega) = \Phi^{-1}(t, \omega)F(t, \Phi(t, \omega)z(y, \omega), \omega),$$

$$z(t_0, \omega) = y_0(\omega). \qquad (7.2.7)$$

From the hypotheses (i)-(iii), (7.2.3) and the fact that

$$\Phi^{-1}(t, \omega) = \left[\frac{(-1)^{i+j}\det(\Phi_{ij}(t, \omega))}{\det \Phi(t, \omega)} \right]^{T},$$

the random function $\Phi^{-}(t, \omega)F(t, y(t, \omega), \omega)$ satisfies the hypotheses corresponding to (ii) and (iii). Here $\Phi_{ij}(t, \omega)$ is the matrix obtained from $\Phi(t, \omega)$ by deleting the ith and the jth columns. Therefore, the IVP (7.2.7) has a solution process $z(t, \omega)$ existing for $t \geq t_0$. Hence

$$z(t, \omega) = y_0(\omega) + \int_{t_0}^{t} \Phi^{-1}(s, \omega)F(s, \Phi(s, \omega)z(s, \omega), \omega)ds. \qquad (7.2.8)$$

From (7.2.6) and (7.2.8), we therefore get

$$y(t, \omega) = \Phi(t, \omega)y_0(\omega) + \int_{t_0}^{t} \Phi(t, \omega)\Phi^{-1}(s, \omega)F(s, y(s, \omega), \omega)ds.$$

The proof is complete.

The result proved above has an extension to the IVP for linear stochastic integro-differential equation. The VPF stated in the next theorem for such equations is the analog of the corresponding Theorem 2.1.2 for linear integrodifferential equation. Hence the proof of the next result can be constructed similarly.

Theorem 7.2.2: *Consider the IVP for linear stochastic integro-differential equation*

$$y' = A(t,\omega)y + \int\limits_{t_0}^{t} k(t,s,\omega)y(s,\omega)ds + F(t,\omega),$$

$$y(t_0,\omega) = y_0(\omega), \tag{7.2.9}$$

where $A(t,\omega)$, $k(t,s,\omega)$ are measurable random matrices those satisfy suitable regularity conditions in order to ensure the existence and uniqueness of sample solution process of (7.2.9) and $F \in C[R^+, R[\Omega, R^n]]$. Then the unique solution of (7.2.9) is given by

$$y(t,\omega) = L(t,\omega)y_0(\omega) + \int\limits_{t_0}^{t} L(t,s,\omega)F(s,\omega)ds, \ t \geq t_0$$

*where $L(t,s,\omega)$ is the solution of the IVP $L_s(t,s,\omega) + L(t,s,\omega)A(s,\omega) +$
$\int\limits_{\sigma}^{t} L(t,\sigma,\omega)k(\sigma,s,\omega)d\sigma = 0$, $L(t,t) = I$, $0 \leq s \leq t < \infty$.*

7.3 NONLINEAR VARIATION OF PARAMETERS

In the present section, let us consider a system of first-order random differential equations

$$x' = f(t,x,\omega), \quad x(t_0,\omega) = x_0(\omega), \tag{7.3.1}$$

where the prime denotes the sample derivative of x and $f \in M[R_+ \times B(z,\rho), R[\Omega, R^n]]$.

As in the corresponding sections of earlier chapters, we need to prove that the solution process $x(t,t_0,x_0,\omega)$ of (7.3.1) is sample differentiable with respect to the initial conditions (t_0,x_0) and that the sample derivatives $\frac{\partial}{\partial x_0}x(t,t_0,x_0,\omega)$ and $\frac{\partial}{\partial t_0}x(t,t_0,x_0,\omega)$ exist and satisfy the equation of variation of (7.3.1) along the solution process (7.3.1). Before we proceed, we prove the following mean value result.

Theorem 7.3.1: *Suppose that $f \in M[J \times D, R[\Omega, R^n]]$ and its sample derivative $\frac{\partial f}{\partial x} = \frac{\partial}{\partial x}f(t,x,\omega)$ exists and is sample continuous in x for each $t \in J$, where D is an open convex set in R^n. Then*

$$f(t,y,\omega) - f(t,x,\omega) = \int\limits_0^1 \left[\tfrac{\partial}{\partial x} f(t, sy + (1-s)x, \omega) \right](y-x)ds.$$

Proof: Set $F(s,\omega,t) = f(t, sy + (1-s)x, \omega)$, $0 \le s \le 1$. The convexity of D implies that $f(s,\omega,t)$ is well defined. Clearly $F(s,\omega,t)$ is sample continuous in s. Further, since $f(t,x,\omega)$ is sample differentiable in x, it follows that $F(s,\omega,t)$ is sample differentiable in $s \in [0,1]$. Moreover, its sample derivative $\frac{\partial F}{\partial s}$ is a sample continuous random variable. Hence

$$\tfrac{\partial}{\partial s} F(s,\omega,t) = \tfrac{\partial}{\partial x} f(t, sy + (1-s)x, \omega)(y-x).$$

Since $F(1,\omega,t) = f(t,y,\omega)$ and $F(0,\omega,t) = f(t,x,\omega)$, the result follows by sample integration of the above relation in the sense of Rieman with respect to s between 0 and 1.

We shall now show that the solution process $x(t,t_0,x_0,\omega)$ of (7.3.1) is sample differentiable with respect to the initial conditions (t_0, x_0) and that the sample derivatives $(\frac{\partial}{\partial x_0})x(t,t_0,x_0,\omega)$ and $(\frac{\partial}{\partial t_0})x(t,t_0,x_0,\omega)$ exist and satisfy the equation of variation of (7.3.1) along the process $x(t,t_0,x_0,\omega)$.

We are now in a position to prove the following result.

Theorem 7.3.2: *Suppose that*

(H_1) *$f \in M[J \times R^n, R[\Omega, R^n]]$ and its sample derivative*

$$\tfrac{\partial f}{\partial x} = \tfrac{\partial}{\partial x} f(t,x,\omega) \qquad (7.3.2)$$

   exists and is sample continuous in x for each $t \in J$;

(H_2) *$K \in IB[J, R[\Omega, R_+]]$ and satisfies*

$$\| f_x(t,x,\omega) \| \le K(t,\omega) \text{ for } (t,x) \in J \times \overline{B}(z,\rho), \qquad (7.3.3)$$

   where $f_x(t,x,\omega) = (\partial/\partial x)f(t,x,\omega)$;

(H_3) *the solution $x(t,\omega) = x(t,t_0,x_0(\omega),\omega)$ of (7.3.1) exists for $t \ge t_0$.*

Then

(a)   *The sample derivative $(\partial/\partial x_{k_0})x(t,t_0,x_0(\omega),\omega)$ exists for all $k = 1,2,3,\dots,n$ and satisfies the systems of linear random differential equations*

$$y'(t,\omega) = f_x(t, x(t,t_0,x_0(\omega),\omega),\omega)y(t,\omega), y(t_0,\omega) = e_k, \qquad (7.3.4)$$

   where $e_k = (e_k^1, e_k^2, \dots, e_k^k, \dots, e_k^n)^T$ is an n-vector such that $e_k^j = 0$ if $j \ne k$ and $e_k^k = 1$ and x_{k0} is the kth component of $x_0 = (x_{10}, x_{k0}, \dots, x_{n0})^T$ and

(b)   *the sample derivative $(\partial/\partial t_0)x(t,t_0,x_0(\omega),\omega)$ exists and satisfies (7.3.4) with*

$$\tfrac{\partial}{\partial t_0} x(t,t_0,x_0(\omega),\omega) = -\,\Phi(t,t_0,x_0(\omega),\omega)f(t_0, x_0(\omega),\omega) \qquad (7.3.5)$$

whenever $f(t_0, x_0(\omega), \omega)$ exists w.p.1, where $\Phi(t, t_0, x_0, \omega)$ is the fundamental matrix solution process of (7.3.4). Moreover, $\Phi(t, t_0, x_0, \omega)$ satisfies the random matrix differential equation

$$X' = f_x(t, x(t), \omega)X, \quad X(t_0, \omega) = \text{unit matrix.} \tag{7.3.6}$$

Proof: First we shall show that (a) holds. From hypotheses, the solutions $x(t, t_0, x_0)$ are unique and continuous with respect to the initial conditions (t_0, x_0). For small λ, $x(t, \lambda, \omega) = x(t, t_0, x_0(\omega) + \lambda e_k, \omega)$ and $x(t, \omega) = x(t, t_0, x_0(\omega), \omega)$ are solution processes of (7.3.1) through $(t_0, x_0 + \lambda e_k)$ and (t_0, x_0), respectively. From the continuous dependence on initial conditions, it is clear that

$$\lim_{\lambda \to 0} x(t, \lambda, \omega) = x(t, t_0, \omega) \text{ uniformly on } J \text{ w.p.1.} \tag{7.3.7}$$

Set

$$\Delta x_\lambda(t, \omega) = \frac{[x(t, \lambda, \omega) - x(t, \omega)]}{\lambda}, \quad \Delta x_\lambda(t_0, \omega) = e_k, \quad \lambda \neq 0, \tag{7.3.8}$$

$x = x(t, \omega)$ and $y = x(t, \lambda, \omega)$. This, together with the application of Theorem 7.3.1, yields

$$\Delta x'_\lambda(t, \omega) = \int_0^1 f_x(t, sy + (1 - s)x, \omega)\Delta x_\lambda(t, \omega)ds. \tag{7.3.9}$$

Define

$$f_x(t, x(t, \omega), \lambda, \omega) = \int_0^1 f_x(t, sx(t, \lambda, \omega) + (1 - s)x(t, \omega), \omega)ds. \tag{7.3.10}$$

Note that the integral is the sample Riemann integral. In view of (H_1), $f_x(t, x, \lambda, \omega)$ is a product-measurable random process which is sample continuous in (x, λ) for fixed t. Furthermore,

$$\| f_x(t, x, \lambda, \omega) \| \leq K(t, \omega) \text{ for } (t, x, \lambda) \in J \times \overline{B}(z, \rho) \times \Lambda,$$

where $\Lambda \subset R$ is an open neighborhood of $0 \in R$. From the sample continuity of $f_x(t, x, \omega)$, we have

$$\lim_{\lambda \to 0} f_x(t, sx(t, \lambda, \omega) + (1 - s)x(t, \omega), \omega) = f_x(t, x(t, \omega), \omega) \tag{7.3.11}$$

uniformly in $s \in [0, 1]$ w.p.1. This, together with (7.3.10), yields

$$\lim_{\lambda \to 0} f_x(t, x(t, \omega), \lambda, \omega) = f_x(t, x(t, \omega), \omega) \text{ w.p.1.} \tag{7.3.12}$$

From (7.3.11) and (7.3.10), relation (7.3.9) reduces to

$$\Delta x'_\lambda(t, \omega) = f_x(t, x(t, \omega), \lambda, \omega)\Delta x_\lambda(t, \omega), \quad \Delta x_\lambda(t_0, \omega) = e_k. \tag{7.3.13}$$

It is obvious that the initial-value problem (7.3.13) satisfies

$$\lim_{\lambda \to 0} \Delta x_\lambda(t, \omega) = y(t, \omega) \text{ uniformly on } J \text{ w.p.1.}, \qquad (7.3.14)$$

where $y(t, \omega)$ is the solution process of (7.3.4). Because of (7.3.8), we note that the limit of $\Delta x_\lambda(t, \omega)$ in (7.3.14) is equivalent to the derivative $(\partial/\partial x_{k0})$ $\times x(t, t_0, x_0(\omega), \omega)$. Hence $(\partial/\partial x_{k0})x(t, t_0, x_0(\omega), \omega)$ exists and is the solution process (7.3.4). This is true for every $k = 1, 2, \ldots, n$. Thus $(\partial/\partial x_0)$ $\times x(t, t_0, x_0(\omega), \omega)$ is the fundamental random matrix solution process of (7.3.4) and satisfies the random matrix solution process of (7.3.4) and satisfies the random matrix equation (7.3.6). $(\partial/\partial x_0)x(t, t_0, x_0(\omega), \omega)$ is denoted by $\Phi(t, t_0, x_0(\omega), \omega)$. This completes the proof of (a) and the last part of (b).

To prove the first part of (b), define

$$\Delta \widehat{x}_\lambda(t, \omega) = \frac{x(t, \lambda, \omega) - x(t, \omega)}{\lambda}, \quad \Delta \widehat{x}_\lambda(t_0, \omega) = \frac{x(t_0, \lambda, \omega) - x_0(\omega)}{\lambda}, \qquad (7.3.15)$$

where $x(t, \lambda, \omega) = x(t, t_0 + \lambda, x_0(\omega), \omega)$ and $x(t, \omega) = x(t, t_0, x_0(\omega), \omega)$ are solution processes of (7.3.1) through $(t_0 + \lambda, x_0)$ and (t_0, x_0), respectively. Again, by imitating the proof of part (a) and by replacing $\Delta x_\lambda(t, \omega)$ by $\Delta \widehat{x}_\lambda(t, \omega)$, one can conclude that the sample derivative $(\partial/\partial t_0)x(t, t_0, x_0(\omega), \omega)$ exists and is the solution of (7.3.4) whenever

$$\lim_{\lambda \to 0} \Delta \widehat{x}_\lambda(t_0, \omega) \text{ exists and is equal to } -f(t_0, x_0(\omega), \omega). \qquad (7.3.16)$$

We shall show that (7.3.16) is true. By uniqueness of solutions, we have

$$\Delta \widehat{x}_\lambda(t_0, \omega) = (x(t_0, t_0 + \lambda, x_0(\omega), \omega) - x(t_0 + \lambda, t_0 + \lambda, x_0(\omega), \omega))/\lambda$$

$$= -\frac{x(t_0 + \lambda, t_0 + \lambda, x_0(\omega), \omega) - x(t_0, t_0 + \lambda, x_0(\omega), \omega)}{\lambda}$$

$$= -\frac{1}{\lambda} \int_{t_0}^{t_0 + \lambda} f(s, x(s, t_0 + \lambda, x_0(\omega), \omega), \omega) ds.$$

This, together with the sample Lebesgue integrability of

$$f(s, x(s, t_0 + \lambda, x_0(\omega), \omega), \omega), \text{ implies that } \lim_{\lambda \to 0} \Delta \widehat{x}_\lambda(t_0, \omega)$$

exists and is equal to $-f(t_0, x_0(\omega, \omega)$ w.p.1. Now by following the proof of the deterministic theorem, we conclude that

$$\frac{\partial}{\partial t_0} x(t, t_0, x_0(\omega), \omega) = -\Phi(t, t_0, x_0(\omega), \omega)f(t_0, x_0(\omega), \omega) \text{ w.p.1.}$$

This completes the proof of the first part of (b). Hence the proof of the theorem is complete.

In the following, we shall present a result similar to Theorem 7.2.1 with respect to

$$y'(t, \omega) = f(t, y(t, \omega), \omega) + F(t, y(t, \omega), \omega), \quad y(t_0, \omega) = x_0(\omega), \qquad (7.3.17)$$

which is a perturbation of a nonlinear system (7.3.1).

Theorem 7.3.3: *Assume that*
(*i*) $f, F \in M[J \times R^n, R[\Omega, R^n]]$ *and* $f(t, x, \omega)$ *and* $F(t, x, \omega)$ *and* $F(t, x, \omega)$
        are a.s. sample continuous in x *for fixed* $t \in J$;
(*ii*) $K \in IB[J, R[\Omega, R_+]]$ *and satisfies*

$$| f(t, x, \omega) | + | F(t, x, \omega) | \leq K(t, \omega) \, for \, (t, x) \in J \times \bar{B} \, (z, \rho);$$

(*iii*) *the initial-value problem* (7.3.1) *has a unique solution process* $x(t, \omega)$, *existing*
        for $t \geq t_0$;
(*iv*) *the sample derivative* $\Phi(t, t_0, x_0, \omega) = (\partial/\partial x_0)x(t, t_0, x_0(\omega), \omega)$ *exists and*
        $\Phi(t, t_0, x_0, \omega)$ *is sample continuous in* x_0 *for fixed* (t, t_0) *and is product-*
        measurable in (t, ω) *for fixed* t_0;
(*v*)   *the inverse* $\Phi^{-1}(t, t_0, x_0, \omega)$ *of random matrix function* $\Phi(t, t_0, x_0, \omega)$ *exists and*
        is product-measurable in (t, ω) *and sample continuous in* x_0 *for fixed* (t, t_0);
(*vi*) $K_1 \in IB[J, R[\Omega, R_+]]$ *and satisfies*

$$\| \Phi^{-1}(t, t_0, x_0, \omega) \| \leq K_1(t, \omega) \, for \, (t, x) \in J \times \bar{B} \, (z, \rho).$$

Then any solution process $y(t, \omega) = y(t, t_0, x_0(\omega), \omega)$ *of* (7.3.17) *satisfies the*
relation

$$y(t, t_0, x_0(\omega), \omega) =$$

$$x\left(t, t_0, x_0(\omega) + \int_{t_0}^{t} \Phi^{-1}(s, t_0, z(s, \omega), \omega)F(s, y(s, \omega), \omega)ds, \omega \right), \quad (7.3.18)$$

as far as $z(t, \omega)$ *exists to the right of* t_0, *w.p.1, where* $z(t, \omega)$ *is a solution process of*
the initial value problem

$$z'(t, \omega) = \Phi^{-1}(t, t_0, z(t, \omega), \omega)F(t, x(t, t_0, z(t, \omega), \omega), \omega),$$

$$z(t_0, \omega) = x_0(\omega). \quad (7.3.19)$$

Proof: In view of the hypotheses, the initial value problems (7.3.1) and (7.3.19)
have solution processes $x(t, t_0, x_0, \omega)$ and $z(t, t_0, x_0, \omega)$, respectively. The method
of variation of parameters requires the determination of a random function $z(t, \omega)$ so
that

$$y(t, t_0, x_0(\omega), \omega) = x(t, t_0, z(t, \omega), \omega), \, z(t_0, \omega) = x_0(\omega), \quad (7.3.20)$$

is a solution process of (7.3.17). The sample differentiation yields

$$f(t, y(t, \omega), \omega) + F(t, y(t, \omega), \omega) = x'(t, t_0, z(t, \omega), \omega)$$

$$+ \frac{\partial}{\partial x_0} \, x(t, t_0, z(t, \omega), \omega)z'(t, \omega).$$

This, together with (iv), (v) and (7.3.20) gives

$$z'(t, w) = \Phi^{-1}(t, t_0, z(t, w), w)F(t, x(t, t_0, z(t, w), w), w),$$

$$z(t_0, w) = x_0(w),$$

which implies

$$z(t, w) = x_0(w) + \int_{t_0}^{t} \Phi^{-1}(s, t_0, z(s, w), w)F(s, x(s, t_0, z(s, w), w), w)ds. \quad (7.3.21)$$

By (7.3.20) and (7.3.21), we have

$$y(t, t_0, x_0(w), w) = x\left(t, t_0, x_0(w) + \int_{t_0}^{t} \Phi^{-1}(s, t_0, z(s, w), w)F(s, y(s, w), w)ds, w\right),$$

which proves the theorem.

Corollary 7.3.1: *Under the hypotheses of Theorem 7.3.3, the following relation is also valid:*

$$y(t, t_0, x_0(w), w) = x(t, t_0, x_0(w), w)$$

$$+ \int_{t_0}^{t} \Phi(t, t_0, z(s, w), w)\Phi^{-1}(s, t_0, z(s, w), w)F(s, y(s, t_0, x_0(w), w), w)ds, \quad (7.3.22)$$

where $z(t, w)$ is any solution process of (7.3.19).

Proof: For $t_0 \leq s \leq t$, we have

$$\frac{d}{ds}x(t, t_0, z(s, w), w) = \frac{\partial}{\partial x_0}x(t, t_0, z(s, w), w)z'(s, w)$$

$$= \Phi(t, t_0, z(s, w), w)\Phi^{-1}(s, t_0, z(s, w), w)F(s, x(s, t_0, z(s, w), w), w),$$

which implies
$$x(t, t_0, z(t, w), w) = x(t, t_0, x_0(w), w)$$

$$+ \int_{t_0}^{t} \Phi(t, t_0, z(s, w), w)\Phi^{-1}(s, t_0, z(s, w), w)F(s, x(s, t_0, z(s, w), w), w)ds.$$

This, together with (7.3.20) yields
$$y(t, t_0, x_0(w), w) = x(t, t_0, x_0(w), w)$$

$$+ \int_{t_0}^{t} \Phi(t, t_0, z(s, w), w)\Phi^{-1}(s, t_0, z(s, w), w)F(s, y(s, t_0, x_0(w), w), w)ds,$$

establishing relation (7.3.22).

Remark 7.3.1: Relations (7.3.18) and (7.3.22) provide two different forms of the variation of parameters formula for the solution process of (7.3.17).

In the following, we give a result which is an analog of a known deterministic nonlinear variation of constants type result due to Alekseev. Furthermore, we shall show that the obtained nonlinear variation of constants formula is equivalent to formula (7.3.22).

Theorem 7.3.4: *Assume that the random functions f and F in (7.3.17) satisfy the hypotheses of Theorem 7.3.2 and Theorem 7.3.3, respectively. Then any solution process $y(t, \omega)$ of (7.3.17) satisfies formulas (7.3.18) and (7.3.22) and in addition the formula*

$$y(t, t_0, x_0(\omega), \omega) = x(t, t_0, x_0(\omega), \omega)$$

$$+ \int_{t_0}^{t} \Phi(t, s, y(s, \omega), \omega) F(s, y(s, \omega), \omega) ds. \tag{7.3.23}$$

Moreover, (7.3.22) and (7.3.23) are equivalent if

$$\Phi(t, s, z(s, \omega), \omega) = \Phi(t, t_0, z(s, \omega), \omega) \Phi^{-1}(s, t_0, z(s, \omega), \omega)$$

Proof: By Theorem 7.3.2, the initial value problem (7.3.1) has a unique solution process $x(t, \omega) = x(t, t_0, x_0(\omega), \omega)$. Furthermore, $(\partial/\partial x_0)x(t, t_0, x_0(\omega), \omega) = \Phi(t, t_0, x_0(\omega), \omega)$ exists and is sample continuous in (t, t_0, x_0) and $\Phi(t, t_0, x_0(\omega), \omega)$ is the fundamental random matrix solution process of (7.3.6), which implies that $\Phi^{-1}(t, t_0, x_0, \omega)$ exists and is sample continuous in (t, t_0, x_0). Thus $f(t, x, \omega)$, $F(t, x, \omega)$, $\Phi(t, t_0, x_0, \omega)$, and $\Phi^{-1}(t, t_0, x_0, \omega)$ satisfy the hypotheses of Theorem 7.3.3. Therefore, the initial value problem (7.3.17) satisfies formulas (7.3.18) and (7.3.22). To show that $y(t, \omega)$ satisfies (7.3.23), consider, for $t_0 \leq s \leq t$, $x(t, s, y(s, \omega), \omega)$ and $y(s, t_0, x_0(\omega), \omega)$ as solution processes of (7.3.1) and (7.3.17) through $(s, y(s, \omega))$ and $(t_0, x_0(\omega))$, respectively. Then

$$\frac{d}{ds}x(t, s, y(s, \omega), \omega) = \frac{\partial}{\partial x_0}x(t, s, y(s, \omega), \omega)y'(s, \omega) + \frac{\partial}{\partial t_0}x(t, s, y(s, \omega), \omega).$$

This, together with (7.3.5) and (7.3.17), yields

$$\frac{d}{ds}x(t, s, y(s, \omega), \omega) = \Phi(t, s, y(s, \omega), \omega) F(s, y(s, \omega), \omega), \tag{7.3.24}$$

which implies

$$x(t, t, y(t, \omega), \omega) = x(t, t_0, x_0(\omega), \omega) + \int_{t_0}^{t} \Phi(t, s, y(s, \omega), \omega) F(s, y(s, \omega), \omega) ds$$

By the uniqueness of the solution process of (7.3.1), we get

$$y(t, t_0, x_0(\omega), \omega) = x(t, t_0, x_0(\omega), \omega)$$

$$+ \int_{t_0}^{t} \Phi(t, s, y(s, w), w) F(s, y(s, w), w) ds$$

proving relation (7.3.23). We next show that formulas (7.3.22) and (7.3.23) are equivalent if and only if

$$\Phi(t, s, y(s, w), w) = \Phi(t, t_0, z(s, w), w) \Phi^{-1}(s, t_0, z(s, w), w).$$

For $t_0 \leq s \leq t$, we note that

$$x(t, s, y(s, w), w) = x(t, t_0, z(s, w), w), \qquad (7.3.25)$$

where $x(t, w)$ and $y(s, w)$ are as defined above and $z(s, w)$ is a solution process of (7.3.19) through (t_0, x_0). Then by sample differentiation with respect to s, (7.3.25) and the uniqueness of the solution of (7.3.1), after substituting for $y'(s, w)$ and $z'(s, w)$, we get

$$\Phi(t, s, y(s, w), w) F(s, y(s, w), w)$$

$$= \Phi(t, t_0, z(s, w), w) \Phi^{-1}(s, t_0, z(s, w), w) F(s, y(s, w), w),$$

which is equivalent to

$$\Phi(t, s, y(s, w), w) = \Phi(t, t_0, z(s, w), w) \Phi^{-1}(s, t_0, z(s, w), w).$$

This, together with (7.3.24) shows that (7.3.22) and (7.3.23) are equivalent if and only if

$$\Phi(t, s, y(s, w), w) = \Phi(t, t_0, z(s, w), w) \Phi^{-1}(s, t_0, z(s, w), w).$$

This completes the proof of the theorem.

The following theorem gives the relationship between the solution process of (7.3.1) and the fundamental random matrix solution process of (7.3.4).

Theorem 7.3.5: *Assume that the hypotheses of Theorem 7.3.1 hold. Furthermore, assume that $x(t, t_0, x_0, w)$ and $x(t, t_0, y_0, w)$ are the solution processes of (7.3.1) through (t_0, x_0) and (t_0, y_0), respectively, existing for $t \geq t_0$, such that x_0, y_0 belong to a convex subset of R^n. Then for $t \geq t_0$,*

$$x(t, t_0, y_0, w) - x(t, t_0, x_0, w) = \int_0^1 \Phi(t, t_0, x_0 + s(y_0 - x_0), w) ds (y_0 - x_0) \ (7.3.26)$$

Proof: Since x_0, y_0 belong to a convex subset of R^n w.p.1, $x(t, t_0, x_0 + s(y_0 - x_0), w)$ is defined for $0 \leq s \leq 1$. Thus

$$\frac{d}{ds} x(t, t_0, x_0 + s(y_0 - x_0), w) = \Phi(t, t_0, x_0 + s(y_0 - x_0), w)(y_0 - x_0),$$

and hence integration from 0 to 1 yields (7.3.26).

Let us now consider the stochastic integro-differential equations (SIDE)

$$y' = f(t, y, \omega) + \int_{t_0}^{t} K(t, s, y(s, \omega), \omega)ds, \, y(t_0, \omega) = y_0(\omega) \qquad (7.3.27)$$

and the corresponding deterministic initial value problem (DIVP)

$$m' = \widehat{f}(t, m) + \int_{t_0}^{t} \widehat{K}(t, s, m(s))ds, \; m(t_0) = E[y_0(\omega)] \qquad (7.3.28)$$

which is obtained by ignoring the random disturbances in the system described in (7.3.27). In our subsequent analysis we will utilize the following random initial value problem (RIVP)

$$x' = \widehat{f}(t, x) + \int_{t_0}^{t} \widehat{K}(t, s, x(s))ds, \; x(t_0, \omega) = x_0(\omega). \qquad (7.3.29)$$

In (7.3.27), (7.3.28) and (7.3.29), $y, m, x \in R^n$, $x_0, y_0 \in R[\Omega, R^n]$; E stands for the expected value of a random variable; (Ω, F, P) is a complete probability space, $f \in M[R_+ \times R^n, R[\Omega, R^n]]$, $K \in M[R_+ \times R_+ \times R^n, R[\Omega, R^n]]$, and $f(t, y, \omega)$, $K(t, s, y, \omega)$ are sample continuous in y for fixed $t, s \in R_+, \widehat{f} \in C[R_+ \times R^n, R^n]$ and $\widehat{K} \in C[R_+ \times R_+ R^n, R^n]$.

We assume that

(H_1) f, K satisfies desired regularity conditions so that the initial value problem (7.3.27) has a sample solution process existing for $t \geq t_0$;

(H_2) $\widehat{f}_x, \widehat{K}_x$ exist and $\widehat{f}_x \in C[R_+ \times R^n, R^n]$, $\widehat{K}_x \in C[R_+ \times R^n \times R^n, R^n]$.

The above conditions imply that $\overline{x}(t) = x(t, t_0, z_0)$ is unique solution process of (7.3.28) or (7.3.29) depending on the choice of z_0, where z_0 is either y_0 or x_0. We need the following results in our subsequent study.

The next result is the analog of the corresponding result proved in Theorem 2.2.1 for integro-differential systems. The proof is similar and hence deleted.

Theorem 7.3.6: *Consider the SIVP for nonlinear integro-differential equations given by*

$$x' = f(t, x, \omega) + \int_{t_0}^{t} g(t, s, x(s), \omega)ds, x(t_0, \omega) = x_0(\omega)$$

and the corresponding DIVP is given by

$$m' = \widehat{f}(t, m) + \int_{t_0}^{t} \widehat{g}(t, s, m(s))ds, m_0 = E[x(t_0, \omega)] = m_0(\omega) \qquad (7.3.30)$$

where $\widehat{f} \in C[R_+ \times R^n, R^n]$, $\widehat{g} \in C[R^+ \times R^+ \times R^n]$ and $\widehat{f}_x, \widehat{g}_x$ exist and are continuous on $R_+ \times R^n$, $R^+ \times R^+ \times R^n$ respectively. Let $x(t, t_0, x_0)$ be the unique *solution of (7.3.30) existing on some interval* $t_0 \le t < a \le \infty$ *and set* $J = [t_0, t_0 + T]$, $t_0 + T < a$.

Define $H(t, t_0, m_0) = \widehat{f}_x(t, m(t))$ and $G(t, s; t_0, m_0) = \widehat{g}_x(t, s, m(s))$.

Then

(*i*) $\Phi(t, t_0, m_0) = \frac{\partial x(t, t_0, m_0)}{\partial x_0}$ exists and is the solution of

$$y'(t) = H(t, t_0, m_0)y(t) + \int_{t_0}^{t} G(t, s; t_0, m_0)y(s)ds$$

with $\Phi(t_0, t_0, m_0) = I$;

(*ii*) $\Psi(t, t_0, m_0) = \frac{\partial x(t, t_0, m_0)}{\partial t_0}$ exists and is the solution of

$$z'(t) = H(t, t_0, m_0)z(t) + \int_{t_0}^{t} G(t, s; t_0, m_0)z(s0ds - \widehat{g}(t, t_0, m_0)$$

such that $\Psi(t, t_0, m_0) = -\widehat{f}(t_0, m_0)$;

(*iii*) The functions $\Phi(t, t_0, m_0)$, $\Psi(t, t_0, m_0)$ satisfy the relation

$$\Psi(t, t_0, m_0) + \Phi(t, t_0, m_0)\widehat{f}(t_0, m_0) + \int_{t_0}^{t} L(t, \sigma; t_0, m_0)\widehat{g}(\sigma, t_0, m_0)d\sigma = 0$$

where $L(t, s; t_0, m_0)$ is the solution of the IVP

$$L_s(t, s; t_0, m_0) + L(t, s; t_0, m_0)H(s, t_0, m_0)$$

$$+ \int_{s}^{t} L(t, \sigma; t_0, m_0)G(\sigma, s, t_0, m_0)d\sigma = 0,$$

$L(t, t; t_0, x_0) = I$ on the interval $t_0 \le s \le t$ and $L(t, t_0; t_0, x_0) = \Phi(t, t_0, x_0)$.

We merely state below the VPF for nonlinear integro-stochastic differential equations. The proof of the next theorem can be written on the lines of similar result proved in Theorem 2.2.2 for integro-differential equations.

Theorem 7.3.7: *Let* $y(t, \omega) = y(t, t_0, y_0(\omega), \omega)$ *and* $x(t, \omega) = x(t, t_0, x_0(\omega))$ *be the sample solution processes of* (7.3.27) *and* (7.3.29) *existing for* $t \ge t_0$ *with* $x_0(\omega) = y_0(\omega)$. *Let the assumptions of Theorem 7.3.6 hold. Set*

$$R(t, y, Ty, \omega) = f(t, y, \omega) - \widehat{f}(t, y) + \int_{t_0}^{t} [K(t, s, y(s, \omega), \omega) - \widehat{K}(t, s, y(s, \omega))]ds.$$

Then

$$y(t, \omega) = x(t, \omega) + \int_{t_0}^{t} \Phi(t, s, y(s, \omega))R(s, y(s, \omega), Ty(s, \omega), \omega)ds$$

$$+ \int_{t_0}^{t} \int_{s}^{t} [\Phi(t, \sigma, y(\sigma, \omega)) - L(t, s, y(s, \omega))]\widehat{K}(\sigma, s, y(s, \omega))d\sigma ds.$$

Remark 7.3.2:

(i)   Observe that in view of (7.3.27) and (7.3.29), the system (7.3.27) can be written as

$$y' = \widehat{f}(t,y) + \int\limits_{t_0}^{t} \widehat{K}(t,s,y(s,\omega))ds + R(t,y,Ty,\omega).$$

(ii)   In case $f(t,y,\omega) = A(t,\omega)y$, $K(t,s,y,\omega) = K(t,s,\omega)y$ then the above VPF becomes

$$y(t,\omega) = x(t,\omega) + \int\limits_{t_0}^{t} \Phi(t,s)R(s,y(s,\omega),Ty(s,\omega),\omega)ds.$$

This form is comparable to the result contained in Theorem 7.2.2.

7.4 VPF IN TERMS OF LYAPUNOV-LIKE FUNCTIONS

The objective of this section is to prove generalized form of VPF for stochastic differential system and stochastic integrodifferential system.   We deal with applications of such VPF s in a subsequent section.

Here we undertake the study of the initial value problem

$$y'(t,\omega) = F(t,y(t,\omega),\omega), y(t_0,\omega) = y_0(\omega) \tag{7.4.1}$$

and the corresponding mean system of differential equations

$$m'(t) = f(t,m(t)), m(t_0) = m_0 = E(y_0(\omega)). \tag{7.4.2}$$

where $f(t,z) = E[F(t,z,\omega)]$. It follows that

$$y'(t,\omega) = f(t,y(t,\omega)) + R(t,y(t,\omega),\omega), y(t_0,\omega) = y_0(\omega), \tag{7.4.3}$$

where

$$R(t,y(t,\omega),\omega) = F(t,y(t,\omega),\omega) - E[F(t,y(t,\omega),\omega].$$

Simultaneously, we also consider the IVP

$$x' = f(t,x), x(t_0,\omega) = y_0(\omega) = x_0(\omega), \tag{7.4.4}$$

where f is as given in (7.4.2).

Assume that all the relations involving random quantities are valid w.p.1. Suppose that the following hypotheses hold.

(H_1) $R \in M[R_+ \times R^n, R[\Omega,R]]$ and R is almost sure sample continuous in x for each $t \in R_+$; $f \in C[R_+ \times R^n, R^n]$, f_x exists and $f_x \in C[R_+ \times R^n, R^{n^2}]$;

(H_2)   The random function F in (7.4.1) satisfies suitable regularity conditions so that (7.4.1) has sample solution process existing for $t \geq t_0$.

This assumption implies that $x(t,\omega) = x(t, t_0, x_0(\omega))$ is a unique solution of (7.4.4) and that $x(t,\omega)$ is sample continuously differentiable with respect to (t_0, x_0).

(H_3) $V(t,x) \in C[R_+ \times R^n, R^m]$ and V_x exists and is continuous for $(t,x) \in R_+ \times R^n$.

The following is the main result of this section.

Theorem 7.4.1: *Let the hypotheses (H_1), (H_2) and (H_3) hold and $x(t,\omega) = x(t_0, x_0(\omega))$ be the sample solution of (7.4.4). Let $y(t,\omega) = y(t, t_0, y_0(\omega), \omega)$ be the sample solution of (7.4.3) for $t \geq 0$. Then*

$$V(t, y(t,\omega))$$
$$= V(t, x(t,\omega)) + \int_{t_0}^{t} V_x(t, x(t, s, y(s, \omega))\Phi(t, s, y(s, \omega))R(s, y(s, \omega), \omega)ds \quad (7.4.5)$$

where $\Phi(t, t_0, y_0(\omega))$ satisfies the random matrix differential equation

$$x' = f_x(t, x(t), \omega)x, \quad x(t_0, \omega) = I.$$

Proof: Let $x(t, s, y(s, \omega))$ be the sample solution process of (7.4.4) through $(s, y(s))$ and $y(s, \omega) = y(s, t_0, y_0(\omega), \omega)$ be the sample solution process of (7.4.3) through (t_0, x_0). From the hypothesis $(H_1) - (H_3)$ and in view of the Theorem 7.3.2, we get

$$\frac{dV}{ds}(t, x(t, s, y(s, \omega)))) =$$

$$= V_x(t, x(t, y(s, \omega)))\left[\frac{d}{ds}x(t, s, y(s, \omega)) + \frac{d}{dy}x(t, s, y(s, \omega))\frac{d}{ds}y(s, \omega)\right]$$

$$= V_x(t, x(t, s, y(s, \omega)))\Phi(t, s, y(s, \omega))R(s, y(s, \omega), \omega) \text{ w.p.1}.$$

Integrating, in the sample sense, both the sides between t_0 and t and noting that $x(t, t, y(t, t_0, y_0(\omega), \omega)) = y(t, t_0, y_0(\omega), \omega)$, we obtain

$$V(t, y(t, \omega)) + V(t, x(t, \omega)) + \int_{t_0}^{t} V_s(t, s(t, s, y(s, \omega)))\Phi(t, s, y(s, \omega))R(s, y(s, \omega), \omega)ds.$$

The proof is complete.

Remarks 7.4.1:
(i)   Let $V(t, x(t, \omega)) = x(t, \omega)$. Then (7.4.5) becomes

$$y(t, \omega) = x(t, \omega) + \int_{t_0}^{t} \Phi(t, s, y(s, \omega))R(s, y(s, \omega), \omega)ds.$$

This result has been proved in Theorem 7.3.4.

(*ii*)   Let $V(t, x(t, \omega)) = |x(t, \omega)|^2$. Then (7.4.5) becomes

$$|y(t, \omega)|^2 = |x(t, \omega)|^2 + 2\int_{t_0}^t x(t, s, y(s, \omega))\Phi(t, s, y(s, \omega))R(s, y(s, \omega, \omega)ds.$$

Theorem 7.4.2: *Let the hypotheses of Theorem 7.4.1 hold. Then*

$$V(t, y(t, \omega) - x(t, \omega)) = V(t, 0) + \int_{t_0}^t V_x[t, x(t, s, y(s, \omega))$$

$$- x(t, s, x(s, \omega))]\Phi(t, s, y(s, \omega))R(s, y(s, \omega), \omega)ds.$$

Proof: Following the proof of Theorem 7.4.1, we get

$$\frac{dv}{ds}(t, x(t, s, y(s, \omega)) - x(t, s, x(s, \omega)))$$

$$= V_x[t, x(t, s, y(s, \omega)) - x(t, s, x(s, \omega))]\Phi(t, s, y(s, \omega))R(s, y(s, \omega), \omega) \text{ w.p.1.} (7.4.6)$$

Integrate (7.4.6) between t_0 and t and use the fact that $x(t, t, y(t, t_0, y_0(\omega), \omega)) = y(t, t_0, y_0(\omega), \omega)$, to obtain the conclusion.

We now proceed to prove the generalized VPF for a system of integro-differential equations with random coefficients. This method gives integral representation of a function of a solution process of (7.3.27) with respect to the solution process of random initial value problem (7.3.29) through $(t_0, y_0(\omega))$. In addition to the hypotheses (H_1) and (H_2) satisfied by the systems (7.3.27)-(7.3.29), we assume the following condition.

(H_3) $V(t, x) \in C[R_+ \times R^n, R^n]$ such that V_x exists and is continuous for $(t, x) \in R_+ \times R^n$.

Theorem 7.4.3: *Let the hypotheses $(H_1), (H_2), (H_3)$ be satisfied and $y(t, \omega) = y(t, t_0, y_0(\omega), \omega)$ and $x(t, \omega) = x(t, t_0, x_0(\omega))$ be the sample solution processes of (7.3.27) and (7.3.29) existing for $t \geq t_0$ with $x_0(\omega) = y_0(\omega)$. Further assume that $V_x(t, x, \omega)$ exists and is sample continuous for $(t, x) \in R_+ \times R^n$. Then*

$$V(t, y(t, \omega), \omega) = V(t_0, x(t, \omega), \omega) + \int_{t_0}^t [V_s(s, x(t, s, y(s, \omega)), \omega)$$

$$+ V_x(s, x(t, s, y(s, \omega)), \omega)\Phi(t, s, y(s, \omega))R(s, y(s, \omega), Ty(s, \omega), \omega)]ds$$

$$+ \int_{t_0}^t \int_s^t [V_x(\sigma, x(t, \sigma, y(\sigma, \omega)), \omega)\Phi(t, \sigma, y(\sigma, \omega))$$

$$- V_x(s, x(t, s, y(s, \omega)), \omega)L(t, s, y(s, \omega), \omega)] \cdot \hat{K}(\sigma, s, y(s, \omega))d\sigma ds.$$

where

$R(t, y, Ty, \omega)$

$$= f(t, y, \omega) - \widehat{f}(t, y) + \int_{t_0}^{t} [K(t, s, y(s, \omega), \omega) - \widehat{K}(t, s, y(s, \omega)))]ds. \quad (7.4.7)$$

Proof: From (7.3.27) and (7.3.29), system (7.3.27) can be rewritten as

$$y' = \widehat{f}(t, y) + \int_{t_0}^{t} \widehat{K}(t, s, y(s, \omega))ds + R(t, y, Ty, \omega) \quad (7.4.8)$$

where $R(t, y, Ty, \omega)$ is as defined in (7.4.7).

Let $x(t, s, y(s, \omega))$ and $x(t, \omega) = x(t, t_0, y_0(\omega))$ be the sample solution processes of (7.3.29) through $(s, y(s, \omega))$ and $(t_0, y_0(\omega))$, respectively, and $y(s, \omega) = y(s, t_0, y_0(\omega))$ be the sample solution process of (7.3.27) through $(t_0, y_0(\omega))$. Now we compute the total sample derivative of $V(s, x(t, s, y(s, \omega)), \omega)$ with respect to s as

$$\frac{d}{ds} V(s, x(t, s, y(s, \omega)), \omega) = V_s(s, x(t, s, y(s, \omega)), \omega)$$

$$+ V_x(s, x(t, s, y(s, \omega)), \omega)[\frac{d}{ds} x(t, s, y(s, \omega))]$$

$$= V_s(s, x(t, s, y(s, \omega)), \omega) + V_x(s, x(t, s, y(s, \omega)), \omega)[-\Phi(t, s, y(s, \omega))\widehat{f}(s, y(s, \omega))$$

$$- \int_{s}^{t} L(t, \sigma; s, y(s, \omega))\widehat{K}(\sigma, s, y(s, \omega))d\sigma$$

$$+ \Phi(t, s, y(s, \omega))\{\widehat{f}(s, y(s, \omega)) + \int_{t_0}^{s} \widehat{K}(s, \xi, y(\xi, \omega))d\xi$$

$$+ R(s, y(s, \omega), Ty(s, \omega), \omega))\}]$$

$$= V_s(s, x(t, s, y(s, \omega)), \omega) + V_x(s, x(t, s, y(s, \omega)), \omega)$$

$$[- \int_{s}^{t} L(t, \sigma; s, y(s, \omega))\widehat{K}(\sigma, s, y(s, \omega))d\sigma$$

$$+ \Phi(t, s, y(s, \omega))\{\int_{t_0}^{s} \widehat{K}(s, \xi, y(\xi, \omega))d\xi$$

$$+ R(s, y(s, \omega), Ty(s, \omega), \omega))\}] \text{ w.p.1.} \quad (7.4.9)$$

Here we have used Theorem 7.3.6 to simplify the expression. Integrating in the sample sense, both sides of (7.4.9) from t_0 to t, and noting $x(t, t, y(t, t_0, y_0(\omega), \omega)) = y(t, t_0, y_0(\omega), \omega)$, we obtain,

$$V(t, y(t, \omega), \omega) = V(t_0, x(t, \omega), \omega) + \int_{t_0}^{t} [V_s(s, x(t, s, y(s, \omega)), \omega)$$

$$+ V_x(s, x(t, s, y(s, \omega)), \omega)\Phi(t, s, y(s, \omega))R(s, y(s, \omega), Ty(s, \omega), \omega)]ds$$

$$- \int_{t_0}^{t} V_x(s, x(t, s, y(s, \omega)), \omega) \int_{s}^{t} L(t, \sigma; s, y(s, \omega))\widehat{K}(\sigma, s, y(s, \omega))d\sigma ds$$

$$+ \int_{t_0}^{t} V_x(s, x(t, s, y(s, \omega)), \omega)\Phi(t, s, y(s, \omega)) \int_{t_0}^{s} \widehat{K}(s, \xi, y(\xi, \omega))d\xi ds \quad (7.4.10)$$

Using Fubini's Theorem, the last term in (7.4.10) can be written as

$$\int_{t_0}^{t} \int_{t_0}^{s} V_x(s, x(t, s, y(s, \omega)), \omega)\Phi(t, s, y(s, \omega))\widehat{K}(s, \xi, y(\xi, \omega))d\xi ds$$

$$= \int_{t_0}^{t} \int_{t_0}^{s} V_x(\sigma, x(t, \sigma, y(\sigma, \omega)), \omega)\Phi(t, \sigma, y(\sigma, \omega))\widehat{K}(\sigma, s, y(s, \omega))d\sigma ds \quad (7.4.11)$$

Using (7.4.11), (7.4.10) can be rearranged as (7.4.7) and hence the theorem follows.

7.5 ITO-TYPE SYSTEMS OF EQUATIONS

In several applications we come across a system of stochastic differential equations of Ito-type

$$dx = f(t, x)dt + \sigma(t, x)dz, x(t_0) = x_0$$

where dx represents a stochastic differential of x, z an m-dimensional normalized Wiener process defined on a complete probability space $(\Omega, \mathcal{F}, P)$, $f(t, x)$ is an average vector, and $\sigma(t, x)$ a diffusion rate matrix function. This section aims at obtaining VPF for stochastic differential equations of Ito-type. In order to present complete theory, one needs to develop Ito-Doob calculus and derive well-known Ito's formula. In the following brief discussion, we assume the formation of such equation, its existence, uniqueness properties and continuous dependence of solutions on the initial data. The literature for such properties is available and is quoted for reference in the bibliography given at the end.

Let $\Omega = (\Omega, \mathcal{F}, P)$ be a complete probability space and $R[\Omega, R^n]$ be the system of all R^n-valued random variables. Let $J = [t_0, t_0 + a]$, $t_0 \geq 0$, $a > 0$. Consider the equation

$$dx = f(t, x)dt + \sigma(t, x)dz, x(t_0) = x_0 \quad (7.5.1)$$

where dx is stochastic differential of x, $x, f \in C[J \times R^n, R^n]$, $\sigma \in C[J \times R^n, R^{nm}]$ and $z(t) \in R[\Omega, R^m]$ is a normalized Wiener process.

For establishing the VPF for nonlinear equation (7.5.1) one needs to show that under appropriate conditions, solution $x(t, t_0, x_0)$ of (7.5.1) possesses derivatives with respect to initial conditions t_0 and x_0. To this end, we state the following result. The proof is on similar lines as the one given for Theorem 2.2.1 and hence omitted.

Theorem 7.5.1: *Suppose that*
(i) *the functions $f(t, x)$ and $\sigma(t, x)$ in (7.5.1) satisfy the relations*

$$| f(t, x) |^2 + | \sigma(t, x) |^2 \leq L^2 (1 + | x |^2),$$

$$| f(t, x) - f(t, y) | + | \sigma(t, x) - \sigma(t, y) | \leq L | x - y |$$

for $(t, x), (t, y) \in J \times R^n$ and where L is a positive constant;
(ii) *$x(t_0)$ is independent of $z(t)$ and $E[| x(t_0 |^2] < \infty$;*
(iii) *$f(t, x)$ and $\sigma(t, x)$ are continuously differentiable with respect to x for fixed $t \in J$ and their derivatives $f_x(t, x)$ and $\sigma_x(t, x)$ are continuous in $(t, x) \in J \times R^n$ and are such that*

$$| f(t, x) | + | \sigma_x(t, x) | \leq c, c > 0 \text{ a constant.}$$

Then (a) the mean square derivative $\frac{\partial x(t, t_0, x_0)}{\partial x_{k0}}$ exists for $k = 1, 2, \ldots, n$ and satisfy Ito-type system of linear differential equations

$$dy = f_x(t, x(t))y dt + \sigma_x(t, x(t))y dz \tag{7.5.2}$$

where $y(t_0) = e_k = (e_k^1, e_k^2, \ldots, e_k^k, \ldots, e_k^n)$ is an n-vector such that $e_k^j = 0$, $j \neq k$ and $e_k^k = 1$ and x_{k0} is the kth component of x_0. Further

$$d\phi = f_x(t, x(t))\phi dt + \sigma_x(t, x(t))\phi dz$$

where $\phi = \phi(t, t_0, x_0) = \frac{\partial x(t, t_0, x_0)}{\partial x_0}$
(b) the mean square derivative $\frac{\partial}{\partial t_0} x(t, t_0, x_0)$ exists and satisfies (7.5.2) with

$$\frac{\partial}{\partial t_0} x(t, t_0, x_0) = -\Phi(t, t_0, x_0) f(t_0, x_0).$$

We now prove the main result of this section.
Let

$$x' = f(t, x), x(t_0) = x_0 \tag{7.5.3}$$

where $f \in C[R_+ \times R^n, R^n]$ be the IVP for deterministic system. The stochastic system of Ito-type

$$dy = f(t, y)dt + \sigma(t, y)dz(t), y(t_0) = x_0, \tag{7.5.4}$$

may be treated as a perturbed system related to (7.5.3). We obtain nonlinear VPF for the solutions of (7.5.1).

Theorem 7.5.2: *Suppose that $f(t, x)$ is twice continuously differentiable with respect to x for each $t \in R_+$ and that $y(t) = y(t, t_0, x_0)$ and $x(t) = x(t, t_0, x_0)$ are solutions of (7.5.1) and (7.5.3) respectively existing for $t \geq t_0$. Then*

$$y(t, t_0, x_0) = x(t, t_0, x_0) + \int_{t_0}^{t} \frac{\partial x}{\partial x_0}(t, s, y(s))\sigma(s, y(s))dz(s)$$

$$+ \frac{1}{2} \int_{t_0}^{t} tr\left(\frac{\partial x}{\partial x_0 \partial x_0}(t, s, y(s))b(s, y(s))\right)ds, \ t \geq t_0 \qquad (7.5.5)$$

where $b(t, y) = \sigma(t, y)\sigma^T(t, y)$.

Proof: Consider the solution $x(t, s, y(s))$ through $(s, y(s))$ of (7.5.3). Under the given hypothesis on f, $x(t, t_0, x_0)$ is continuously differentiable with respect to t_0 and twice differentiable with respect to x_0. Apply Ito's formula to $x(t, s, y(s))$ to get

$$dx(t, s, y(s)) = \frac{\partial x}{\partial t_0}(t, s, y(s))ds + \frac{\partial x}{\partial x_0}(t, s, y(s)dy(s)$$

$$+ \frac{1}{2}\left(tr\left(\frac{\partial^2}{\partial x_0 \partial x_0}x(t, s, y(s))\sigma(s, y(s))\sigma^T(s, y(s))\right)\right)ds. \qquad (7.5.6)$$

It is known that

$$\frac{\partial x}{\partial t_0}(t, t_0, x_0) = -\frac{\partial}{\partial x_0}x(t, t_0, x_0)f(t_0, x_0).$$

From (7.5.6), we therefore get

$$dx(t, s, y(s)) = \frac{\partial x}{\partial x_0}(t, s, y(s))\sigma(s, y(s))dz(s)$$

$$+ \frac{1}{2}\left(tr\left(\frac{\partial^2}{\partial x_0 \partial x_0} x(t, s, y(s))\sigma(s, y(s))\right)\right)ds.$$

Integration of this relation between t_0 and t, yields the conclusion (7.5.5).

Corollary 7.5.1: *Let $f(t, x) = Ax$, where A is an $n \times n$ matrix. Then (7.5.5) takes the form*

$$y(t, t_0, x_0) = x(t, t_0, x_0) + \int_{t_0}^{t} e^{A(t-s)}\sigma(s, y(s))dz(s), \ t \geq t_0.$$

Now consider the Ito system

$$dx = \sigma(t, x)dz(t), \ x(t_0) = x_0 \qquad (7.5.7)$$

to be an unperturbed system and (7.5.4) be viewed as a perturbed system. Here $f(t, x)dt$ if the deterministic perturbation.

Theorem 7.5.3: *Assume that $\sigma(t, x)$ is twice continuously differentiable with respect to x for each $t \in R_+$ and that $x(t) = x(t, t_0, x_0)$ and that $x(t) = x(t, t_0, x_0)$ and $y(t) = y(t, t_0, x_0)$ are solutions of (7.5.7) and (7.5.4) respectively, existing for $t \geq t_0$. Then*

$$y(t, t_0, x_0) = x\left[t, t_0, x_0 + \int_{t_0}^{t} \psi^{-1}(s, t_0, u(s))f(s, y(s, t_0, x_0))ds\right], \; t \geq t_0$$

where $\frac{\partial}{\partial x_0}x(t, t_0, x_0) = \psi(t, t_0, x_0)$ and $u(t)$ is a solution of

$$u'(t) = \psi^{-1}(t, t_0, u(t))f(t, x(t, t_0, u(t)))$$

$$u(t_0) = x_0.$$

Proof: Let $x(t, t_0, x_0)$ be the solution of (7.5.7) existing for $t \geq t_0$. The technique of VPF involves finding $u(t)$ such that

$$y(t, t_0, x_0) = x(t, t_0, u(t)), u(t_0) = u_0,$$

is a solution of (7.5.4). Since $x(t, t_0, x_0)$ is twice differentiable with respect to x_0 and that $\psi^{-1}(t, t_0, x_0)$ exists, we obtain by Ito's formula

$$dy = dx + \psi du + d\psi du + \frac{1}{2}tr\left(\frac{\partial^2}{\partial x_0 \partial x_0}x(t, t_0, u(t))(du)^T(du)\right).$$

Let us suppose that $u(t)$ satisfies

$$du = g(t, u)dt + G(t, u)dz(t), \; u(t_0) = x_0$$

and determine g and G. Hence, we get

$$f(t, y)dt = \psi du + \sigma x(t, x)\psi Gdt$$

$$+ \frac{1}{2}tr\left(\frac{\partial^2}{\partial x_0 \partial x_0}x(t, t_0, u)G(t, u)^T G(t, u)\right)dt.$$

As a result, we find that $G \equiv 0$ and

$$g(t, u) = \psi^{-1}(t, t_0, u(t))f(t, y).$$

It now follows that

$$y(t, t_0, x_0) = x\left[t, t_0, x_0 + \int_{t_0}^{t} \Psi^{-1}(s, t_0, u(s))f(s, y(s, t_0, x_0))ds\right] \; t \geq t_0.$$

Remark: Let $\sigma(t, x) = \lambda(t)x$ in (7.5.7). Then

$$x(t, t_0, x_0) = x_0 \exp\left[-\frac{1}{2} \int_{t_0}^{t} \lambda^2(s)ds + \int_{t_0}^{t} \lambda(s)dz(s) \right].$$

Hence

$$\psi(t, t_0) = \exp\left[-\frac{1}{2} \int_{t_0}^{t} \lambda^2(s)ds + \int_{t_0}^{t} \lambda(s)dz(s) \right],$$

thus any solution $y(t)$ of (7.5.4) takes the form

$$y(t) = \psi(t, t_0)\left[x_0 + \int_{t_0}^{t} \exp\left(\frac{1}{2} \int_{t_0}^{s} \lambda^2(\tau)d\tau - \int_{t_0}^{s} \lambda(\tau)dz(\tau) \right) f(s, y(s))ds \right].$$

7.6 BOUNDEDNESS OF SOLUTIONS

The present section is devoted to the study of boundedness of the solution process w.p.1. of (7.4.1). Such a result will help us to understand the need for various assumptions made in the subsequent discussions. Based on these assumptions, we obtain estimates of the difference between the solution of the mean of (7.4.2) and the solution of (7.4.1). Let F in (7.4.1) satisfy the hypothesis (H_2) given in Section 7.4.

Definition 7.6.1: The differential system (7.4.1) is said to be bounded w.p.1., if for every $\alpha > 0$, $t_0 \in J$, there exists a positive function $\beta = \beta(t_0, \alpha)$ which is continuous in t_0 for each α such that the inequality $|y_0(\omega)| < \alpha$, w.p.1. implies $|y(t, \omega)| < \beta$, w.p.1., $t \geq t_0$.

The comparison random differential equation is

$$u'(t, \omega) = g(t, u(t, \omega), \omega), \ u(t_0, \omega) = u_0(\omega),$$

where $g \in M[J \times R^m, R[\Omega, R^m]]$ is such that $g(t, u, \omega)$ satisfies Caratheodory conditions in (t, u) w.p.1. and $g(t, u, \omega)$ is quasimonotonic, nondecreasing in u for fixed t w.p.1. This differential system is said to be bounded w.p.1., if given $\alpha > 0$, $t_0 \in J$, then exists positive function $\beta = \beta(t_0, \epsilon)$ such that $\sum_{i=1}^{m} u_{i0}(\omega) \leq \alpha$, w.p.1. implies that $\sum_{i=1}^{m} u_i(t, \omega) < \beta$, $t \geq t_0$, w.p.1.

We state the following comparison result.

Lemma 7.6.1: *Suppose that*
(a) $g \in M[J \times R^m, R[\Omega, R^m]]$ *and* $g(t, u, \omega)$ *is sample continuous and quasimonotone nondecreasing in u for fixed $t \in J$,*
(b) $V \in C[J \times R^n, R[\Omega, R^m]]$ *satisfies a local Lipschitz condition in y w.p.1. and*

for $(t, y) \in J \times R^n$

$$D_{7.4.1}^+ V(t, y(t, \omega)) \leq g(t, v(t, y(t, \omega))),$$

(c) $$\sum_{i=1}^{n} V_i(t, y(t, \omega)) \geq b(\mid y \mid)$$

where $b \in K$ *on the interval* $0 \leq u < \infty$ *and* $b(u) \to \infty$ *as* $u \to \infty$.
Then if the comparison differential system is bounded, then the system (7.4.1) *is bounded.*

Based on the above lemma and the VPF obtained in Section 7.4, we obtain, under suitable conditions, estimates for the difference between solution of (7.4.1) and the solution of the mean.

In addition to the hypotheses (H_1)-(H_3) given in Section 7.4, we assume the following additional conditions:

(H_4) $$\mid V_x[t, x(t, s, y(s, \omega))] \times \Phi(t, s, y(s, \omega))R(s, y(s, \omega)\omega) \mid$$

$$\leq c(\mid y(s, \omega) - m(s) \mid)g(s, \omega)$$

where V, V_x, Φ, R are defined in Section 7.4, $C \in C[R_+, R_+]$ and nondecreasing on R_+, $g \in M[R_+, R[\Omega, R_+]]$ and is sample Lebesgue integrable;

(H_5) $$b(\mid x \mid) \leq V(t, x) \leq a(\mid x \mid)$$

where $a, b \in C[R_+, R_+]$, a is differentiable on R_+; b^{-1}, inverse of function b, exists and is nondecreasing and continuous on R_+.

The following theorem provides us the estimates of solutions of (7.4.1).

Theorem 7.6.1: *Let the hypotheses* (H_1)-(H_5) *hold. Then*

$$b(\mid y(t, \omega) - m(t) \mid) \leq H^{-1}\left[\int_{t_0}^{t} g(s, \omega)ds + H(N(t, \omega))\right], \quad w.p.1. \quad (7.6.1)$$

where

$$N(t, \omega) = a(\mid x(t, \omega) - m(t) \mid)$$

$$\frac{dH(s)}{ds} = \frac{1}{h(s)}, \quad h(s) = C(b^{-1}(s)).$$

Further, if H^{-1} *is a concave function, then we have*

$$E[b(\mid y(t, \omega)) - m(t) \mid)]$$

$$\leq H^{-1}\left[E\left(\int\limits_{t_0}^{t}g(s,\omega)ds\right) + E(H(N(t,\omega)))\right]. \qquad (7.6.2)$$

Proof: From the Corollary 7.4.2, the hypothesis (H_4) and (H_5), we have

$b(\mid y(t,\omega) - m(t) \mid)$

$$\leq a(\mid x(t,\omega) - m(t) \mid) + \int\limits_{t_0}^{t}C(\mid y(s,\omega) - m(s) \mid)g(s,\omega)ds. \qquad (7.6.3)$$

Set $r(t,\omega) = \int\limits_{t_0}^{t}C(\mid y(s,\omega) - m(s) \mid)g(t,\omega)ds$. Then

$$r'(t,\omega) = C(\mid y(t,\omega) - m(t) \mid)g(t,\omega) \qquad (7.6.4)$$

From (7.6.3), we have

$$b(\mid y(t,\omega) - m(t) \mid) \leq N(t,\omega) + r(t,\omega). \qquad (7.6.5)$$

In view of (7.6.4) and (7.6.5), we then have

$$r'(t,\omega) \leq h(N(t,\omega) + r(t,\omega))g(t,\omega). \qquad (7.6.6)$$

Now set $u(t,\omega) = N(t,\omega) + r(t,\omega)$, $u(t_0,\omega) = N(t_0,\omega)$. Hence it follows that

$$u'(t,\omega) \leq N'(t,\omega) + h(u(t,\omega))g(t,\omega). \qquad (7.6.7)$$

Consider the following related differential equation

$$V'(t,\omega) = N'(t,\omega) + h(V(t,\omega))g(t,\omega), V(t_0,\omega) = u(t_0,\omega). \qquad (7.6.8)$$

Applying Bihari's integral inequality, we get

$$V(t,\omega) \leq H^{-1}\left[\int\limits_{t_0}^{t}g(s,\omega)ds + H(N(t,\omega))\right]. \qquad (7.6.9)$$

Further, in view of (7.6.7), (7.6.8), (7.6.9) and the comparison theorem, we obtain

$$r(t,\omega) \leq -N(t,\omega) + H^{-1}\left[\int\limits_{t_0}^{t}g(s,\omega)ds + H(N(t,\omega))\right]. \qquad (7.6.10)$$

Now (7.6.5) and (7.6.10) together yield

$$b(\mid y(t,\omega) - m(t) \mid) \leq H^{-1}\left[\int_{t_0}^{t} g(s,\omega)ds + H(H(t,\omega))\right]. \qquad (7.6.11)$$

This is the conclusion of (7.6.1).

To prove (7.6.2), let H^{-1} be concave. Taking expectation on both sides of (7.6.11) and using Jenson's inequality, we get

$$E(b(\mid y(t,\omega) - m(t) \mid)) + H^{-1}\left[E\int_{t_0}^{t} g(s,\omega)ds + EH(N(t,\omega)))\right].$$

The proof is complete.

Corollary 7.6.1: *Suppose that the assumptions in Theorem 7.6.1 are true, and* $\mid \Phi(t,s,y(s,\omega)) \mid \ \leq K, \ k > 0.$ *Then*

$$b(\mid y(t,\omega) - m(t) \mid) \leq H^{-1}\left[H(\bar{N})(t_0,\omega) + \int_{t_0}^{t} g(s,\omega)ds\right] \qquad (7.6.12)$$

where

$$\bar{N}(t_0,\omega) = a(K \mid x_0(\omega) - m_0 \mid).$$

Further, if H^{-1} is concave, then

$$E(b(\mid y(t,\omega) - m(t) \mid)) \leq H^{-1}\left[E(H(\bar{N}(t_0,\omega))) + E\left(\int_{t_0}^{t} g(s,\omega)ds\right)\right] (7.6.13)$$

Proof: From Theorem 7.6.1, we have

$$b(\mid y(t,\omega) - m(t) \mid) \leq a(\mid x(t,\omega) - m(t) \mid)$$

$$+ \int_{t_0}^{t} C(\mid y(s,\omega) - m(s))g(s,\omega)ds.$$

By using the result of Theorem 7.3.5, we get

$$x(t,t_0,x(\omega)) - x(t,t_0,m(t_0))$$

$$= \int_{0}^{1} \Phi(t,t_0,m_0 + s(x_0(\omega) - m_0))ds \cdot (x_0(\omega) - m_0).$$

In view of the hypothesis of this corollary, we get

$$| x(t, t_0, x(\omega)) - x(t, t_0, m(t_0)) | \leq K | x_0(\omega) - m_0 | . \qquad (7.6.14)$$

The inequality (7.6.14) together with the conclusion of Theorem 7.6.1 yields (7.6.12). The conclusion (7.6.13) also follows similarly from Theorem 7.6.1.

In order to illustrate the nature of assumption (H_4), we choose now

$$V(t, x) = | x |^2, \quad | \Phi(t, s, y(s, \omega)) | \leq K. \qquad (7.6.15)$$

Note that (H_4) is feasible if the solution processes of (7.4.1) are bounded w.p.1. From (7.6.15), observe that

$$| V_x(t, x) | \leq 2 | x | . \qquad (7.6.16)$$

From the boundedness assumption, the relation (7.6.15) and (7.6.16), we get g satisfying

$$| V_x(t, x(t, s, y(s, \omega)) - x(t, s, m(s))) | \Phi(t, s, y(s, \omega)) R(s, y(s, \omega), \omega) |$$

$$\leq 2 | x(t, s, y(s, \omega)) - x(t, s, m(s)) | g(s, \omega) \qquad (7.6.17)$$

where $g(s, \omega) \in M[R_+, R[\Omega, R]]$. The application of Corollary 7.6.1 yields

$$b(| y(t, \omega) - m(t) |) \leq H^{-1} \left[K | x_0(\omega) - m_0 | + \int_{t_0}^{t} g(s, \omega) ds \right]$$

$$= \left[K(| x_0(\omega) - m_0 |) + \int_{t_0}^{t} g(s, \omega) ds \right]^2 \qquad (7.6.18)$$

since $H(s) = s^{1/2}$. Note that $b(u) = a(u) = u^2$. Hence (7.6.18) takes the for

$$| y(t, \omega) - m(t) |^2 \leq \left[K | x_0(\omega) - m_0 | + \int_{t_0}^{t} g(s, \omega) ds \right]^2 . \qquad (7.6.19)$$

Taking the expectations on both sides of (7.6.19), on obtains

$$E | y(t, \omega) - m(t) |^2 \leq E \left[K(| x_0(\omega) - m_0 |) + \int_{t_0}^{t} g(s, \omega) ds \right]^2$$

$$= E \left[K^2 | x_0(\omega) - m_0 |^2 + 2K | x_0(\omega) - m_0 | \int_{t_0}^{t} g(s, \omega) ds \right.$$

$$\left. + \left(\int_{t_0}^{t} g(s, \omega) ds \right)^2 \right] . \qquad (7.6.20)$$

Remarks 7.6.1: (*i*) Taking square root and then expectation on both sides of (7.6.19), we get

$$E \mid y(t,\omega) - m(t) \mid \ \leq E[K \mid x_0(\omega) - m_0 \mid] + E\left[\int_{t_0}^{t} g(s,\omega)ds\right].$$

(*ii*) When $\mid x_0(\omega) - m_0 \mid$ and $g(s,\omega)$ are independent random processes, from (7.6.20) we get

$$E \mid y(t,\omega) - m(t) \mid^2 \ \leq K^2 E(\mid x_0(\omega) - m_0 \mid^2 + 2KE(\mid x_0(\omega) - m_0 \mid)$$

$$+ E\left(\int_{t_0}^{t} g(s,\omega)ds\right)$$

$$+ E\left[\int_{t_0}^{t} \int_{t_0}^{t} (g(s,\omega)g(u,\omega))dsdu\right]. \tag{7.6.21}$$

(*iii*) If $g(s,\omega)$ is any stationary Gaussian process and $Eg(s,\omega)$ is a constant, say equal to zero, then $E(g(s,\omega)g(u,\omega))$ depends on $s-u$ and then (7.6.21) takes the form

$$E \mid y(t,\omega) - m(t) \mid^2 \ \leq K^2 E(\mid x_0 - m_0 \mid)^2 + \int_{t_0}^{t} \int_{t_0}^{t} C(u-s)duds,$$

where $E(g(s,\omega)g(u,\omega)) = C(u-s)$.

Note that in the above considerations, the condition (H_4) contained on the right-side of the product of the two random functions. Let us consider that the right-hand side of (H_4) consists of a product of a deterministic function and a random function.

Corollary 7.6.2: *Suppose that the hypotheses of Theorem 7.6.1 hold and in place of (H_4) assume that*

(H_6) $\mid V_x(t, x(t, s, y(s,\omega))) - x(t, s, m(s)))\Phi(t, s, y(s,\omega))) \cdot R(s, y(s,\omega), \omega) \mid \ \leq$ $w(s)g(s,\omega)$ *where* $w(s) \in C[R_+, R_+]$. *Then*

$$E(b \mid y(t,\omega) - m(t) \mid) \leq E(a(K \mid x_0(\omega) - m_0 \mid))$$

$$+ \left[\int_{t_0}^{t} w^2(s)ds\right]^{1/2} E\left[\int_{t_0}^{t} g^2(s,\omega)ds\right]^{1/2}.$$

The proof of this corollary is based on Theorem 7.6.1, Corollary 7.6.2 and Hölder's inequality.

The flexibility of the hypothesis (H_6) can be seen from the following example.

Let $V(t, x) = \|x\|$ and $\mid \Phi(t, s, y(s,\omega)) \mid \ \leq K$. Clearly $\mid V_x(t, x) \mid \ \leq 1$. Further

$$V_x(t, x(t, s, y(s, \omega))) - x(t, s, m(s))\Big)\Phi(t, s, y(s, \omega))R(s, y(s, \omega), \omega)$$

$$\leq g(s, \omega)$$

where $g \in M[R_+, R[\Omega, R]]$. Choose $b(u) = a(u) = u$. Then the application of the corollary yields

$$|\, y(t, \omega) - m(t)\,| \leq K\,|\, x_0(\omega) - m_0\,| + \int_{t_0}^{t} g(s, \omega).$$

Since $H(s) = s$, taking expectations on both sides, we obtain

$$E(y(t, \omega) - m(t)) \leq KE(\,|\, x_0(\omega) - m_0\,|\,) + E\left(\int_{t_0}^{t} g(s, \omega)ds\right).$$

We present below a simple illustrative example.
Let

$$y'(t, \omega) = -a(t, \omega)y^3(t, \omega), y(t_0, \omega) = y_0(\omega), \qquad (7.6.22)$$

where $a(t, \omega) \in M[R, R[\Omega, R_+]]$ and satisfies appropriate conditions for the existence of solution process for $t \geq t_0$. Let

$$y'(t, \omega) = -E(a(\cdot, \omega))y^3(t, \omega) + [E(a(t, \omega))) - a(t, \omega)]y^3(t, \omega)$$

$$y(t_0, \omega) = y_0(\omega) \qquad (7.6.23)$$

$$m'(t) = -E(a(t, \omega))m^3(t), m(t_0) = E(x_0(\omega)) \qquad (7.6.24)$$

$$x' = -E(a(t, \omega))x^3, x(t_0, \omega) = x_0(\omega). \qquad (7.6.25)$$

Assume that the solution of (7.6.22) is bounded w.p.1. That implies $y^2(t, \omega) \leq \alpha^2$ w.p.1., $\alpha \neq 0$. From (7.6.25) we get

$$x(t, \omega) = \frac{x_0(\omega)}{[1 + 2x_0^2(\omega)\int_{t_0}^{t} Ea(s, \omega)ds]^{1/2}}. \qquad (7.6.26)$$

From (7.6.26), we obtain

$$\Phi(t, t_0, x_0(\omega)) = \frac{1}{[1 + 2x_0^2(\omega)\int_{t_0}^{t} a(s, \omega)ds]^{3/2}}. \qquad (7.6.27)$$

From (7.6.23) and (7.6.27), the inequality

$$|\, \Phi(t, s, y(s, \omega))R(s, y(s, \omega), \omega)\,|$$

$$\leq |\, a(s, \omega) - Ea(s, \omega)\,| \setminus [\alpha^2 + 2\int_{s}^{t} Ea(u, \omega)du]^{3/2}. \qquad (7.6.28)$$

holds because $y^2 \leq \alpha^{-2}$, w.p.1. Further

$$| \Phi(t, s, y(s, \omega)) R(s, y(s, \omega), \omega) | \leq K_1 | a(s, \omega) - Ea(s, \omega) | \quad (7.6.29)$$

where $K_1 > 0$ is a constant.

Let

$$V(t, x) = x, b(z) = a(z) = z. \quad (7.6.30)$$

Here (7.6.30) and (7.6.28) satisfy the assumption (H_6). Further
$| \Phi(t, s, y(x, \omega)) | \leq 1$ and $H(s) = s$. From the relations (7.6.28), (7.6.30), we get

$$E | y(t, \omega) - m(t) |$$

$$\leq E | x_0(\omega) - m_0 | + E \left\{ \int_{t_0}^{t} \frac{|a(s, \omega) - Ea(s, \omega)| ds}{[\alpha^2 + 2 \int_{s}^{t} Ea(u, \omega) du]^{3/2}} \right\}. \quad (7.6.31)$$

In case we employ (7.6.29), we get the inequality

$$E | y(t, \omega) - m(t) | \leq E | x_0(\omega) - m_0 | \leq E | x_0(\omega) - m_0 |$$

$$+ E \left\{ \int_{t_0}^{t} | a(s, \omega) - Ea(s, \omega) | ds \right\}.$$

One may choose $V(t, x) = x^2$, $V(t, x) = x^p$ or consider that $| x_0(\omega) - m_0 |$ and $| a(s, \omega) - Ea(s, \omega) |$ are independent processes. Thus one can compute a variety of error estimates of solutions and mean of solutions of given differential equations.

7.7 NOTES

For preliminaries and notation for stochastic differential equations, see Doob [1], Ito [1], Ladde and Lakshmikantham [1], Bharucha Reid [1], Tsokos and Padgett [1].
     The result on linear VPF in Section 7.2 taken from Ladde and Lakshmikantham [2]. Theorem 7.2.2 for stochastic integro-differential equation is due to Ladde and Sathananthan [2]. The result on differentiability with respect to initial data is taken from Ladde and Lakshmikantham [1]. The result in Theorem 7.3.6, 7.3.7 is the work of Ladde and Sathananthan [2]. The contents of Section 7.4 are adopted from Ladde and Sambandham [1] and Ladde and Sathananthan [ ]. The VPF for Ito-type system of equations is due to Kulkarni and Ladde [1]. The nonlinear VPF is taken from Ladde and Lakshmikantham [1]. The results have been extended to Ito-type of integro-differential equations by Ladde and Sathananthan [1, 3] and Ladde, Sambandham, Sathananthan [4]. The application for boundedness of solutions is adopted from Ladde and Sambandham [1]. The extension of VPF for difference equations has been proved by Ladde and Sambandham [2, 3]. Also refer to Fai and Chaghey [1, 2].

8. Miscellaneous Equations

8.0 INTRODUCTION

In the earlier chapters, we considered specific areas of differential equations to establish the linear and nonlinear variation of parameters method and study some significant applications of this technique to discover qualitative properties of solutions. There are several other areas of dynamic systems in which VPFs are proved and profitably employed. For example, complex differential equations, partial differential equations, singular differential equations, operator equations, renewal equations, dynamic systems on time-scale and several others.

The aim of the present chapter is to provide the VPF in few of these miscellaneous topics and present applications.

In Section 8.1, we consider differential equations with piecewise constant delays. The equations of this type are of relatively recent origin and represent models in disease dynamics. Essentially, so far, linear equations are explored while nonlinear equations are not yet studied significantly. Here we establish VPF for linear differential equations and give a suitable application of this technique to stability problems.

In Section 8.2, we consider complex differential equations. Here as before, we present linear variation of parameters and obtain solution of perturbed linear complex differential equations. In order to consider nonlinear variation, we need to show that solutions of nonlinear differential equations are differentiable with respect to initial conditions. After proving this result we obtain nonlinear VPF for complex differential equations.

In the next section (Section 8.3), we consider hyperbolic partial differential equations and obtain VPF for a particular type of linear equations. This formula is then used to establish existence of solution of nonlinear equations of hyperbolic type. The monotone iterative method is employed to obtain the existence theory. This theory has natural extension to higher order hyperbolic partial differential equations.

Section 8.4 is devoted to the study of operator equations. Here we prove a VPF for linear case by using functional analytic methods. It is shown that the VPF for linear Volterra integral equation is a particular case of this more general setting in abstract form. Finally, in Section 8.5, we consider dynamic systems on time scales which include at the same time continuous and discrete dynamics. This area is of very recent origin and has much promise. We discuss linar and nonlinear VPFs.

Section 8.6 containing notes include reference to extensive literature in these areas of dynamics systems.

8.1 EQUATIONS WITH PIECEWISE CONSTANT DELAY

We consider, in the present section, differential equations involving piecewise constant delays. These equations are of the type

$$x'(t) = f(t, x(t), x(h(t))$$

where the lagging argument is of the form $h(t) = [t], [t - h], [t - 1]$ etc; $[t]$ denoting the largest integral in $t \in R$. Depending upon the nature of lagging argument one needs to place appropriate initial data to find solutions of an initial value problem (IVP).

Consider the scalar IVP

$$x'(t) = ax(t) + a_0 x([t]) + a_1 x([t - 1]),$$

$$x(-1) = c_{-1}, x(0) = c_0 \tag{8.1.1}$$

with constant coefficients, $[t]$ designates the greatest integer function. This equation is closely related to impulse and loaded equation. One may write (8.1.1) as

$$x'(t) = ax(t) + \sum_{i=-\infty}^{\infty} [a_0 x(i) + a_1 x(i - 1)](H(t - i) - H(t - i - 1)), \tag{8.1.2}$$

where $H(t) = 1$ for $t > 0$ and $H(t) = 0$ for $t < 0$. If distributional derivates are admitted, then differentiating (8.1.2) yields

$$x''(t) = ax'(t) + \sum_{i=-\infty}^{\infty} [a_0 x(i) + a_1 x(i - 1)](\delta(t - i) - \delta(t - i - 1)),$$

where δ denotes the delta functional. This impulse equation contains the values of the unknown solution for the integral values of t.

Denote $b_0 = e^a + a^{-1}a_0(e^a - 1), b_1 = a^{-1}a_1(e^a - 1)$ and let λ_1 and λ_2 be the roots of the equation

$$\lambda^2 - b_0\lambda - b_1 = 0.$$

We have the following result.

Theorem 8.1.1: *The problem* (8.1.1) *has a unique solution*

$$x(t) = c_{[t]}e^{a(t-[t])} + a^{-1}(a_0 c_{[t]} + a_1 c_{[t]-1})(e^{a(t-[t])} - 10 \tag{8.1.3}$$

where

$$c_{[t]} = [\lambda^{[t]+1}(c_0 - \lambda_2 c_{-1}) + (\lambda_1 c_{-1} - c_0)\lambda_2^{[t]+1}]/\lambda_1 - \lambda_2.$$

Proof: Assume that $x_n(t)$ is a solution of (8.1.1) on the interval $n \le t < n + 1$, with the conditions $x(n) = c_n, x(n - 1) = c_{n-1}$. Then we have

$$x'_n(t) = ax(t) + a_0 c_n + a_1 c_{n-1}$$

having general solution on the given interval as

$$x_n(t) = e^{a(t-n)}c - a^{-1}(a_0 c_n + a_1 c_{n-1}),$$

c being an arbitrary constant. For $t = n$, we get

$$c_n = c - a^{-1}(a_0 c_n + a_1 c_{n-1}).$$

Hence $c = (1 + a^{-1}a_0)c_n + a^{-1}a_1 c_{n-1}$ and

$$x_n(t) = (a^{-1}a_1 c_{n-1} + (1 + a^{-1}a_0)c_n)e^{a(t-n)} - a^{-1}(a_0 c_n + a_1 c_{n-1}).$$

We repeat the above argument for $n - 1 \le t < n$ and note that $x_{n-1}(n) = x_n(n)$ for $n = 1, 2, \ldots$ so that finally we get (8.1.3).

Many properties like boundedness, stability, asymptotic behavior can be inferred from (8.1.3) for appropriate choices of the constants a, a_0 and a_1. Here we aim at obtaining VPF for a system of equations involving piecewise constant delays. To begin with we obtain representation of a solution of a linear system.

Let $C(J)$ denote the space of continuous functions mapping $J = [0, \infty)$ into R^n. For an $n \times n$ matrix $M = (m_{ij})$, define the norm $|M| = \max_J \sum_{i=1}^{n} |m_{ij}|$ and let I denote the $n \times n$ identity matrix.

We consider the following systems of equations.

$$x'(t) = A(t)x(t), \tag{8.1.4}$$

$$y'(t) = A(t)y(t) + B(t)y([t]), \tag{8.1.5}$$

$$z'(t) = A(t)z(t) + B(t)z([t]) + c(t), \tag{8.1.6}$$

for $t \ge 0$, with initial conditions

$$x(0) = y(0) = z(0) = c_0 \tag{8.1.7}$$

and with the following property:

(H) A, B are $n \times n$ matrices with entries as real-valued continuous functions of $t \in J$, c is an n-column vector with entries as real-valued continuous functions for $t \in J$, x, y, z are n-vectors, and c_0 is a real constant n-column vector. Let ϕ be the fundamental matrix (FM) of (8.1.4) such that $\phi(0) = E$.

The method of iteration is one of the powerful tools used to obtain solutions of differential equations. We employ this method below to obtain the solution of the initial value problem (IVP) (8.1.5), (8.1.7).

Theorem 8.1.2: *Let the property (H) hold. Then there exists a unique solution to the IVP (8.1.5), (8.1.7) for $t \in J$ and it is given by*

$$y(t) = \lim_k \left\{ \phi(t, 0) + \int_0^t \phi(t, t_1)B(t_1)\phi([t_1], 0)dt_1 \right.$$

$$+ \int_0^t \int_0^{[t_1]} \phi(t, t_1) B(t_1) \phi([t_1], t_2) B(t_2) \phi([t_2], 0) dt_2 dt_1 + \ldots$$

$$+ \int_0^t \int_0^{[t_1]} \cdots \int_0^{[t_{k-1}]} \phi(t, t_1) B(t_1) \phi([t_1], t_2) B(t_2) \ldots$$

$$\times B(t_k) \phi([t_k], 0) dt_k \ldots dt_2 dt_1 \Big\} c_0. \qquad (8.1.8)$$

Proof: Let Δ be any compact interval in J, such that $0 \in \Delta$. The space $C(\Delta)$ of continuous functions from Δ into R^n, with supnorm is complete. Consider the space $C_\lambda(\Delta)$, $\lambda \geq 0$, of continuous functions from Δ into R^n with norm

$$|y|_\lambda = \sup_{t \in \Delta} \left\{ |y(t)| \exp\left(-\int_0^t |\phi(t, s) B(s)| ds \right) \right\}.$$

Clearly, $C_0(\Delta) = C(\Delta)$. It is seen that the norms $|y|_\lambda$ are all equivalent for $\lambda \geq 0$ so that $C_\lambda(\Delta)$ is also a complete space. Define $T: C_\lambda(\Delta) \to C_\lambda(\Delta)$ as

$$(Ty)(t) = \phi(t, 0) c_0 + \int_0^t \phi(t, s) B(s) y([s]) ds.$$

It can be shown that

$$|Ty_1 - Ty_2|_\lambda \leq \tfrac{1}{\lambda} |y_1 - y_2|_\lambda.$$

That is, for $\lambda > 1$, T is a contraction map on $C_\lambda(\Delta)$. Hence, by the Banach fixed point theorem there exists a unique y in $C(\Delta)$ such that $(Ty)(t) = y(t)$.
    Defining

$$y_0(t) = c_0$$

$$y_k(t) = \phi(t, 0) c_0 + \int_0^t \phi(t, s) B(s) y_{k-1}([s]) ds, \quad k = 1, 2, \ldots$$

and by using the method of successive approximation, the result (8.1.8) follows. The proof is complete.
    The closed form solution of the IVP (8.1.5), (8.1.7) is given by

$$y(t) = \left(\phi(t, [t]) + \int_{[t]}^t \phi(t, s) B(s) ds \right) c_{[t]}, \quad t \geq 0,$$

where

$$c_{[t]} = \prod_{k=[t]}^1 \left(\phi(k, k-1) + \int_{k-1}^k \phi(k, s) B(s) ds \right) c_0. \qquad (8.1.9)$$

In order to study the perturbation effects on (8.1.5), we now treat (8.1.4) as our basic equation. The following definition is now in order.

Definition 8.1.1: The function

$$\psi(t) = \left\{ \phi(t,[t]) + \int\limits_{[t]}^{t} \phi(t,s)B(s)ds \right\}$$

$$\times \prod\limits_{k=[t]}^{1} \left\{ \phi(k,k-1) + \int\limits_{k-1}^{k} \phi(k,s)B(s)ds \right\}, \quad t \in J \qquad (8.1.10)$$

satisfying the matrix IVP, $Y'(t) = A(t)Y(t) + B(t)Y([t]), Y(0) = E$ is called the FM for (8.1.5).

We use the notation $\psi(t,k) = \psi(t)\psi^{-1}(k), k = 0, 1, \ldots, [t], t \in J$. Observe from (8.1.9) and (8.1.10) that the solution $y(t)$ of (8.1.5) is given by $y(t) = \psi(t)c_0, t \in J$.

Let $x(t)$, $y(t)$, $z(t)$, $t \in J$, be the solutions of (8.1.4), (8.1.5), (8.1.6), respectively, satisfying (8.1.7). It is natural to expect that x, y, and z are related to each other. This relationship is established below through the method of VPF.

Theorem 8.1.3: *Let $y(t)$ be the solution of (8.1.5), (8.1.7). Let ϕ and ψ be the FMs of (8.1.4) and (8.1.5), respectively. Then the unique solution of (8.1.6), (8.1.7) for $t \in J$, is given by*

$$z(t) = y(t) + \sum\limits_{k=1}^{[t]} \int\limits_{k-1}^{k} \psi(t,k)\phi(k,s)c(s)ds + \int\limits_{[t]}^{t} \phi(t,s)c(s)ds. \quad (8.1.11)$$

Proof: Let $\tilde{z}$ (t represent the integral terms on the right-hand side of (8.1.11). It is enough to prove that $\tilde{z}$ (t) is a solution of (8.1.6). Differentiate $\tilde{z}$ (t) and use (8.1.4) and (8.1.5) to get

$$\tilde{z}\,'(t) = \sum\limits_{k=1}^{[t]} \int\limits_{k-1}^{k} \{A(t)\psi(t,k) + B(t)\psi([t],k)\}\phi(k,s)c(s)ds$$

$$+ c(t) + \int\limits_{[t]}^{t} A(t)\phi(t,s)c(s)ds$$

$$= A(t)\tilde{z}\,(t) + B(t)\tilde{z}\,([t]) + c(t).$$

The proof is complete.

We can arrive at the same result (8.1.11) by taking (8.1.4) as the basic equation and using the usual variation of parameters formula to get the solution of (8.1.6). This has been achieved in the next theorem.

Theorem 8.1.4: *Let ϕ and ψ be the FMs of (8.1.4) and (8.1.5), respectively, then*

$$z(n) = y(n) + \sum_{k=1}^{n} \int_{k-1}^{k} \psi(n, k)\phi(k, s)c(s)ds, \qquad (8.1.2)$$

where $n \geq 1$ is an integer

$$\psi(n, k) = \phi(n, k) + \sum_{r=k+1}^{n} \int_{r-1}^{r} \psi(r - 1, k)\phi(n, s)B(s)ds,$$

$$for \ n > k \qquad (8.1.13)$$

and

$$\psi(n, k) = I \ for \ n = k, n = 1, 2, \ldots, [t].$$

Proof: Let for $t \geq 0$, $y(t)$ and $z(t)$ be solutions of (8.1.5) and (8.1.6), respectively. By the VPF we have, for $t \in J$,

$$y(t) = c_0 + \int_0^t \phi(t, s)B(s)y([s])ds, \qquad (8.1.14)$$

$$z(t) = c_0 + \int_0^t \phi(t, s)B(s)z([s])ds + \int_0^t \phi(t, s)c(s)ds. \qquad (8.1.15)$$

Clearly, (8.1.12) holds for $n = 1$, since $\psi(1, 1) = I$. Assume that the result (8.1.12) is true for $n = 1, 2, \ldots, m$. For $t = m + 1$, we have from (8.1.15)

$$z(m + 1) = c_0 + \sum_{k=1}^{m+1} \int_{k-1}^{k} \phi(m + 1, s)\{B(s)z(k - 1) + c(s)\}ds.$$

In similar form we can write $y(m + 1)$ from (8.1.14) in terms of $y(0)$, $y(1), \ldots, y(m)$. Since (8.1.12) holds for $n = 1, 2, \ldots, m$, we get the value of $z(m + 1)$ in terms of $y(0), y(1), \ldots, y(m)$. From the expression of $z(m + 1)$, the sum of the terms containing $y(0), y(1), \ldots, y(m)$ can be replaced by $y(m + 1)$, then we get

$$z(m + 1) = y(m + 1) + \sum_{k=1}^{m+1} \int_{k-1}^{k} \phi(m + 1, s)$$

$$\times B(s) \left\{ \sum_{p=1}^{k-1} \int_{p-1}^{p} \psi(k - 1, p)\phi(p, r)c(r)dr \right\} ds$$

$$+ \sum_{k=1}^{m+1} \int_{k-1}^{k} \phi(m + 1, k)\phi(k, s)c(s)ds.$$

By changing the order of integration in the second term on the right side, we obtain

$$z(m+1) = y(m+1) + \sum_{k=1}^{m+1} \int_{k-1}^{k} \left\{ \phi(m+1,k) \right.$$

$$\left. + \sum_{p=k+1}^{m+1} \int_{p-1}^{p} \psi(p-1,k)\phi(m+1,r)B(r)dr \right\} \tag{8.1.16}$$

$$\times \phi(k,s)c(s)ds.$$

Now use (8.1.13) in (8.1.16) to see the result (8.1.12) is true for $n = m+1$. Hence the theorem.

Example 8.1.1: We verify the relation (8.1.11) for the one dimensional case. Assume that $y_n(t)$ and $z_n(t)$ are the solutions of (8.1.12) and (8.1.13), respectively, in the interval $[n, n+1)$, with scalar functions $A(t) \equiv a(t)$, $B(t) = b(t)$, for $t \in J$. Clearly $\phi(t) = \exp(\int_0^t a(s)ds)$ and $\psi(t)$ is given by (8.1.10). Let $z_n(n) = d_n$ for $n = 0, 1, 2, \ldots$. It is easy to verify that

$$z_n(t) = d_n \exp\left(\int_n^t a(r)dr \right) + \int_n^t \exp\left(\int_s^t a(r)dr \right) \{r(s)d_n + c(s)\}ds \tag{8.1.17}$$

We can similarly write $z_{n-1}(t)$. Since $\lim_{t \to n} z_{n-1}(t) = z_n(n) = d_n$, we obtain for $n = 1, 2, 3, \ldots,$

$$d_n = d_{n-1}\left\{ \exp\left(\int_{n-1}^n a(r)dr \right) + \int_{n-1}^n \exp\left(\int_s^n a(r)dr \right) b(s)ds \right\}$$

$$+ \int_{n-1}^n \exp\left(\int_s^n a(r)dr \right) c(s)ds. \tag{8.1.18}$$

Use the recurrence relation (8.1.18) to obtain d_n in terms of d_0. Substitute d_n in (8.1.17) and use the fact that $d_0 = c_0$ to obtain

$$z_n(t) = y_n(t) + \sum_{k=1}^n \int_{k-1}^k \psi(t,k)\phi(k,s)c(s)ds + \int_n^t \phi(t,s)c(s)ds. \tag{8.1.19}$$

If we take $n = [t]$, then (8.1.19) is true for any t and hence, we write $z_n(t) = z(t)$, $y_n(t) = y(t)$, $t \in J$. Now (8.1.19) is in the form of (8.1.11).

Consider the systems

$$y'(t) = Ay(t) + By([t]) \tag{8.1.20}$$

$$z'(t) = Az(t) + Bz([t]) + f(t, z(t)), \tag{8.1.21}$$

where A, B are $n \times n$ constant matrices and A is nonsingular and $f \in C[J \times R^n, R^n]$. Further, assume that f satisfies the condition

$$| f(t, z) | \leq \alpha(t) | z |, \ t \in J \tag{8.1.22}$$

where $\alpha: J \to R^+$ is continuous and is such that

$$\int_0^\infty \alpha(s)ds < \infty. \tag{8.1.23}$$

We now prove the following theorem.

Theorem 8.1.5: *Assume the FMs ϕ and ψ of (8.1.20) with $B = 0$ and (8.1.20) respectively, satisfy the conditions*

$$| \psi(t) | \leq N, \ | \phi(t, s) | \leq M_0,) \leq s \leq t < \infty$$

$$| \psi(t, k)\phi(k, s) | \leq M_k \text{ for } k = 1, 2, \dots, [t], \tag{8.1.24}$$

$$0 \leq s \leq k < \infty.$$

Take $M = \max \{M_0, M_1, \dots, M_{[t]}\}$, where N and M_k, $k = 0, 1, \dots, [t]$, are constants. Let f satisfy (8.1.22) and (8.1.23). Then all the solutions $z(t)$ of (8.1.21) exist for $t \geq 0$ and there exists a constant $K > 0$ such that

$$| z(t) | \leq K | c_0 |, \ t \in J. \tag{8.1.25}$$

Further, if $y(t)$ is the solution of (8.1.20) with $y(0) = c_0$ and $y(t)$ satisfies $\lim_{t \to \infty} y(t) = 0$, then $\lim_{t \to \infty} z(t) = 0$.

Proof: Using the VPF, we have

$$z(t) = y(t) + \sum_{k=1}^{[t]} \int_{k-1}^k \psi(t, k)\phi(k, s)f(s, z(s))ds$$

$$+ \int_{[t]}^t \phi(t, s)f(s, z(s))ds, \ t \in J.$$

The conditions (8.1.22) and (8.1.24) yield for $t \in J$

$$| z(t) | \leq N | c_0 | + \int_0^t \alpha(s) | z(s) | ds.$$

Application of Gronwall's integral inequality and the condition (8.1.23) leads us to the conclusion (8.1.25). The condition $\lim_{t \to \infty} y(t) = 0$ implies that, given any $\epsilon > 0$, there exists a $T(\epsilon) > 0$ such that $| y(t) | < \epsilon$ for all $t \geq T(\epsilon)$. Proceeding as before, for $t \geq T(\epsilon)$, we get

$$| z(t) | \leq \epsilon + M \int_0^t \alpha(s) | z(s) | ds.$$

Hence $| z(t) | \leq K_1 \epsilon$ for some constant K_1 where $K_1 \leq \exp\{M \int_0^\infty \alpha(s)ds\}$ which is independent of ϵ and T. This implies $\lim_{t \to \infty} z(t) = 0$.

Theorem 8.1.6: *Let $c_0 \geq 0$ be a constant and $u, a, b \in C[J, R^+]$. If the inequality*

$$u(t) \leq c_0 + \int_0^t (a(s)u(s) + b(S)u([s]))ds, \ t \in J \qquad (8.1.26)$$

holds, then for $t \in J$, we have

$$u(t) \leq c_0 \prod_{k=1}^{[t]} \left\{ \exp\left(\int_{k-1}^k a(r)dr \right) + \int_{k-1}^k \exp\left(\int_s^k a(r)dr \right) b(s)ds \right\}$$

$$\times \left\{ \exp\left(\int_{[t]}^t a(r)dr \right) + \int_{[t]}^t \exp\left(\int_s^t a(r)dr \right) b(s)ds \right\}. \qquad (8.1.27)$$

Proof: In the interval $n \leq t < n+1$

$$u(t) \leq u(n) \left(1 + \int_n^t b(s)ds \right) + \int_n^t a(s)u(s)ds. \qquad (8.1.28)$$

Then it is known that

$$u(t) \leq u(n) \left\{ \exp\left(\int_n^t a(r)dr \right) + \int_s^t \exp\left(\int_s^t a(r)dr \right) b(s)ds \right\}, \ n = 0, 1, \ldots.$$

Applying this inequality successively for $u(n)$, $u(n-1), \ldots, u(1)$, we get the desired conclusion (8.1.27).

Remark 8.1.1: Observe that the right-hand side of the inequality (8.1.27) is in fact a solution of the related delay differential equation

$$y'(t) = a(t)y(t) + b(t)y([t]), \ y(0) = c_0.$$

In this sense, (8.1.27) is the best estimate. When $b = 0$ in (8.1.26), (8.1.27) reduces to $u(t) \leq c_0 \exp(at), \ t \geq 0$.

Consider the perturbed system

$$z'(t) = Az(t) + Bz([t]) + f(t, z(t), z([t])), \qquad (8.1.29)$$

where A, B are $n \times n$ constant matrices, A is nonsingular, and $f \in C[J \times R^n \times R^n, R^n]$. Assume

$$| f(t, z(t), z([t])) | \leq \alpha(t) | z(t) | + \beta(t) | z([t]) | , \qquad (8.1.30)$$

where α, β are nonnegative continuous functions on J both satisfy (8.1.23).

Theorem 8.1.7: *Assume the FMs ϕ and ψ of (8.1.20) with $B = 0$ and (8.1.20) respectively, satisfy (8.1.24). Let the conditions (8.1.30) and (8.1.23) hold. If $y(t)$ is the solution of (8.1.20) with $y(0) = c_0$ and $y(t)$ satisfies $\lim_{t \to \infty} y(t) = 0$, then $\lim_{t \to \infty} z(t) = 0$.*

Proof: Proceeding as in the proof of Theorem 8.1.5, and using the conditions (8.1.30) and (8.1.23), we get

$$| z(t) | \leq \epsilon + M \int_0^t (\alpha(s) | z(s) | + \beta(s) | z([s]) |) ds.$$

Now use inequality (8.1.27) to get $| z(t) | \leq \epsilon K_1$ where

$$K_1 = \prod_{k=1}^{[t]} \left\{ \exp\left(M \int_{k-1}^k \alpha(r) dr \right) + \int_{k-1}^k M \exp\left(M \int_s^k \alpha(r) dr \right) \beta(s) ds \right\}$$

$$\times \left\{ \exp\left(M \int_{[t]}^t \alpha(r) dr \right) + \int_{[t]}^t M \exp\left(M \int_s^t \alpha(r) dr \right) \beta(s) ds \right\}.$$

Clearly it can be seen that

$$K_1 \leq \exp\left(M \int_0^\infty \alpha(r) dr \right) \prod_{k=1}^\infty \left(1 + \int_{k-1}^k M \beta(s) ds \right).$$

In view of (8.1.23) we can see that K_1 is finite and hence, $\lim_{t \to \infty} z(t) = 0$. The proof is complete.

To extend the foregoing arguments further, we consider the differential equations with piecewise constants delays of the form

$$x'(t) = a(t)x(t) \qquad (8.1.31)$$

$$y'(t) = a(t)y(t) + c(t)y\left(2\left[\frac{t+1}{2}\right]\right) \qquad (8.1.32)$$

and

$$z'(t) = a(t)z(t) + c(t)z\left(2\left[\frac{t+1}{2}\right]\right) + f(t) \qquad (8.1.33)$$

with initial conditions

$$x(0) = y(0) = z(0) = c_0. \qquad (8.1.34)$$

where a, c, f are real-valued continuous functions of t defined on $[0, \infty)$. Observe that the argument deviation $t - 2\left[\frac{t+1}{2}\right]$ is negative for $2n - 1 \le t < 2n$ and positive for $2n < t < 2n + 1$ for each integer n. Hence the two equations (8.1.32), (8.1.33) are of advanced type on $[2n - 1, 2n)$ and of retarded type on $(2n, 2n + 1)$. It is known that if

$$\int\limits_{2n-1}^{2n} u^{-1}(t)c(t)dt \neq u^{-1}(2n), \quad n = 1, 2, \ldots,$$

where u^{-1} is the reciprocal of u and $u(t) = \exp(\int_0^t a(s)ds)$ then the above equations possess unique solutions. In case $a(t) = a$ and $c(t) = c$ are constant functions, the solution of the unperturbed equation is given by

$$y(t) = \lambda\left(t - 2\left[\frac{t+1}{2}\right]\right) \left(\frac{\lambda(1)}{\lambda(-1)}\right)^{\left[\frac{t+1}{2}\right]} c_0$$

if $\lambda(-1) \neq 0$ and

$$\lambda(t) = \exp(at)(1 + \tfrac{c}{a}) - \tfrac{c}{a}.$$

Let ϕ denote the solution of the IVP

$$x' = a(t)x, \; x(0) = 1$$

and let ψ denote the solution of (8.1.32) such that $\psi(0) = 1$.

Let $x(t), y(t), z(t)$ be solutions of (8.1.31), (8.1.32), and (8.1.33) respectively. It is natural to expect that solutions x, y, z are related to each other. This relationship is established below through the method of VPF.

Theorem 8.1.8: *The unique solution of* (8.1.33), (8.1.34) *is given by*

$$z(t) = y(t) + \sum_{k=0}^{\left[\frac{t+1}{2}\right]-1} \frac{1}{\lambda(1)} \int\limits_{2k}^{2k+1} \psi(t, 2k)\phi(2k+1, s)f(s)ds$$

$$- \sum_{k=1}^{\left[\frac{t+1}{2}\right]} \frac{1}{\lambda(-1)} \int\limits_{2k}^{2k-1} \psi(t, 2k)\phi(2k-1, s)f(s)ds$$

$$+ \int\limits_{2\left[\frac{t+1}{2}\right]}^{t} \phi(t, s)f(s)ds$$

where ϕ and ψ are as defined above and $y(t)$ is the solution of (8.1.32), (8.1.34).

The proof of the above theorem can be formulated by following Theorems 8.1.3, 8.1.4.

8.2 COMPLEX DIFFERENTIAL EQUATIONS

The present section includes the results on complex differential equations. We prove below VPFs for linear and nonlinear equations. The study of complex equations becomes difficult when solutions are studied in the neighborhood of singular points. Hence, below, we assume smooth conditions to avoid complexities.

An nth order differential equation is linear and homogeneous if it is the sum of the terms of the form $Pr(z)w^{(n-k)}(z)$ equated to zero where $Pr(z)$ is a polynomial function in z of finite order and having the form

$$L(w) = P_0(z)w^{(n)}(z) + P_1(z)w^{(n-1)}(z) + \ldots + P_n(z)w(z) = 0. \quad (8.2.1)$$

If w_1 and w_2 are two solutions of (1) then

$$L(c_1 w_1 + c_2 w_2) = c_1 L(w_1) + c_2 L(w_2) = 0.$$

Along with (8.2.1) the initial data

$$w(z_0) = w_0, w'(z_0) = w_1, \ldots, w^{(n-1)}(z_0) = w_{n-1}. \quad (8.2.2)$$

The IVP (8.2.1) and (8.2.2) has a unique solution when $z = z_0$ does not belong to fixed singularities of the equation. Further the solution of the equation

$$L(w) = P(z) \quad (8.2.3)$$

has the form $w(z) + v(z)$ where $w(z)$ is the general solution of (8.2.1) and $v(z)$ is the particular solution of (8.2.3).

Based on these considerations, let us study the first order case. The method of quadrature can be employed to solve the equation

$$w'(z) = F_0(z) + F_1(z)w(z). \quad (8.2.4)$$

Set $w = uv$ so that $(uv)' = uv' + vu' = F_0 + F_1 uv$ and choose $v' = F_1 v$ and $vu' = F_0$, so that we obtain

$$u(z) = c + \int F_0(t)\exp[-\int F_1(s)ds]dt.$$

Let $w(z, z_0, w_0)$ denote the solution of (8.2.4) such that $w(z_0, z_0, w_0) = w_0$, then

$$w(z, z_0, w_0) = \exp\left[\int_{z_0}^{z} F_1(s)ds\right]\left[w_0 + \int_{z_0}^{z} F_0(t)\exp\left\{-\int_{z_0}^{t} F_1(s)ds\right\}dt\right] \quad (8.2.5)$$

If the mappings $z \to F_0(z)$ and $z \to F_1(z)$ are holomorphic in a common domain of holomorphy D containing z_0, then all the steps given above are justified. Here the integration may be performed along any path in the domain D and consequently solution (8.2.5) is locally holomorphic. If the domain D is simply connected then (8.2.5) is holomorphic in the large. The solution possesses no movable singularities and the only fixed singularities are those of F_0 and F_1 and the point at infinity.

The Bernoulli equation in complex domain is of the form

$$y'(z) = f(z)y(z) + g(z)[y(z)]^n, \; n \neq 0, 1, \tag{8.2.6}$$

where the coefficients are holomorphic functions of z in the domain D and can be solved by setting $v(z) = [y(z)]^{1-n}$. For $n = 2$, we have solution

$$y(z) = \left[v_0 \exp\left[-\int_{z_0}^{z} f(s)ds \right] - \int_{z_0}^{z} g(s) \exp\left[\int_{s}^{z} f(t)dt \right] ds \right]^{-1}.$$

The solution $y(z)$ has the form

$$y(z) = \frac{1}{v_0(c(z) + D(z))}.$$

This expression is a Mobius transformation on v_0. More generally, let a, b, c, d be any four arbitrary constants with $ad - bc \neq 0$, set

$$y \frac{a + bv}{c + dv}$$

and substitute y in (8.2.6) for $n = 2$. Then

$$(cb - ad)v' = [acf(z) + g(z)a^2] + [bcf(z) + adf(z) + 2abg(z)]v$$

$$+ [bdf(z) + b^2g(z)]v^2$$

which is again a Riccati equation and can be resolved by the method of VP.
    Consider the nonlinear complex differential equation

$$w' = F(z, w), w(z_0) = w_0; \tag{8.2.7}$$

where F is holomorphic in D: $| z - z_0 | \leq a$, $| w - w_0 | \leq b$, $a > 0$, $b > 0$ in the complex space c^2. Let $w(z, z_0, w_0)$ denote a solution of (8.2.7). Since F is holomorphic in D, it has partial derivative $F_w(z, w)$ which is also holomorphic in D. Assume that there exists $B > 0$ such that

$$| F_w(z, w) | \leq B, (z, w) \in D. \tag{8.2.8}$$

The relation (8.2.8) implies that F is Lipschitz in w and that

$$F(z, w_1) - F(z, w_2) = F_w(z, w_1)(w_1 - w_2) + o(| w_1 - w_2 |). \tag{8.2.9}$$

The equation related to (8.2.7) is

$$w'(z) = F_w(z, w(z, z_0, w_0))w(z), \; w(z_0) = 1; \tag{8.2.10}$$

is called variational equation and possesses unique solution

$$w(z, z_0, 1) = \exp\left\{ \int_{z_0}^{z} F_w[s, w(s, z_0, w_0] \, ds \right\}.$$

Since D is convex, we may consider the integral along a straight line. We assume the following

(H_1) Let $w(z, z_0, w_0)$, $w(z, z_0, w_1)$ be two solutions of the equation (8.2.7) when $(z_0, w_0), (z_0, w_1)$ are in D. Let $D_0 \subset D$ be a disc having center at $z = z_0$ and in which the two solutions are holomorphic and bounded.

We now prove the following result.

Theorem 8.2.1: *Let (H_1) hold. Then there exists a disc D^* in the w-plane such that if w_0 and w_1 are both in D^* and z lies in D_0, then*

$$| w(z, z_0, w_1) - w(z, z_0, w_0) | \leq | w_1 - w_0 | \exp[B \, | z - z_0 |], \quad (8.2.11)$$

where B is as given in (8.2.8).

Proof: We have (after using Lipschitz criteria)

$$| w(z, z_0, w_1) - w(z, z_0, w_0) |$$

$$\leq | w_1 - w_0 | + B \, | \int_{z_0}^{z} | w_1(s) - w_0(s) | \, dr | \quad (8.2.12)$$

where r is the arc length on the rectilinear path of integration. set $u(t) = | w(z_0 + te^{i\theta}, z_0, w_1) - w(z_0 + te^{i\theta}, z_0, w_0 |$, $\theta = \arg(z - z_0)$ and $h(t) = | w_1 - w_0 |$. Then the inequality (8.2.12) reduces to

$$u(t) \leq h(t) + B \int_0^t u(s) ds$$

$$\leq h(t) \exp(B \, | t |) \text{ (by Gronwall's inequality)}$$

which is (8.2.11) once we note that $| z_t - z_0 | = | t |$.

In order to prove the analog of Theorem 8.2.1 for z_0, we assume the following hypothesis.

(H_2) Let $| F(z, w) | \leq \widetilde{B}$, $(z, w) \in D$. Let $w(z, z_1, w_0) = w_1(z)$ and $w(z, z_2, w_0) = w_2(z)$ be two solutions of equation (8.2.7) when (z_1, w_0), (z_2, w_0) are in D. Let $D_0 \subset D$ be a disc having center at z_0, $z, z_1, z_2 \in D_0$, and $w_1(z), w_2(z)$ are holomorphic and bounded.

Theorem 8.2.2: *Let the hypothesis (H_2) hold. Then*

$$| w(z, z_1, w_0) - w(z, z_2, w_0) | \leq \widetilde{B} \, | z_1 - z_2 | \exp[B\delta_2] \quad (8.2.13)$$

where $\delta z = \min[\mid z - z_1 \mid, \mid z - z_2 \mid]$.

Proof: Suppose that z is nearer to z_1 than z_2. Then

$$w_1(z) = w_0 + \int\limits_{z_1}^{z} F[s, w_1(s)]ds$$

$$w_2(z) = w_0 + \int\limits_{z_1}^{z} F[s, w_2(s)]ds - \int\limits_{z_1}^{z_2} F[s, w_2(s)]ds.$$

Here, it has been assumed that the integral of $F[s, w_2(s)]$ along the perimeter of the triangle with vertices at z, z_1, z_2, is equal to zero. Observe that

$$\mid \int\limits_{z_1}^{z_2} F[s, w_2(s)]ds \mid \; \leq \widetilde{B} \mid z_1 - z_2 \mid .$$

Employing Lipschitz's condition (8.2.8) we get

$$\mid w_2(z) - w_1(z) \mid \; \leq \widetilde{B} \mid z_2 - z_1 \mid + B \mid \int\limits_{z_1}^{z} \mid w_2(s) - w_1(s) \mid ds \mid$$

$$\leq \widetilde{B} \mid z_2 - z_1 \mid \exp[B \mid z - z_1 \mid].$$

Since $\delta(z) = \mid z - z_1 \mid$, the result (8.3.13) follows.

We now show that solutions $w(z, z_0, w_0)$ of (8.2.7), under the given hypotheses, are differentiable with respect to initial data namely z_0 and w_0 we have the following results due to I. Bendixson.

Theorem 8.2.3: *The solution $w(z, z_0, w_0)$ is differentiable with respect to w_0 and hence locally holomorphic. Further*

$$w(z, z_0, w_1) - w(z, z_0, w_0) = W(z, z_0, 1)(w_1 - w_0) + o(\mid w_1 - w_0 \mid), \quad (8.2.14)$$

and

$$\frac{\partial w}{\partial w_0}(z, z_0, w_0) = \exp\left[\int\limits_{z_0}^{z} F_w(s, w(s, z_0, w_0))ds\right]. \quad (8.2.15)$$

Proof: We need to show that

$$\Delta z = w_1(z) - w_0(z) - W(z)(w_1 - w_0) = 0(\mid w_1 - w_0 \mid) \quad (8.2.16)$$

where $w_1(z) = w(z, z_0, w_1), w_0(z) = w(z, z_0, w_0)$ and $w(z)$ is the solution of (8.2.10). Hence

$$\Delta z = \int\limits_{z_0}^{z} [F(s, w, (s)) - F(s, w_0(s)) - (w_1 - w_0)F_w(s, w_0(s))w(s)]ds.$$

$$= \int_{z_0}^{z} F_w(s, w_0(s))[w_1(s) - w_0(s) - (w_1 - w_0)w(s)]ds + o(\,|\,w_1 - w_0\,|\,).$$

Hence

$$|\,\Delta(z)\,|\, \leq o(\,|\,w_1 - w_0\,|\,) + B\,|\int_{z_0}^{z} |\,\Delta(s)\,|\,ds\,|$$

$$\leq o(\,|\,w_1 - w_0\,|\,)\exp[B\,|\,z - z_0\,|\,]$$

which yields (8.2.14) and (8.2.15). Observe that in (8.2.14) the path of limit $w_1 \to w_0$ is immaterial. Further

$$\lim_{w_1 \to w_0} \frac{w(z, z_0, w_1) - w(z, z_0, w_0)}{w_1 - w_0} = W(z, z_0, 1)$$

implying that w is analytic.

Theorem 8.2.4: *Let* (H_2) *hold. Let* $W(z) = W(z, z_0, 1)$ *be the solution of* (8.2.10). *Then* $w(z, z_0, w_0)$ *is a differentiable function of the complex variable* z_0 *and·*

$$\frac{\partial}{\partial z_0} w(z, z_0, w_0) = - F(z_0, w_0)W(z). \qquad (8.2.17)$$

Proof: Consider two solutions of (8.2.7), namely $w_0(z) = w(z, z_0, w_0)$ and $w_1(z) = w(z, z_1, w_0)$, where z, z_0, z_1 are in the disc D_0. Set

$$D(z) = w_1(z) - w_0(z) + F(z_0, w_0)W(z)(z_1 - z_0).$$

It is enough to show that $D(z) = o(\,|\,z_1 - z_2\,|\,)$ for any z_1 in D_0. Observe that

$$w_1(z) - w_0(z) = \int_{z_1}^{z} F(s, w_1(s))ds - \int_{z_0}^{z} F(s, w_0(s))ds$$

$$= \int_{z_0}^{z} \{F(s, w_1(s)) - F(s, w_0(s))\}ds - \int_{z_0}^{z_1} F(s, w_1(s))ds$$

$$= \int_{z_0}^{z} F_w(s, w_0(s))[w_1(s) - w_0(s)] + E_1 - \int_{z_0}^{z_1} F(s, w_1(s))ds$$

where E_1 denotes the error term and it small because of (8.2.9) and (H_2). In fact, $E_1 = o(z_1 - z_2)$. Note that

$$w(z) = 1 + \int_{z_0}^{z} F_w(s, w_0(s))w(s)ds.$$

Hence

$$-\int_{z_0}^{z_1} F(s, w_0(s))ds + F(z_0, w_0)(z_1 - z_0)$$

$$= -\int_{z_0}^{z_1} \{F(s, w_0(s)) - F(z_0, w_0)\}ds$$

the modulus value of which does not exceed

$$B \mid \int_{z_0}^{z_1} \mid w_0(s) - w_0 \mid \mid ds \mid \mid \leq B \mid \int_{z_0}^{z_1} \mid dz \mid \mid \int_{z_0}^{s} \mid F(t, w_0(t)) \mid \mid dt \mid \mid$$

$$\leq \tfrac{1}{2} B^2 \mid z_1 - z_0 \mid^2.$$

Hence, finally it follows that

$$D(z) = \int_{z_0}^{z} F_w(s, w_0(s))D(s) + E_2$$

where $E_2 = o(\mid z_1 - z_2 \mid)$.

This fact implies that $w(z, z_0, w_0)$ is differentiable with respect to z_0 and yields (8.2.17). The proof is complete.

In order to develop the VPF for complex differential equations, consider the perturbed complex equation

$$p' = F(z, p) + \mathcal{F}(z, p), p(z_0) = w_0 \qquad (8.2.18)$$

where F is given in (8.2.7) and $\mathcal{F}$ is holomorphic in D. Let $w(z, z_0, w_0)$, $p(z, z_0, w_0)$ denote solutions of (8.2.7) and (8.2.18) respectively existing on a common region of holomorphy D_0. Let $w(z, z_0, w_0)$ satisfy appropriate conditions as given in Theorems 8.2.1, 8.2.2 so that w is differentiable with respect to z_0 and w_0 in D_0. We have now an analog of Theorem 1.2 for complex differential equations as given below.

Theorem 8.2.5: *Let* $F, \mathcal{F}$ *satisfy the foregoing conditions and solutions* $w(z, z_0, w_0)$, $p(z, z_0, w_0)$ *exist on* D_0, *a common region of holomorphy. Let* $W(z, z_0, 1)$ *be the solution of* (8.2.10) *existing in* D_0. *Then*

$$p(z, z_0, w_0) = w(z, z_0, w_0) + \int_{z_0}^{z} w(s, z_0, p(s)(\mathcal{F}(s, p(s, z_0, w_0)))ds. \quad (8.2.19)$$

for all $z \in D_0$.

Proof: Consider the solution $w(z, z_0, p(z, z_0, w_0))$ of (8.2.7) through $(z_0, p(z, z_0, w_0))$. Set

$$v(s) = w(z, s, p(s, z_0, w_0)). \qquad (8.2.20)$$

Differentiate (8.2.20) with respect to s to get

$$\frac{d}{ds}v(s) = \frac{dw(s,z_0,p(s,z_0,w_0))}{ds} + \frac{dw(s,z_0,p(s,z_0,w_0))}{dp} \, p'(s,z_0,w_0). \qquad (8.2.21)$$

Substitute the values of $\frac{dw}{ds}$ and $\frac{dw}{dp}$ from (8.2.17) and (8.2.14) in (8.2.21) to get

$$\frac{d}{ds}v(s) = -W(z,s,p'(s,z_0,w_0))F(s,p(s,z_0,w_0))$$

$$+ W(z,s,p(s,z_0,w_0))[F(s,p(s,z_0,w_0)) + \mathcal{F}(s,p(s,z_0,w_0))]$$

$$= W(z,s,p(s,z_0,w_0))\mathcal{F}(s,p(s,z_0,w_0)).$$

Integration between z_0 and z yields the conclusion (8.2.19) once we recognize from (8.2.20) that $v(z) = p(z,z_0,w_0)$ and $v(z_0) = w(z,z_0,w_0)$. The proof is complete.

Remark 8.2.1:   If $F(z,w) + \mathcal{F}(z,w) = F_0(z) + F_1(z)w(z)$ in (8.2.18), then treating F_0 as the perturbation term, we get

$$w(z,z_0,w_0) = \exp\left[\int_{z_0}^{z} F_1(s)ds\right]w_0$$

and

$$w(z,z_0,1) = \int_{z_0}^{z} F_1(s)ds.$$

With these substitutions, the relation (8.2.19) becomes

$$P(z,z_0,w_0) = \exp\left[\int_{z_0}^{z} F_1(s)d\right]w_0$$

$$+ \int_{z_0}^{z}\exp\left[\int_{t}^{z} F_1(s)ds\right]F_0(t)dt.$$

This is the result embodied in (8.2.5).

8.3 HYPERBOLIC PARTIAL DIFFERENTIAL EQUATIONS

In the present section we organize the known results on VPF for linear hyperbolic partial differential equations and provide an application of this technique in which monotone iterative method is employed to prove the existence of extremely solutions for nonlinear hyperbolic equations. The advantage of this method is that the iterates are solutions of linear initial value problems which are explicitly computable.

Let $a,b \in R$, $a > 0$, $b > 0$ and I_a, I_b denote the intervals $[0,a]$ and $[0,b]$, respectively, Iab the rectangle $I_a \times I_b$. By $v \in C^{1,2}[I_{ab}, R]$ we mean that v is a continuous function on I_{ab} and its partial derivatives v_x, v_y, v_{xy} exist and are continuous on I_{ab}.

Let $\sigma \in C^1[I_a, R], \tau \in C^1[I_b, R], u_0 \in R$. Consider the equation

$$u_{xy} = -M^2 u + M u_x + M u_y + h(x, y), \quad (x, y) \in I_{ab},$$

$$u(x, 0) = \sigma(x), x \in I_a; \quad u(0, y) = \tau(y), y \in I_b, \qquad (8.1.3)$$

$$\sigma(0) = \tau(0) = u_0$$

where $M \geq 0$ is a constant and $h \in C[I_{ab}, R]$.

Theorem 8.3.1: (Variation of Parameters Formula) *Any solution $u(x, y)$ on I_{ab} of (8.3.1) is given by the formula*

$$u(x, y) = \sigma(x)e^{My} + \tau(y)e^{Mx} - u_0 e^{M(x+y)}$$

$$+ e^{M(x+y)} \int_0^x \int_0^y h(s, t)e^{-M(s+t)} dt\, ds. \qquad (8.3.2)$$

Proof: We find u of the form

$$u(x, y) = z(x, y)e^{M(x+y)},$$

for a suitable function $z \in C^{1,2}[I_{ab}, R]$. Then using (8.3.1), we get

$$u_x = [z_x + M z]e^{M(x+y)},$$

$$u_y = [z_y + M z]e^{M(x+y)},$$

$$u_{xy} = [z_{xy} + M^2 z + M(z_x + z_y)]e^{M(x+y)},$$

and

$$[-M^2 z + M(z_x + z_y + 2M z)]e^{M(x+y)} + h(x, y)$$

$$= [z_{xy} + M^2 z + M(z_x + z_y)]e^{M(x+y)},$$

which implies $z_{xy} = h(x, y)e^{-M(x+y)}$. Hence, it follows that

$$z(x, y) = z(x, 0) + z(0, y) - z(0, 0) + \int_0^x \int_0^y h(s, t)e^{-M(s+t)} dt\, ds$$

which yields

$$u(x, y) = u(x, 0)e^{My} + u(0, y)e^{Mx} - u(0, 0)e^{M(x+y)}$$

$$+ e^{M(x+y)} \int_0^x \int_0^y h(s,t)e^{-M(s+t)} dt\, ds$$

which is (8.3.2). This completes the proof.

Based on Theorem 8.3.1, we prove the following inequality.

Theorem 8.3.2: *Let* $m \in C^{1,2}[I_{ab}, R]$. *Then*

$$m_{xy}(x,y) \geq - M^2 m(x,y) + M m_x(x,y) + M m_y(x,y), \tag{8.3.3}$$

for $(x,y) \in I_{ab}$ *and* $M \geq 0$, *a constant,*

$$m(0,0) = 0;\ m_x(x,0) \geq 0, x \in I_a;\ and\ m_y(0,y) \geq 0,\ y \in I_b, \tag{8.3.4}$$

imply

$$m(x,y) \geq 0, m_x(x,y) \geq 0\ and\ m_y(x,y) \geq 0\ for\ (x,y) \in I_{ab}. \tag{8.3.5}$$

Proof: By Theorem 8.3.1, we have, using (8.3.3),

$$m(x,y) = m(x,0)e^{My} + m(0,y)e^{Mx} - m(0,0)e^{M(x+y)}$$

$$+ e^{M(x+y)} \int_0^x \int_0^y h(s,t)e^{-M(s+t)} dt\, ds, \tag{8.3.6}$$

where $h(x,y) \geq 0$ is a continuous function on I_{ab}. Since $h(x,y) \geq 0$ on I_{ab} and $M \geq 0$, we obtain, from (8.3.6) and two other equations resulting from differentiation with respect to x and y, the following inequalities:

$$m(x,y) \geq m(x,0)e^{My} + m(0,y)e^{Mx} - m(0,0)e^{M(x+y)}, \tag{8.3.7}$$

$$m_x(x,y) \geq m_x(x,0)e^{My} + M m(0,y)e^{Mx} = M m(0,0)e^{M(x+y)}, \tag{8.3.8}$$

$$m_y(x,y) \geq m_y(0,y)e^{Mx} + M m(x,0)e^{My} - M m(0,0)e^{M(x+y)}. \tag{8.3.9}$$

Conclusion (8.3.5) now readily follows from (8.3.4), (8.3.7)-(8.3.9) and the fact that $M \geq 0$, completing the proof.

We aim at establishing the existence of the extremal solutions of the initial value problem

$$u_{xy} = f(x,y,u,u_x,u_y),\ (x,y) \in I_{ab},$$

$$u(x,0) = \sigma(x),\ x \in I_a,$$

$$u(0,y) = \tau(y),\ y \in I_b,$$

$$\sigma(0) = \tau(0) = u_0 \tag{8.3.10}$$

where $f \in C[I_{ab} \times R^3, R]$, $\sigma \in C^1[I_a, R]$, $\tau \in C^1[I_b, R]$. The existence proof is based on monotone iterative technique which provides a constructive method for obtaining the extremal solutions of the initial value problem (8.3.10).

Definition 8.3.1: A function $z \in C^{1,2}[I_{ab}, R]$ is said to be an upper solution of (8.3.10) on I_{ab}, if

$$z_{xy} \geq f(x, y, z, z_x, z_y), \quad (x, y) \in I_{ab},$$

$$z_x(x, 0) \geq \sigma'(x), \quad x \in I_a,$$

$$z_y(0, y) \geq \tau'(y), \quad y \in I_b,$$

$$z(0, 0) \geq u_0,$$

and a lower solution of (8.3.10) on I_{ab} if the reversed inequalities hold.

Theorem 8.3.3: *Suppose that*
(i) $v, w \in C^{1,2}[I_{ab}, R]$, $(v, v_x, v_y) \leq (w, w_x, w_y)$ *on* I_{ab}, *and* v, w *are, respectively, the lower and upper solutions of* (8.3.10) *such that* $v(0, 0) = u_0 = w(0, 0)$;
(ii) f *satisfies*

$$f(x, y, z, p, q) - f(x, y, \overline{z}, \overline{p}, q) \geq -M^2(z - \overline{z}) + M(p - \overline{p}) + M(q - \text{(8.3.11)}$$

whenever $(v, v_x, v_y) \leq (\overline{z}, \overline{p}, \overline{q}) \leq (z, p, q) \leq (w, w_x, w_y)$ *on* I_{ab}, *where* $M \geq 0$ *is a constant.*
  Then, there exist monotone sequences $\{v^n\}$, $\{p^n\}$, $\{q^n\}$, $\{w^n\}$, $\{\widetilde{p}^n\}$ *and* $\{\widetilde{q}^n\}$ *such that* $\lim_n v^n = \rho$, $\lim_n q^n = \rho_y$, $\lim_n w^n = r$, $\lim_n \widetilde{p}^n = r_x$ *and* $\lim_n \widetilde{q}^n = r_y$ *uniformly and monotonically on* I_{ab}, *where* ρ *and* r *are, respectively, the minimal and the maximal solutions of* (8.3.10) *and satisfy*

$$(v, v_x, v_y) \leq (\rho, \rho_x, \rho_y) \leq (r, r_x, r_y) \leq (w, w_x, w_y) \quad \text{on} \quad I_{ab}.$$

Proof: For any $\eta \in C^{1,2}[I_{ab}, R]$ such that

$$(v, v_x, v_y) \leq (\eta, \eta_x, \eta_y) \leq (w, w_x, w_y),$$

consider the initial value problem

$$u_{xy} = -M^2 u + M u_x + M u_y + h(x, y), \quad (x, y) \in I_{ab},$$

$$u(x, 0) = \sigma(x), \quad x \in I_a,$$

$$u(0, y) = \tau(y), \quad y \in I_b,$$

$$\sigma(0) = \tau(0) = u_0, \qquad (8.3.12)$$

where $h(x, y) = f(x, y, \eta(x, y), \eta_x(x, y), \eta_y(x, y)) + M^2 \eta(x, y) - M \eta_x(x, y) -$

$M\eta_y(x,y)$, $M \geq 0$ being the constant appearing in (8.3.11). It is clear that (8.3.12) has a unique solution u on I_{ab} for every such η. Define a mapping A on $\langle v, w \rangle$ by $A\eta = u$. We prove that

(a) (i) $(v, v_x, v_y) \leq (Av, (Av)_x, (Av)_y)$,

     (ii) $(w, w_x, w_y) \geq (Aw, (Aw)_x, (Aw)_y)$;

(b) for any $\eta^1, \eta^2, (v, v_x, v_y) \leq (\eta^1, \eta_x^1, \eta_y^1) \leq (\eta^2, \eta_x^2, \eta_y^2) \leq (w, w_x, w_y)$ implies
     $(A\eta^1, (A\eta^1)_x, (A\eta^1)_y) \leq (A\eta^2, (A\eta^2)_x, (A\eta^2 0_y))$.

To prove (a), set $v^1 = Av$ and let $m = v^1 - v$. Then, from the facts that v is the lower solution of (8.3.10) and v^1 is the (unique) solution of (8.3.12) with $\eta = v$, we obtain $m_{xy} \geq -M^2 m + M m_x + M m_y$. Also there hold $m(0,0) = 0$, $m_x(x,0) \geq 0$ for $x \in I_a$ and $m_y(0,y) \geq 0$ for $y \in I_b$. An application of Theorem 8.3.2 therefore yields (a)(i). Similar proof holds for (a)(ii). To prove (b), set $u^1 = A\eta^1$, $u^2 = A\eta^2$ and $m = u^2 - u^1$. Then, as in the proof of (a), we obtain the inequalities (8.3.3) and (8.3.4). Therefore, invoking Theorem 8.3.2 once again, the assertion (b) follows. Now, define the sequences $\{v^n\}$, $\{p^n\}$, $\{q^n\}$, $\{w^n\}$, $\{\widetilde{p}^{\,n}\}$ and $\{\widetilde{q}^{\,n}\}$ as follows:

$$v^0 = v, \, p^0 = v_x, \, q^0 = v_y,$$

$$w^0 = w, \, \widetilde{p}^{\,0} = w_x, \, \widetilde{q}^{\,0} = w_y;$$

and, for $n \geq 1$,

$$v^n = Av^{n-1}, \, p^n = (Av^{n-1})_x, \, q^n = (Av^{n-1})_y,$$

$$w^n = Aw^{n-1}, \, \widetilde{p}^{\,n} = (Aw^{n-1})_x, \, \widetilde{q}^{\,n} = (Aw^{n-1})_y.$$

By repeated applications of Theorem 8.3.2, it is easy to see that the six sequences defined above satisfy

$$v = v^0 \leq v^1 \leq \ldots \leq v^n \leq \ldots \leq \ldots \leq w^n \leq \ldots \leq w^1 \leq w^0 = w,$$

$$v_x = p^0 \leq p^1 \leq \ldots \leq p^n \leq \ldots \leq \ldots \leq \widetilde{p}^{\,n} \leq \ldots \leq \widetilde{p}^{\,1} \leq \widetilde{p}^{\,0} = w_x,$$

$$v_y = q^0 \leq q^1 \leq \ldots \leq q^n \leq \ldots \leq \ldots \leq \widetilde{q}^{\,n} \leq \ldots \leq \widetilde{q}^{\,1} \leq \widetilde{q}^{\,0} = w_y.$$

Now, using standard arguments, it is easy to conclude that $\lim_n v^n = \rho$ and $\lim_n w^n = r$ uniformly and monotonically on i_{ab}. Also, it is not hard to see that ρ and r are both solutions of (8.3.10). To prove that they are, respectively, the minimal and maximal solutions of (8.3.10), we once again appeal to Theorem 8.3.2. Let u be any solution of (8.3.10) such that $(v, v_x, v_y) \leq (u, u_x, u_y) \leq (w, w_x, w_y)$ on I_{ab}. We assert that

$$(v, v_x, v_y) \leq (\rho, \rho_x, \rho_y) \leq (u, u_x, u_y) \leq (r, r_x, r_y) \leq (w, w_x, w_y) \quad (8.3.13)$$

on I_{ab}. Suppose that, for some $n \geq 1$,

$$(v^n, p^n, q^n) \leq (u, u_x, u_y) \leq (w^n, \widetilde{p}^{\,n}, \widetilde{q}^{\,n}), \quad\quad\quad (8.3.14)$$

and let $m = w^{n+1} - u$. Then $m_{xy} \geq -M^2 m + M m_x + M m_y$, together with the facts $m(0,0) = 0$, $m_x(x,0) \geq 0$ for $x \in I_a$ and $m(0,y) \geq 0$ for $y \in I_b$, implies, by Theorem 8.3.2, that $(u, u_x, u_y) \leq (w^{n+1}, \widetilde{p}^{\,n+1}, \widetilde{q}^{\,n+1})$. Similarly, it follows that $(v^{n+1}, p^{n+1}, q^{n+1}) \leq (u, u_x, u_y)$, whence we conclude, by induction, that (8.3.14) holds for all n. Letting $n \to \infty$ in (8.3.14) yields the assertion (8.3.13) and the proof of the theorem is complete.

Remark 8.3.1: The special case $M = 0$, which, by assumption (ii) of Theorem 8.3.1 implies that $f(x, y, z, p, q)$ is monotone nondecreasing in z, p and q is included in Theorem 8.3.3. We remark that, in this case, we do not need to assume $v(0,0) = w(0,0)$.

Corollary 8.3.1: *Suppose that in Theorem* 8.3.3, *f satisfies Lipschitz condition, i.e.,*

$$| f(x, y, z, p, q) - f(x, y, \overline{z}, \overline{p}, \overline{q}) |$$

$$\leq L(| z - \overline{z} | + | p - \overline{p} | + | q - \overline{q} |)$$

for $(x, y) \in I_{ab}$, $z, \overline{z}, p, \overline{p}, q, \overline{q} \in R$, *where $L \geq 0$ is a constant. Then*

$$\rho(x, y) = r(x, y), (x, y) \in I_{ab}$$

and consequently there exists a unique solution of the IVP (8.3.10) *in the section* (v, w).

In place of (8.3.10), one may consider higher order partial differential equation

$$\frac{\partial^{n+m} u}{\partial x^n \partial y^m} = f(x, y, \langle u \rangle)$$

together with the initial conditions

$$\frac{\partial^j u(x,0)}{\partial y^j} = \alpha_j(x), 0 \leq j \leq m - 1 \tag{8.3.15}$$

$$\frac{\partial^i u(0,y)}{\partial x^i} = \beta_i(y), 0 \leq i \leq n - 1 \tag{8.3.16}$$

where $\langle u \rangle$ stands for

$$\left(u, \frac{\partial u}{\partial x}, \ldots, \frac{\partial^n u}{\partial x^n}, \frac{\partial u}{\partial y}, \ldots, \frac{\partial^{n+1} u}{\partial x^n \partial y}, \ldots, \frac{\partial^{m-1} u}{\partial y^{m-1}}, \ldots, \right.$$

$$\left. \frac{\partial^{n+m-1} u}{\partial x^n \partial y^{m-1}}, \frac{\partial^m u}{\partial y^m}, \ldots, \frac{\partial^{n-1+m} u}{\partial x^{n-1} \partial y^m} \right);$$

$f \in C[I_{ab} \times R^{nm+n+m}, R]$, $\alpha_j \in C^{(n)}[I_a, R]$, $\beta_j \in C^{(m)}[I_b, R]$ and $\alpha_j^{(i)}(0) = \beta_i^{(j)}(0)$; $0 \leq j \leq m - 1, 0 \leq i \leq n - 1$.

In order to establish the existence of extremal solutions of the above initial value problem one needs to prove a VPF for higher order linear hyperbolic partial differential equation. We present below this generalization. It is stated without proof since essentially the proof is similar to Theorem 8.3.1. This result is then employed to develop monotone iterative method for the above initial value problem. For this purpose we need to extend the proof of Theorem 8.3.3 to cover the general case. These details are omitted here.

Theorem 8.3.4: *Any solution (only solution)* $u(x, y)$ *on* I_{ab} *of the linear differential equation*

$$\frac{\partial^{n+m} u}{\partial x^n \partial y^m} = -\sum_{i=0}^{n} \sum_{j=0}^{m}{}' \binom{n}{i} \binom{m}{j} (-M_1)^{n-i}(-M_2)^{m-j} \frac{\partial^{i+j} u}{\partial x^i \partial y^j} + h(x, y)$$

together with (8.3.15), (8.3.16) *can be written as*

$$u(x, y) = \varphi(x, y) + \psi(x, y)$$

where M_1 *and* M_2 *are constants, and the functions* $\varphi(x, y)$, $\psi(x, y)$ *are defined as follows*

$$\varphi(x, y) = e^{M_1 x} \sum_{i=0}^{n-1} \frac{x^i}{i!} \left(\sum_{k=0}^{n-i-1} \frac{(-xM_1)^k}{k!} \right) \beta_i(y)$$

$$+ e^{M_2 y} \sum_{j=0}^{m-1} \frac{y^j}{j!} \left(\sum_{l=0}^{n-i-1} \frac{(-yM_2)^l}{l!} \right) \alpha_j(x)$$

$$- e^{M_1 x + M_2 y} \sum_{i=0}^{n-1} \sum_{j=0}^{m-1} \frac{x^i}{i!} \frac{y^j}{j!} \left(\sum_{k=0}^{n-i-1} \frac{(-xM_1)^k}{k!} \right) \left(\sum_{l=0}^{n-j-1} \frac{(-yM_2)^l}{l!} \right) \alpha_j^{(i)}(0),$$

$$\psi(x, y) = e^{M_1 x + M_2 y} \frac{1}{(n-1)!(m-1)!}$$

$$\times \int_0^x \int_0^y (x-s)^{n-1}(y-t)^{m-1} e^{-M_1 s - M_2 t} h(s, t) dt \, ds.$$

8.4 OPERATOR EQUATIONS

The purpose of this section is to study the VPF for operator equations using functional analytic techniques.

Let $\mathcal{F}$ be a given Fréchet space. Consider the abstract equations

$$y = f + Ty \tag{8.4.1}$$

$$x = f + T[x + g(x)] \tag{8.4.2}$$

satisfying

(H_1) $T: \mathcal{F} \to \mathcal{F}$ is a continuous linear map such that if $I = $ identity map of F-space $\mathcal{F}$, then $I - T: \mathcal{F} \to \mathcal{F}$ is one to one and onto.

(H_2) $f \in \mathcal{F}$ and g maps $\mathcal{F}$ into $\mathcal{F}$.

Theorem 8.4.1: *Suppose $\mathcal{F}$ and T satisfy the hypothesis (H_1) and (H_2). Then the nonlinear equation (8.4.2) is equivalent to*

$$x = y - Rg(x) \qquad (8.4.3)$$

where $R = I - (I - T)^{-1}: \mathcal{F} \to \mathcal{F}$ is continuous and $y = f - Rf$ is the solution of (8.4.1).

Proof: Since $(I - T)^{-1}$ exists then $R = I - (I - T)^{-1}$ is a well-defined linear map of $\mathcal{F}$ into $\mathcal{F}$. Since $I - R = (I - T)^{-1}$ then $y = (I - R)f = (I - R)^{-1}f$ solves (8.4.1). Since $I - T$ is continuous on $\mathcal{F}$, it is closed. Thus $(I - T)^{-1}$ is a closed linear map defined on all of $\mathcal{F}$. By the closed graph theorem $I - R$ and also R is continuous.

From (8.4.2) we have

$$x - Tx = (I - T)x = f + Tg(x).$$

Hence

$$x = (I - T)^{-1}f + (I - T)^{-1}Tg(x)$$

$$= (I - R)f + (I - T)^{-1}Tg(x)$$

$$= y + (I - T)^{-1}Tg(x).$$

Note that $(I - T)^{-1}T = -R$, it follows that

$$x = y - Rg(x)$$

which is (8.4.3). Since the calculations are reversible it follows that (8.4.3) implies (8.4.2). The proof is complete.

As an illustration, we show that the above VPF includes, as a special case, the VPF for Volterra integral equations.

We state below the VPF for linear Volterra integral system.

Theorem 8.4.2: *Consider the linear Volterra integral system*

$$y(t) = f(t) + \int\limits_0^t a(t, s)y(s)ds, \ t \in R_+ \qquad (8.4.4)$$

where a is an $n \times n$ continuous matrix defined on R_+ and $f \in C[R_+, R^n]$. Let $r(t, s)$ be the solution of the resolvent equation

$$r(t, s) = -a(t, s) + \int\limits_s^t r(t, u)a(u, s)ds, \ (t, s) \in R_+ \times R_+. \qquad (8.4.5)$$

Then solution $y(t)$ of (8.4.4) is given by

$$y(t) = f(t) - \int\limits_0^t r(t, u)f(u)du, \ t \in R^+. \qquad (8.4.6)$$

Further, let the perturbed equation to (8.4.4) be

$$x(t) = f(t) + \int_0^t a(t,s)[x(s) + G(s,x(s))]ds, \tag{8.4.7}$$

where $G \in C[R_+ \times R^n, R^n]$. Then the solution $x(t)$ of (8.4.7) is given by

$$x(t) = y(t) - \int_0^r r(t,s)G(s,x(s))ds \tag{8.4.8}$$

where $y(t)$ is the solution of (8.4.4) given by (8.4.6) and $r(t,s)$ is the solution of (8.4.5). The relation (8.4.8) represents the VPF for (8.4.7).

To show that (8.4.8) is a particular case of (8.4.3), consider the equations (8.4.4) and (8.4.7). Let $\mathcal{F}$ denote the solution space and $f \in \mathcal{F}$. In this setting, assumption (H_1) implies that for any f, the linear expression (8.4.4) has a unique solution y in $\mathcal{F}$. The map T is defined by

$$T\phi(t) = \int_0^t a(t,s)\phi(s)ds, \ t \geq 0.$$

Here the operator R reduces to

$$R\phi(t) = \int_0^t r(t,s)\phi(s)ds$$

where $r(t,s)$ is the resolvent of $a(t,s)$. For this final conclusion, one needs to assume certain admissibility information about the equation (8.4.4).

8.5 DYNAMIC SYSTEMS ON MEASURE CHAINS

It is now known that from a modeling point of view, it is most realistic to consider dynamic systems on measure chains or time scales, which incorporate both the theory of difference equations and differential equations. See Lakshmikantham, Sivasundaram, and Kaymakcalan [1].

In this section, we shall develop nonlinear variation of parameters formula for dynamic systems on measure chains parallel to the well-known Alekseev's formula in the theory of differential equations. For this purpose, we first need to investigate continuity and differentiability of solutions of nonlinear dynamic systems on measure chains and obtain the relation between the derivatives relative to the initial data. Curiously enough, such a relation differs from the usual expected expression and hence the discussion becomes more interesting. Once this relation is obtained, one can use suitable chain rule to derive the nonlinear variation of parameters for dynamic systems on measure chains. Of course, when the measure chain is R, obtained formula reduces to the Alekseev's formula and when the measure chain is a discrete set, we deduce the corresponding nonlinear variation of parameters formula which is new. Naturally, as is to be expected, when the dynamic system is linear, the variation of parameters formula obtained reduces to the respective linear variation of parameter formula.

Before we proceed to discuss NVP formula, we need to describe calculus on measure chains.

We begin with the notion of a measure chain in several steps. Let $\mathbb{T}$ (for "time") be some set.

Axiom 1: There is a relation " $\leq$ " on $\mathbb{T}$, which is reflexive ($t \leq t$ for all $t \in \mathbb{T}$), transitive ($r \leq s$ and $s \leq t \Rightarrow r \leq t$ for all $r, s, t \in \mathbb{T}$), antisymmetric ($r \leq s$ and $s \leq r \Rightarrow r = s$ for all $r, s \in \mathbb{T}$) and total ($r \leq s$ or $s \leq r$ for all $r, s \in \mathbb{T}$). The pair($\mathbb{T}$, $\leq$) is said to be a linearly ordered set or in short a chain.

Axiom 2: The chain ($\mathbb{T}$, $\leq$) is conditionally complete, that is, any nonvoid subset of $\mathbb{T}$ which is bounded above has a least upper bound (in $\mathbb{T}$).

A set satisfying Axioms 1 and 2 is called a conditionally complete chain. Any closed subset of R (with Euclidean topology) is an example of a conditionally complete chain. R and $hZ(h > 0)$ are conditionally complete and Q is not.

The set of so-called open intervals

$$]t_1, t_2[\; := \{t \in \mathbb{T} : t_1 < t < t_2\}, t_1, t_2 \in \mathbb{T} \cup \{-\infty, +\infty\}$$

generates a topology on $\mathbb{T}$, the order topology, with which $\mathbb{T}$ will always be equipped in the sequel. It can be shown by straight forward means that a subset of $\mathbb{T}$ is compact if and only if it is bounded and closed, thus $\mathbb{T}$ is a locally compact topological space.

The mapping $\sigma : \mathbb{T} \to \mathbb{T}$, defined by $\sigma(t) : -\inf\{s \in \mathbb{T} : s > t\}$ is called jump operator. Accordingly, we define the backward jump operator

$$\rho : \mathbb{T} \to \mathbb{T}, \text{ defined by } \rho(t) : = \sup\{s \in \mathbb{T} : s < t\}.$$

A nonmaximal element $t \in \mathbb{T}$ is said to be right-scattered, if $\sigma(t) > t$, and right-dense, $\sigma(t) = t$. We call a nonminimal element $t \in \mathbb{T}$ left-scattered, if $\rho(t) < t$, and left-dense, if $\rho(t) = t$.

Axiom 3: On the conditionally complete chain $\mathbb{T}$, there exists a mapping $\mu : \mathbb{T} \times \mathbb{T} \to R$ with the following properties (for all $r, s, t \in \mathbb{T}$)
(i) $\mu(r, s) + \mu(s, t) = \mu(r, t)$ (cocyclic property)
(ii) if $s > t \Rightarrow (\mu(s, t) > 0$ (strong isotony)
(iii) μ is continuous. (continuity)

Such a function is called growth calibration. It can be shown that each growth calibration is a measure on $\mathbb{T}$. The triple ($\mathbb{T}$, $\leq$, μ) satisfying axioms 1, 2 and 3 is called a (strong) measure chain and the function $\mu^*(t) = \mu(\sigma(t), t)$ being defined on $\mathbb{T}^k$ is called the graininess of the measure chain, where $\mathbb{T}^k = \{t \in \mathbb{T} : t$ nonmaximal or left dense$\}$.

We call a triple ($\mathbb{T}, \mu, X$) a dynamical triple, if ($\mathbb{T}, \mu$) is a measure chain and X is a Banach space. Let $f : \mathbb{T} \to X$. At $t \in \mathbb{T}$, f has the derivative $f_t^\Delta \in X$, if for each $\epsilon > 0$ there exists a neighborhood U of t such that for all $s \in U$

$$| f(\sigma(t)) - f(s) - f_t^{\Delta} \mu(\sigma(t), s) | \leq \epsilon | \mu(\sigma(t), s) | .$$

f is called differentiable in $t \in \mathbb{T}$ if f has exactly one derivative f_t^{Δ} in t.

Let $t \in \mathbb{T}$ be left-dense. If for a mapping $g: \mathbb{T} \to X$, the limit

$$\lim_{s \to t} g(s) = \lim_{s \to t, s < t} g(s): = g(t^-)$$

exists, then we call it the left-sided limit of g at t. Accordingly we define the notion of the right-sided limit $g(t^*)$ for right-dense point t. The mapping g is called rd-continuous, if it is continuous in each right-dense or maximal $t \in \mathbb{T}$ and the left-sided limit $g(t^-)$ exists in each left dense t.

The following implication is immediate

$$\text{continuous} \Rightarrow \text{rd-continuous.}$$

The converse is not true (in general).

The mapping $f: \mathbb{T}^k \times X \to X$ is called rd-continuous, if it is continuous at each (t, x) with right-dense or maximal t, and the limits

$$f(t^-, x): = \lim_{(s,y) \to (t,x), s < t} f(s, y)$$

and $\lim_{y \to x} f(t, y)$ exist at each (t, x) with left-dense t. By $C_{rd}[\mathbb{T}^k \times X, X]$, we denote the set of rd-continuous mappings from $\mathbb{T}^k \times X$ to X. See Lakshmikantham, Sivasundaram and Kaymakcalan [1] for details of used results.

Let $(\mathbb{T}, \mu, X)$ a dynamical triple, and $B(X)$ be a Banach algebra with unity of the continuous endomorphims on a Banach space X.

A mapping $A: \mathbb{T}^k \to B(X)$ is called regressive, if for each $t \in \mathbb{T}^k$ the mapping $A(t)\mu^*(t) + id: X \to X$ is invertible. This is the case e.g., if $| A(t)\mu^*(t) | < 1$ for all $t \in \mathbb{T}$. Obviously, in case $\mathbb{T} = R$ any A is regressive (since $\mu^* = 0$) and in case $\mathbb{T} = Z$, A is regressive if $| A(t) | < 1$ (since $\mu^* \equiv 1$).

Suppose $A: \mathbb{T}^k \to B(X)$ is rd-continuous and regressive and $F: \mathbb{T}^k \times X \to X$ is rd-continuous, then a mapping $x: \mathbb{T} \to X$ is called a solution of the dynamic equation

$$x^{\Delta} = A(t)x + F(t, x) \tag{8.5.1}$$

if $x^{\Delta}(t) = A(t)x(t) + F(t, x(t))$ for all $t \in \mathbb{T}^k$.

If a solution $x(\cdot)$ of (8.5.1) in addition satisfies the condition $x(\tau) = \eta$ for a pair $(\tau, \eta) \in \mathbb{T} \times X$, it is called a solution of the initial value problem (IVP)

$$x^{\Delta} = A(t)x + F(t, x), x(\tau) = \eta. \tag{8.5.2}$$

Consider the IVP, in the Banach algebra $B(X)$,

$$x^{\Delta} = A(t)x, x(\tau) = I, \tag{8.5.3}$$

(where I is the unity ov $B(X)$), and note that it admits exactly one solution $\Phi_A(\tau): = x(\cdot; \tau, I)$.

We call it principal solution. The corresponding transition function is defined to be $\Phi_A(t, \tau) := \Phi_A(\tau)(t)$. In the particular case, when $A: \mathbb{T} \to B(X)$ is constant, we call te transition function exponential function $(e_L(t, \tau))$.

Theorem 8.5.1: *We consider the IVP* (8.5.2) *with rd-continuous and regressive right-hand side. Then the solution of* (8.5.2) *is given by*

$$x(t) = \Phi_A(t, \tau)\eta + \int_\tau^t \Phi_A(t, \sigma(s))F(s, x(s))\Delta s.$$

Proof: By defining $x(t) = \Phi_A(t, \tau)z(t)$ and substituting in (8.5.2), we obtain using product rule

$$\Phi_A(\sigma(t), \tau)z^\Delta + \Phi_A^\Delta(t, \tau)z = A(t)\Phi_A(t, \tau)z + F(t, x),$$

which yields

$$z^\Delta = \Phi_A^{-1}(\sigma(t), \tau)F(t, x(t))$$

$$= \Phi_A(\tau, \sigma(t))F(t, x(t)).$$

Thus $z(t) = \eta + \int_\tau^t \Phi_A(t, \sigma(s))F(s, x(s))\Delta s$.

Multiplying this equation by $\Phi_A(t, \tau)$ gives

$$x(t) = \Phi_A(t, \tau)\eta + \int_\tau^t \Phi_A(t, \tau)\Phi_A(\tau, \sigma(s))F(s, x(s))\Delta s$$

$$= \Phi_A(t, \tau)\eta + \int_\tau^t \Phi_A(t, \sigma(s))F(s, x(s))\Delta s.$$

We need the following results before we proceed further.

Lemma 8.5.2: *Let $V: \mathbb{T} \times R^n \to R^n$ and $x: \mathbb{T} \to R^n$ be two functions such that x^Δ, V_t^Δ, V_x exist. Then*

$$[V(t, x(t))]_t^\Delta = V_t^\Delta(t, x(t)) + V_x(\sigma(t), x(t))x_t^\Delta.$$

Proof: The proof of this lemma follows directly.

Lemma 8.5.3: *Let $f \in C_{rd}[\mathbb{T}^k \times D, R^n]$, where D is an open, convex set in R^n, and let f_x exist and be rd-continuous. Then,*

$$f(t, x_2) - f(t, x_1) = \int_0^1 f_x(t, sx_2 + (1 - s)x_1)\Delta s(x_2 - x_1).$$

Proof: Define

$$F(t, s) = f(t, sy + (1 - s)x), \ 0 \leq s \leq 1.$$

The convexity of D implies that $F(t, s)$ is well defined. It is obvious that $F(t, s)$ is rd-continuous. Since $f_x(t, x)$ exists, it follows that $f_s^\Delta(t, s)$ exists. Moreover, its derivative $F_s^\Delta(t, s)$ is rd-continuous. Hence

$$F_s^\Delta(t, s)) = f_x(t, sx_2 + (1 - s)x_1) \cdot (x_2 - x_1) \qquad (8.5.4)$$

since $F(t, 1) = f(t, x_2)$ and $F(t, 0) = f(t, x_1)$, the result follows by integrating (8.5.4), in the sense of Cauchy with respect to s from 0 to 1. Which completes the proof.

Theorem 8.5.4: *Let* $\mathbb{T} = [\tau, s]$ *be some compact measurable chain and L be a negative constant with $L\mu(s, \tau) < 1$. The right-hand side in the IVP*

$$x^\Delta = f(t, x), x(\tau) = \eta, \qquad (8.5.5)$$

is rd-continuous and satisfies the Lipschitz condition

$$\mid f(t, x_1) - f(t, x_2) \mid \ \leq L \mid x_1 - x_2 \mid, \ t \in \mathbb{T}^k, \ x_1, x_2 \in R^n.$$

Then the solutions $x(\cdot, \tau, \eta)$ of (8.5.5) are unique and continuous with respect to the initial values (τ, η).

Proof: Since the uniqueness of solutions is valid, we have to prove the continuity part only.

Let $x(t, \tau, \eta), x(t, \tau_1, \eta_1)$ be the solutions through (τ, η), (τ_1, η_1) respectively. Define $v(t) = \mid x(t, \tau, \eta) - x(t, \tau_1, \eta_1) \mid$, then $v(t) = \mid \eta - x(\tau, \tau_1, \eta_1) \mid$.

We claim that

$$v(t) \leq m(t), \ \text{for all } t \geq \tau$$

where $m(t) = e_L(t, \tau)v(\tau)$ is a solution of IVP in R,

$$u^\Delta = Lu, u(\tau) = v(\tau).$$

We apply the induction principle to the following

$$A(t)v(t) \leq m(t), \ \text{for all } t \geq \tau.$$

(*i*)   Obviously, $A(\tau)$ is true.
(*ii*)   Let t be right-scattered and $A(t)$ be true.

We need to show that $A(\sigma(t))$ is true. We have,

$$v(\sigma(t)) - m(\sigma(t)) = [v^\Delta(t) - m^\Delta(t)]\mu^*(t) + v(t) - m(t). \qquad (8.5.6)$$

Since

$$v_t^\Delta(t) = \frac{v(\sigma(t)) - v(t)}{\mu^*(t)}$$

$$= \frac{|x(\sigma(t),\tau,\eta) - x(\sigma(t),\tau_1,\eta_1)| - |x(t,\tau,\eta) - x(t,\tau_1,\eta_1)|}{\mu^*(t)}$$

$$\leq \;|\, x^\Delta(t,\tau,\eta) - x^\Delta(t,\tau_1,\eta_1)\,|$$

$$= \;|\, f(t,x(t,\tau,\eta)) - f(t,x(t,\tau_1,\eta_1))\,|$$

$$\leq Lv(t).$$

Therefore, in view of (8.5.6), we obtain

$$v(\sigma(t)) - m(\sigma(t)) = L[v(t) - m(t)]\mu^*(t) + v(t) - m(t) \leq 0,$$

since $A(t)$ is valid.

Hence $A(\sigma(t))$ is true.

(iii) Let t be right-dense and U be a neighborhood of t. Suppose that $A(t)$ is true. As in the theory of ordinary differential equations, one can get

$$v(s) \leq m(s), s > t, s \in U.$$

Hence $A(s)$ is satisfied for all $s \in U$, $s > t$.

(iv) Let t be left-dense such that $A(s)$ is true for all s with $s < t$. By continuity of v and m, it follows that

$$v(t) = \lim_{s \to t^-} v(s) \leq \lim_{s \to t^-} m(s) = m(t).$$

Hence $A(t)$ is valid.

Thus $A(t)$ is true for all $t \geq \tau$.

Let $t \in \mathbb{T}$ be fixed and U be a compact neighborhood of τ containing t. We define

$$M: = \max\{e_L(t,s): s \in U\} > 0.$$

For $\epsilon > 0$, there exists a neighborhood of τ such that for all $s \in V$,

$$|\, x(s,\tau,\eta) - x(\tau,\tau,\eta)\,| \leq \tfrac{\epsilon}{2M}.$$

For $s \in U \cap V$ and $y \in N_{\frac{\epsilon}{2M}}$ (a neighborhood of η),

$$|\, x(t,s,y) - x(t,\tau,\eta)\,| \leq e_L(t,s)\,|\,y - x(s,\tau,\eta)\,|$$

$$\leq e_L(t,s)[\,|\,y - \eta\,| + |\,x(s,\tau,\eta) - x(\tau,\tau,\eta)\,|\,]$$

$$\leq M\left[\tfrac{\epsilon}{2M} + \tfrac{\epsilon}{2M}\right] = \epsilon.$$

The proof is complete.

Theorem 8.5.5:   *Let* $\mathbb{T} = [\tau, s]$ *be some compact measure chain and L be a nonnegative constant with* $L\mu(s, \tau) < 1$. *Let* $f \in C_{rd}[\mathbb{T}^k \times D \times A, R^n]$, *D is an open set in* R^n, *and A is an open parameter* μ_0-*set in* R^m, *and for* $\mu = \mu_0$, *let* $x_0(t) = x(t, \tau, \eta, \mu_0)$ *be a solution of*

$$x^\Delta = f(t, x, \mu_0), \; x(\tau) = \eta, \tag{8.5.7}$$

existing for $t \geq \tau$. *Assume further that*

$$\mid f(t, x, \mu) - f(t, x, \mu_0) \mid \; < \eta, \textit{provided } \mu \in U_{\mu_0} \textit{ (a neighborhood of } \mu_0),$$

and

$$\mid f(t, x, \mu) - f(t, x, \mu_0) \mid \; \leq L \mid x_1 - x_2, t \in \mathbb{T}^k.$$

Then, given $\epsilon > 0$, *there exists a neighborhood* U_{μ_0} *of* μ_0 *such that for every* $\mu_1 \in U_{\mu_0}$, *the IVP*

$$x^\Delta = f(t, x, \mu_1), x(\tau) = \eta, \tag{8.5.8}$$

admits a unique solution $x(t) = x(t, \tau, \eta, \mu_1)$ *satisfying*

$$\mid x(t) - x_0(t) \mid \; < \epsilon, t \geq \tau.$$

Proof: The uniqueness of solution is obvious. Consider the IVP in R,

$$u^\Delta = Lu + \eta, u(\tau) = 0 \text{ (here } \eta = \eta(\epsilon), \epsilon > 0).$$

By using variation of parameter formula

$$u(t, \tau, 0, \eta) = \int_\tau^t e_L(t, \sigma(s))\eta\Delta s$$

$$\leq \eta \sup_{s \, \in \, [\tau, t]^k} e_L(t, \sigma(s))(t - \tau).$$

Let $\epsilon > 0$ and let V be a compact neighborhood of t. Then

$$u(t, \tau, 0, \eta) < \epsilon, t \in V.$$

Now, define

$$m(t) = \mid x(t) - x_0(t) \mid,$$

where $x(t)$, $x_0(t)$ are the solutions of (8.5.8), (8.5.7) respectively. Then, as in the proof of Theorem 8.5.4, we can prove that

$$m(t) \leq u(t, \tau, 0, \eta), t \in \mathbb{T} \text{ whenever } \mu_1 \in U_{\mu_0}.$$

Hence

$$| \, x(t) - x_0(t) \, | \, < \epsilon, t \in V, \text{ provided that } \mu_1 \in U_{\mu_0}$$

which completes the proof of the theorem.

Theorem 8.5.6: *Let* $\mathbb{T} = [\tau, s]$ *be some compact measure chain. Assume that* $f \in C_{rd}[\mathbb{T}^k \times R^n, R^n]$, *and possesses rd-continuous partial derivatives* f_x *on* $\mathbb{T}^k \times R^n$. *Let L be a nonnegative constant with* $L\mu(s, \tau) < 1$ *and* $| \, f_x(t, x) \, | \, \leq L$ *on* $\mathbb{T}^k \times R^n$. *Let the solution* $x_0(t) = x(t, \tau, \eta)$ *of*

$$x^\Delta = f(t, x), x(\tau) = \eta,$$

exists for $t \geq \tau$.
  Then
(i) $\Phi(t, \tau, \eta) = x_\eta(t, \tau, \eta)$ *exists and is the solution of*

$$y^\Delta = H(t, \tau, \eta)y \tag{8.5.9}$$

  where

$$H(t, \tau, \eta) = \lim_{h \to 0} \int_0^1 f_x(t, px(t, \tau, \eta) - (1 - p)x(t, \tau, \eta + h))\Delta p$$

  such that $\Phi(\tau, \tau, \eta)$ *is the unit matrix;*
(ii) $\Psi(t, \tau, \eta) = x_\tau^\Delta(t, \tau, \eta)$ *exists, is the solution of*

$$z^\Delta = H(t, \sigma(\tau), \tau, \eta)z$$

  such that

$$\Psi(\sigma(\tau), \tau, \eta) = -f(\tau, \eta),$$

  where

$$H(t, \sigma(\tau), \tau, \eta) = \int_0^1 f_x(t, px(t, \sigma(\tau), \eta) - (1 - p)x(t, \tau, \eta))\Delta p;$$

(iii)   *the function* $\Phi(t, \tau, \eta), \Psi(t, \tau, \eta)$ *satisfy the relation*

$$\Psi(t, \tau, \eta) = -\Phi(t, \sigma(\tau, \eta)f(\tau, \eta)$$

$$+ \int_{\sigma(\tau)}^t \Phi(t, \sigma(s), \eta)[H(s, \sigma(\tau), \tau, \eta) - H(s, \sigma(\tau), \tau, \eta)]\Psi(s, \tau, \eta)\Delta s.$$

Proof: Under the assumptions of f, it is clear, that the solutions $x(t, \tau, \eta)$ exist, are unique and continuous in (t, τ, η) on some interval.

First we shall prove (ii). Let $\epsilon > 0$ and let U be a neighborhood of τ. By Lemma 8.5.3, we have for $s \in U$,

$$[x(t, \sigma(\tau), \eta) - x(t, s, \eta)]_t^\Delta = f(t, x(t, \sigma(\tau), \eta)) - f(t, x(t, s, \eta))$$

$$= \int_0^1 f_x(t, px(t, \sigma(\tau), \eta) + (1 - p)x(t, s, \eta))dp[x(t, \sigma(\tau), \eta) - x(t, s, \eta)].$$

Thus we get IVP of the type

$$y_t^\Delta(t) = H(t, \sigma(\tau), s, \eta)y(t), \qquad (8.5.11)$$

$$y(\sigma(\tau)) = -\frac{1}{\mu(\sigma(\tau), s)} \int_s^{\sigma(\tau)} f(\xi, x(\xi, \tau, \eta))\Delta\xi, \text{ (here } \mu(\sigma(\tau), s) = \sigma(\tau) - s)$$

where

$$H(t, \sigma(\tau), s, \eta) = \int_0^1 f_x(t, px(t, \sigma(\tau), \eta) + (1 - p)x(t, s, \eta))dp.$$

If τ is right-dense, that is, $\sigma(\tau) = \tau$, then we get desired results as in the theory of ordinary differential equation.

If τ is right-scattered, then consider

$$\frac{x(t, \sigma(\tau), \eta) - x(t, s, \eta)}{\mu(\sigma(\tau), s)},$$

we know that

$$\left| \frac{x(t, \sigma(\tau), \eta) - x(t, s, \eta)}{\mu(\sigma(\tau), s)} - \frac{x(t, \sigma(\tau), \eta) - x(t, \tau, \eta)}{\mu^\cdot(t)} \right| \leq \epsilon,$$

since the mapping $g(\cdot) := \frac{x(t, \sigma(\tau), \eta) - x(t, \cdot, \eta)}{\mu(\sigma(\tau), \cdot)}$ is defined and continuous in a neighborhood of τ.

Now,

$$\left| x(t, \sigma(t), \eta) - x(t, s, \eta) - \frac{x(t, \sigma(\tau), \eta) - x(t, \tau, \eta)}{\mu^\cdot(t)}\mu(\sigma(\tau), s) \right| \leq \epsilon \, |\mu(\sigma(\tau), s)|.$$

It further implies that $x_\tau^\Delta = \frac{\partial x(t, \tau, \eta)}{\partial \tau} = \frac{x(t, \sigma(\tau), \eta) - x(t, \tau, \eta)}{\mu^\cdot(\tau)}$ exists.

We now show that x_τ^Δ is the solution of $(8.5.10)$.

By using Lemma 8.5.3, we get

$$\frac{\partial x(\sigma(\tau), \tau, \eta)}{\partial \tau} = \frac{x(\sigma(\tau), \sigma(\tau), \eta) - x(\sigma(\tau), \tau, \eta)}{\mu^\cdot(\tau)}$$

$$= \frac{\eta - x(\sigma(\tau), \tau, \eta)}{\mu^\cdot(\tau)}$$

$$= -\frac{1}{\mu^*(\tau)} \int_\tau^{\sigma(\tau)} x_\tau^\Delta(\xi, \tau, \eta) \Delta\xi$$

$$= -\frac{1}{\mu^*(\tau)} \int_\tau^{\sigma(\tau)} f(\xi, \tau, \eta)) \Delta\xi$$

$$= -f(\tau, \eta).$$

The proof of (i) follows immediately from the theory of ordinary differential equation, using the rd-continuous nature of f_x.

Finally, to prove (iii), we proceed as follows. Since

$$\Phi(t, \sigma(\tau, \eta)) = \frac{\partial x(t, \sigma(\tau, \eta))}{\partial \eta}$$

exists and is the solution of

$$y^\Delta = H(t, \sigma(\tau), \eta) y,$$

where

$$H(t, \sigma(\tau), \eta) = \lim_{h \to 0} \int_0^1 f_x(t, px(t, \sigma(\tau), \eta) - (1-p)x(t, \sigma(\tau), \eta+h)) \Delta p$$

such that

$$\Phi(\sigma(\tau), \sigma(\tau, \eta))$$

is the unit matrix. But

$$\Psi_t^\Delta = H(t, \sigma(\tau), \eta) \Psi + [H(t, \sigma(\tau), \tau, \eta) - H(t, \sigma(\tau), \eta)] \Psi.$$

Hence, by variation of parameter formula (Theorem 8.5.1), we have

$$\Psi(t, \tau, \eta) = -\Phi(t, \sigma(\tau), \eta) f(\tau, \eta)$$

$$+ \int_{\sigma(\tau)}^t \Phi(t, \sigma(s), \eta) [H(s, \sigma(\tau), \tau, \eta) - H(s, \sigma(\tau), \eta)] \Psi(s, \tau, \eta) \Delta s.$$

This completes the proof.

Finally, we are now in a position to prove the nonlinear variation of parameter formula.

Theorem 8.5.7: *Let* $\mathbb{T} = [\tau, s]$ *be some compact measure chain. Assume that* $f, F \in C_{rd}[\mathbb{T}^k \times R^n, R^n]$, *and let* f_x *exists and be rd-continuous on* $\mathbb{T}^k \times R^n$.

Let L be a nonnegative constant with $L\mu(s,\tau) < 1$ and $|f_x(t,x)| \leq L$ on $\mathbb{T}^k \times R^n$. If $x(t,\tau,\eta)$ is the solution

$$x^\Delta = f(t,x), x(\tau) = \eta,$$

exists for $t \geq \tau$, any solution $y(t,\tau,\eta)$ for

$$y^\Delta = f(t,y) + F(t,y), \text{ with } y(\tau) = \eta, \text{ satisfies the integral equation}$$

$$y(t,\tau,\eta) = x(t,\tau,\eta) + \int_\tau^t \Phi(t,\sigma(s),y(s))F(s,y(s))\Delta s$$

$$+ \int_\tau^t \int_{\sigma(s)}^t \Phi(t,\sigma(p),y(s))[H(p,\sigma(s),s,y(s))$$

$$- H(p,\sigma(s),y(s))]\Phi(p,s,y(s))\Delta p\Delta s.$$

Proof: Set $y(t) = y(t,\tau,\eta)$. Then, by using Lemma 8.5.2 and Theorem 8.5.6, we get

$$[x(t,s,y(s))]_s^\Delta = x_s^\Delta(t,s,y(s)) + x_y(t,\sigma(s),y(s))y_s^\Delta$$

$$= -\Phi(t,\sigma(s),y(s))f(s,y(s)) + \int_{\sigma(s)}^t \Phi(t,\sigma(p),y(s))[H(p,\sigma(s),s,y(s))$$

$$- H(p,\sigma(s),y(s))]\Psi(p,s,y(s))\Delta p + \Phi(t,\sigma(s),y(s))y_s^\Delta.$$

$$= \Phi(t,\sigma(s),y(s))F(s,y(s)) + \int_{\sigma(s)}^t \Phi(t,\sigma(p),y(s))[H(p,\sigma(s),s,y(s))$$

$$- H(p,\sigma(s),y(s))]\Psi(p,s,y(s))\Delta p.$$

Since $x(t,t,y(t,\tau,\eta)) = y(t,\tau,\eta)$, by integrating above equation we obtain

$$y(t,\tau,\eta) = x(t,\tau,\eta) + \int_\tau^t \Phi(t,\sigma(s),y(s))F(s,y(s))\Delta s$$

$$+ \int_\tau^t \int_{\sigma(s)}^t \Phi(t,\sigma(p),y(s))[H(p,\sigma(s),s,y(s))$$

$$- H(p,\sigma(s),y(s))]\Psi(p,s,y(s))\Delta p\Delta s.$$

Remarks: It is easy to see from the definitions of $H(p,\sigma(s),s,y(s))$ and $H(p,\sigma(s),y(s))$ that they are identical if the measure chain is R, and consequently, the foregoing Variation of Parameter formula reduces to the usual Alekseev's formula. Moreover, when the dynamic systems is linear, the last term in the expression of the foregoing formula disappears, and one gets the corresponding linear Variation of Parameters formula for both continuous and discrete systems respectively.

8.6 NOTES

Equations with piecewise constant delay equations is relatively new area and mathematical model representing such equations was first given by Busenberg and Cooke [1]. Theorem 8.1.1 is taken from Cooke and Wiener [1]. The method of iteration of Theorem 8.1.2 and the VPF in Theorems 8.1.3, 8.1.4 is due to Jayasree and Deo [1]. The applications given in Theorems 81.5 to 8.1.7 is taken from Jayasree and Deo [1] while Theorem 8.1.8 is adopted from Jayasree and Deo [2]. For additional application, refer to Cooke and Wiener [1, 2], Shah and Wiener [1], Wiener and Aftabizadeh [1] and Aftabizadeh and Wiener [1].

The contents of Section 8.2 are taken from the classical book by Hille [1]. See also another book by Hille [2]. The differentiability of solutions with respect to initial conditions has been given by Bendixson [1]. See also Gronwall [1].

The Theorems 8.3.1-8.3.3 are due to Lakshmikantham and Pandit [1]. A variant of the VPF in Theorem 8.3.1 for hyperbolic equations is given in Blakley and Pandit [1]. Theorem 8.3.4 is taken from Agarwal [1]. Refer to the work of Sheng and Agarwal [1, 4] wherein periodic solutions and monotone method for higher-order hyperbolic partial differential equation have been extensively studied. Also refer to the papers of Hale and Perissinotto [1], Pachpatte [3], Lakshmikantham and Oguztoreli [1], Devsahayam [1], Pazy [1]. The Riemann's method for hyperbolic PDE is a VPF and is given in Copson [1].

The content of Section 8.4 appears in the monograph by Miller [2]. For additional details of linear operators, refer to Dunford and Schwartz [1]. For Volterra-type integral equations, refer to Corduneanu [1]. Also refer to Kunhert [1], Chen [1], Knapp and Wanner [1], Wanner and Rietberger [1].

Several areas of dynamic systems where the VPF is established and applied fruitfully, are not included in this chapter. For example, perturbations of the nonlinear renewal equation has been studied by Brauer [4], singular measure differential equations considered by Zhihong, Xiangcai, Yongqing [1]. Methods and applications of interval analysis, a monograph by Moore [1] include some interesting results.

Another useful area not covered here is the boundary value problems for differential equations. One of the immediate applications of VPF in BVPs is the formulation of Green's function. We have already proved in previous chapters, theorems on differentiability of solution with respect to initial conditions. Similar results namely differentiability of solutions with respect to boundary conditions have been proved by Spencer [1], Hale and Ladiera [1], Ehme and Henderson [1], Ehme [1] and Murthy and Sivasundaram [1].

The results of Section 8.5 are taken from Lakshmikantham, Sivasundaram and Kaymakcalam [1]. See the recent monograph of Lakshmikantham, Sivasundaram and Kaymakcalan [1] on *Dynamic Systems and Measure Chains*" for further results in this direction.

Bibliography

Aftabizadeh, A.R.
[1] Variation of constants, vector Lyapunov functions and comparison theorem, *Appl. Math. and Computation* **7** (1980), 341-352.

Aftabizadeh, A.R. and Wiener, J.
[1] Differential inequalities for delay differential equations with piecewise constant argument, *Appl. Math. and Computation* **24** (1987), 183-194.

Agarwal, R.P.
[1] The method of upper, lower solutions and monotone iterative scheme for higher order hyperbolic partial differential equations, *J. Austral. Math. Soc.* (Ser A) **47** (1989), 153-170.
[2] *Difference Equations and Inequalities: Theory, Methods, and Applications*, Marcel Dekker, Inc., New York 1992.

Agarwal, R.P. and Lalli, B.S.
[1] Discrete polynomial interpolation, Green's functions, maximum principles, error bounds and boundary value problems, *Computers and Mathematics with Applications*, (to appear).

Agarwal, R.P. and Usmani, R.A.
[1] Monotone convergence of iterative methods for right focal point boundary value problem, *J. Math. Anal. Appl.*, (to appear).

Alekseev, V.M.
[1] An estimate for the perturbations of solutions of ordinary differential equations, *Vestnik Moskov, Univ. Ser I. Mat. Meh.* **2** (1961), 26-36. (Russian)

Aulbach, B. and Hilger, S.
[1] *A Unified Approach to Continuous and Discrete Dynamics*, Qualitative Theory of Differential Equations **53**, Szeged, Hungary 1988.
[2] *Linear Dynamic Processes with Inhomogeneous Time Scales*, Nonlinear Dynamics and Quantum Dynamical Systems, Academie-Verlag, Berlin 1990.

Bainov, D.D. and Simeonov, P.S.
[1] *Systems with Impulsive Effects: Stability, Theory and Applications*, Ellis Horwood, Chicester, UK 1989.

Balchandran, K. and Lalitha, D.
[1] Controllability of nonlinear Volterra integrodifferential systems, *J. Appl. Math. Stoch. Analy.* **5** (1992), 139-146.

Barnett, S.
[1] Matrix differential equations and Kronecker product, *SIAM J. Appl. Math.* **24** (1973), 1-5.

Barnett, S. and Cameron, R.G.
[1] *Introduction to Mathematical Control Theory*, Oxford University Press, Oxford, UK 1985.

Beesack, P.R.
[1] On some variation of parameter methods for integrodifferential, integral, and quasilinear partial integrodifferential equations, *J. Appl. Math. Comput.* **22** (1987), 189-215.

Bellen, A.
[1]   Monotone methods for periodic solutions of second order functional differential equations, *Numer. Math.* **42** (1983), 15-30.
Bellman, R.
[1]   *Introduction to Matrix Analysis*, 2nd Ed., McGraw-Hill, New York 1970.
[2]   Topics in pharmacokinetics III: Repeated dosage and impulse control, *Math. Biosci.* **12** (1971), 1-2.
Bellman, R. and Cooke, K.L.
[1]   *Differential Difference Equations*, Academic Press, New York 1963.
Bendixson, I.
[1]   Demonstration de l'existence de l'integral d'une equation aux derivees partielles lineaires, *Bull. Soc. Math. France* **24** (1896), 220-225.
Bernfeld, S. and Lakshmikantham, V.
[1]   Perturbations of functional differential equations with nonuniform stability behavior, *J. Math. Analy. Appl.* **46** (1974), 249-260.
[2]   Monotone methods for nonlinear BVPs in Banach spaces, *J. Nonlinear Analysis: TMA* **3** (1979), 303-316.
Bernfeld, S. and Lord, M.E.
[1]   A nonlinear variation of constants method for integrodifferential and integral equations, *J. Appl. Math. Computation* **22** (1977), 189-215.
Bharucha-Reid, A.T.
[1]   *Probabilistic Analysis and Related Topics*, Academic Press, New York 1983.
[2]   *Random Integral Equations*, Academic Press, New York 1972.
Blakley, R.D. and Pandit, S.G.
[1]   On sharp linear comparison result and application to nonlinear Cauchy problem, *Dynamic Systems and Applications* **3** (1994), 135-140.
Brauer, F.
[1]   Perturbations of nonlinear systems of differential equations, *J. Math. Analy. Appl.* **14** (1966), 198-206.
[2]   Perturbations of nonlinear systems of differential equations II, *J. Math. Analy. Appl.* **17** (1967), 418-434.
[3]   A nonlinear variation of constants formula for Volterra equations, *Math. Systems Theory* **6** (1972), 226-234.
[4]   Perturbations of the nonlinear renewal equations, *Advances in Math* **22** (1976), 32-51.
Brauer, F. and Strauss, A.
[1]   Perturbation of nonlinear systems of differential equations III, *J. Math. Analy. Appl.* **31** (1970), 37-48.
Brunner, H.
[1]   The application of the variation of constants formula in the numerical analysis of integral and integrodifferential equations, *Utilitas Mathematica* **19** (1981), 255-290.
Burton, T.A.
[1]   Periodicity and limiting equations in Volterra systems, *Bulletino UMI, Analisis Funzionale e Appl. Sene VI*, Vol. **IV-C**:1 (1985), 31-39.
Busenberg, S. and Cooke, K.L.
[1]   Models of vertically transmitted diseases with sequential continuous dynamics, *Nonlinear Phenomena in Mathematical Sciences* (ed. by V. Lakshmikantham), Academic Press, New York (1982), 179-187.

Chandra, J. and Lakshmikantham, V.
[1]   Explicit bounds on solutions of a class of nonlinear differential inequalities, *J. Math. Phy. Sci.* **VI** (1972), 83-88.

Chen, K.T.
[1]   On a generalization of Picard's approximation, *J. Diff. Equations* **2** (1966), 438-448.

Coddington, E. and Levinson, N.
[1]   *Theory of Ordinary Differential Equations*, McGraw-Hill, New York 1955.

Cooke, K.L. and Wiener, J.
[1]   Retarded differential equations with piecewise constant delays, *J. Math. Analy. Appl.* **99** (1984), 265-297.
[2]   An equation alternatively of retarded and advanced type, *Proc. Amer. Math. Soc.* **99** (1987), 726-732.

Coppel, W.A.
[1]   *Stability and Asymptotic Behavior of Differential Equations*, D.C. Heath and Company 1965.

Corduneanu, C.
[1]   *Almost Periodic Functions*, Interscience Publishers, New York 1968.
[2]   *Principles of Differential and Integral Equations*, Allyn and Becon, Boston 1971.

Daoyi, Xu
[1]   Inequality approach to the stability analysis of delay differential systems, *J. Sichuan Normal Univ.* **16** (1993), 22-28.

Dauer, J.P.
[1]   Controllability of nonlinear systems with restraint controls, *J. Optim. Theory Appl.* **14** (1974), 251-262.

Davison, E.J. and Macki, M.C.
[1]   The numerical solutions of matrix Riccati differential equation, *IEEE Trans. Automat. Control* **18** (1973), 71-73.

Deimling, K.
[1]   *Ordinary Differential Equations in Banach Spaces*, Springer-Verlag, Lecture Notes **596**, Berlin 1977.

Delfour, M.C. and Mitter, S.K.
[1]   Hereditary differential equations with constant delays I. General case, *J. Diff. Eqns.* **12** (1972), 213-235.

Deo, S.G. and Torres, E.F.
[1]   Generalized variation of constant formula for nonlinear functional differential equations, *Applied Math. Computation* **24** (1987), 263-274.

Deo, S.G. and Sivasundaram, S.
[1]   Controllability of nonlinear integrodifferential systems, (preprint).

Devsahayam, M.P.
[1]   Monotone iterative scheme for nonlinear hyperbolic boundary value problem, *Applicable Analy.* **20** (1985), 49-55.

Doob, J.L.
[1]   *Stochastic Processes*, John Wiley and Sons, New York 1953.

Dunford, N. and Schwartz, J.T.
[1]   *Linear Operators. Part I*, Interscience, New York 1958.

Ehme, J. and Henderson, J.
[1]   Differentiation of solutions of boundary value problems with respect to boundary conditions, *Appl. Analy.* **46** (1992), 175-194.

Elaydi, S.
[1]   Stability of Volterra difference equations of convolution type, *Dynamical Systems* (ed. by Liao Shan-Tao, et. al), World Scientific (1993), 66-73.
[2]   Periodicity and stability of linear Volterra difference systems, *J. Math. Analy. Appl.* **181** (1994), 483-492.
El'sgol'tz, L.E.
[1]   *Introduction to the Theory of Differential Equations with Deviating Arguments*, Holden-Day 1966.
Evans, R.B.
[1]   Asymptotic behavior of perturbed autonomous linear functional differential equations, *Proc. Amer. Math. Soc.* **48** (1975).
[2]   Asymptotic equivalence of linear functional differential equations, *J. Math. Analy. Appl.* **51** (1975), 223-228.
Fai, Ma and Caughey, T.K.
[1]   On the stability of linear and nonlinear stochastic transformations, *Int. J. Control* **34** (1981), 501-511.
[2]   On the stability of stochastic difference systems, *Int. J. Nonlinear Mechanics* **16** (1981), 139-153.
Fausett, D.W. and Köksal, S.
[1]   Variation of parameters formula and Lipschitz stability of nonlinear matrix differential equation, (to appear).
Gear, G.W. and Tu, T.K.
[1]   The effect of variable mesh size on the stability of multistep methods, *SIAM JNA* **1** (1974), 1025-1043.
Grobner, W.
[1]   *Die Lie-Reihen und Ihre Anwendungen*, D. Verlag der Wiss, Berlin 1960, 1967.
Grobner, W. and Knapp, H. (eds.)
[1]   *Contributions to the Method of Lie Series*, B.I. Htb. **802/802a**, Mannheim 1967.
Gronwall, T.H.
[1]   Note on the derivative with respect to a parameter of the solutions of a system of differential equations, *Ann. Math.* **20**:2 (1918), 292-296.
Grossman, S.L. and Miller, R.K.
[1]   Perturbation theory for Volterra integrodifferential system, *J. Diff. Eqns.* **8** (1970), 457-474.
[2]   Nonlinear Volterra integrodifferential systems with L^1-kernel, *J. Diff. Eqns.* **13** (1973), 551-556.
Guo, D.
[1]   Initial value problems for integrodifferential equations of Volterra type in Banach spaces, *J. Appl. Math. and Stoch. Analy.* **7** (1994), 13-23.
Gyllenberg, M. and Webb, G.F.
[1]   Asynchronous exponential growth of semigroup of nonlinear operators, *J. Math. Analy. Appl.* **167** (1992), 443-467.
Halaney, A.
[1]   *Differential Equations: Stability Oscillations, Time Lags*, Academic Press, New York 1966.
Halaney, A. and Wexler, D.
[1]   *Qualitative Theory of Impulsive Systems*, Editura Academiei Republici Socialiste Romania, Bucuresti 1968. (`Mir', Moscow 1971).

Hale, J.K.
[1] *Functional Differential Equations*, Springer-Verlag, New York 1971.
[2] Critical cases for neutral functional differential equations, *J. Diff. Eqns.* **10** (1971), 59-82.

Hale, J.K. and Ladeira, L.A.C.
[1] Differentiability with respect to delays, *J. Diff. Eqns.* **92** (1991), 14-26.

Hale, J.K. and Mayer, K.
[1] A class of functional equations of neutral type, *Mem. Amer. Math. Soc.* **76** (1967).

Hale, J.K. and Perissinotto Jr., A.
[1] Global attractor and convergence for one-dimensional semilinear thermoelasticity, *Dynamic Systems and Applications* **2** (1993), 1-10.

Hartman, P.
[1] *Ordinary Differential Equations*, 2nd Edition, Birkhäuser, Boston 1982.

Hastings, S.P.
[1] Variation of parameters for nonlinear differential-difference equations, *Proc. Amer. Math. Soc.* **19** (1968), 1211-1216.

Hilger, S.
[1] Analysis on measure chains. A unified approach to continuous and discrete calculus, *Res. in Mathematics* **18** (1990), 18-56.

Hille, E.
[1] *Lectures on Ordinary Differential Equations*, Addison-Wesley Publishing Co., London 1969.

Hille, E. and Phillips, R.S.
[1] Functional analysis and semigroups, *Colloq. Amer. Math. Soc.* **31** (1957).

Horn, R.A. and Johnson, C.R.
[1] *Topics in Matrix Analysis*, Cambridge Univ. Press, Cambridge, UK 1991.

Hu, S., Lakshmikantham, V. and Rao, M.R.M.
[1] Nonlinear variation of parameters formula for integrodifferential equations of Volterra type, *J. Math. Analy. Appl.* **129** (1988), 223-230.

Ito, K.
[1] On stochastic differential equations, *Mem. Amer. Math. Soc.* **4** (1951).

Ize, A.F. and Molfetta, N.A.
[1] Asymptotically autonomous neutral functional differential equations with time dependent lag, *J. Math. Analy. Appl.* **51** (1975), 299-335.

Ize, A.F. and Ventura, A.
[1] Asymptotic behavior of a perturbed neutral functional differential equations related to the solution of the unperturbed linear system, *J. of Math. Analy. Appl.* **111** (1984), 57-91.
[2] An extension of the Alekseev variation of constant formula for neutral nonlinear perturbed equation with an application to the relative asymptotic equivalence, *J. Math. Analy. Appl.* **122** (1987), 16-35.

Jaysree, K.N. and Deo, S.G.
[1] Variation of parameters formula for the equations of Cooke and Wiener, *Proc. Amer. Math. Soc.* **112** (1991), 75-80.
[2] On piecewise constant delay differential equations, *J. Math. Analy. Appl.* **169** (1992), 55-69.

Jiongyu, W.
[1] The nonlinear variation of constant formula in RFDE's and its applications, *Ann. of Diff. Eqns.* **5** (1989), 75-85.

Kamala, P.S. and Lakshmikantham, V.

[1]   Asymptotic self invariant sets and functional differential equations in Banach spaces, *Annali di Matematica pura ed applicata* **XCVI** (1973), 217-231.

Kartsatos, A.G.

[1]   Global controllability of perturbed quasilinear systems, *Problems of Control and Information Theory* **3** (1974), 137-145.

[2]   The Leray-Schauder Theorem and the existence of solutions to boundary value problems on infinite interval, *Indiana Univ. Math. J.* **23** (1974), 1021-1029.

[3]   *Advanced Ordinary Differential Equations*, Mancorp Publishing, Tampa, USA 1993.

Kato, T.

[1]   Nonlinear semigroups and evolution equations, *J. Math. Soc. Japan* **19** (1967), 508-520.

[2]   Integration of the equation of evolution in a Banach space, *J. Math. Soc. Japan* **5** (1953), 208-234.

Kaymakcalan, B.

[1]   Linear and nonlinear variation of parameters formulae for dynamic systems on time scales, *Nonlinear Differential Equations*, (to appear).

Kaymakcalan, B. and Leela, S.

[1]   A survey of dynamic systems on time scales, *Nonlinear Times and Digest* **1** (1994), 37-60.

Kaymakcalan, B. and Rangarajan, L.

[1]   Variation of Lyapunov's method for dynamic systems on time-scale, (preprint).

Kelley, W.G. and Peterson, A.C.

[1]   *Difference Equations: An Introduction with Applications*, Academic Press, New York 1991.

Knapp, H. and Wanner, G.

[1]   Numerical solutions of differential equations by Grobner's method of the series, *MRC Techn. Sum. Rep.* **880**, Madison, Wisconsin (June 1968).

Konstantinore, M.M., Petkov, P.Hr. and Christos, N.D.

[1]   Perturbation analysis of the continuous and discrete matrix Riccati equations, *Proc. ACC*, Seattle, WA **1** (1986), 636-639.

Konstantinov, M.M., Christov, N.D. and Petkov, P.Hr.

[1]   Perturbation analysis of linear control problems, (preprint).

Kruger-Thiemer, E.

[1]   Formal theory of drug dosage regiments, *I. Jour. Theo. Biology* **13** (1966).

Kuhnert, K.

[1]   Generalization of Grobner's integral equation and an iterated equation of chen., Chap. III of Final Tech. Rep. Oct. 1970, *European Research Office Contract No. JA37-70-C-0249*, Dept. of Math. University of Innsbruck.

Kulev, G.K. and Bainov, D.D.

[1]   Continuous dependence and differentiability of solutions of impulsive systems of integro-differential equations with respect to initial data and parameter, *Rendiconti di Matematica*, Serie VII **13** (1933), 105-124.

Kulkarni, R.M. and Ladde, G.S.

[1]   Stochastic perturbations of nonlinear systems of differential equations, *J. Math. Physical Sci.* **10** (1976), 33-45.

Ladas, G., Ladde, G.S. and Lakshmikantham, V.
[1]   On some fundamental properties of solutions of differential equations in a Banach space, unpublished manuscript.
Ladde, G.S.
[1]   Variation of comparison theorem and perturbation of nonlinear systems, *Proc. Amer. Math. Soc.* **52** (1975), 181-187.
Ladde, G.S. and Lakshmikantham, V.
[1]   *Random Differential Inequalities*, Academic Press, New York 1980.
Ladde, G.S., Lakshmikantham, V. and Leela, S.
[1]   A new technique in perturbation theory, *Rocky Mountain J. Math.* **6** (1977), 133-140.
Ladde, G.S., Lakshmikantham, V. and Vatsala, A.S.
[1]   *Monotone Iterative Techniques for Nonlinear Differential Equations*, Pitman, Boston 1985.
Ladde, G.S. and Sambandham, M.
[1]   Error estimates of solutions and means of solutions of stochastic differential systems, *J. Math. Phys.* **24** (1983), 815-822.
[2]   Random difference inequalities, *Trends in Theory and Practice of Nonlinear Analysis* (ed. by V. Lakshmikantham), Elsevier Science Publishers, North Holland (1985), 231-240.
[3]   Variation of constants formula and error estimates to stochastic difference equations, *J. Math. Phys. Sci.* **22** (1988), 557-584.
Ladde, G.S., Sambandham, M. and Sathananthan, S.
[1]   Comparison theorems and its applications, *General Inequalities* (ed. by W. Walter), Birkhäuser Verlag, Basel **6** (1992), 321-342.
Ladde, G.S. and Sathananthan, S.
[1]   Ito-type systems of stochastic integro-differential equations, *Integral Methods in Science and Engineering* (ed. by A. Haji-Sheikh), North Holland Publishers **90** (1991), 75-89.
[2]   Stochastic integro-differential equations with random parameters I, *Dynamic Systems and Applications* **1** (1992), 369-390.
[3]   Error estimates and stability of Ito-type systems of nonlinear stochastic integro-differential equations, *J. Appl. Analy.* **43** (1992), 163-189.
Lakshmikantham, V.
[1]   A variation of constants formula and Bellman-Gronwall-Reid inequalities, *J. Math. Analy. Appl.* **41** (1973), 199-204.
Lakshmikantham, V., Bainov, D.D. and Simeonov, P.S.
[1]   *Theory of Impulsive Differential Equations*, World Scientific, Singapore 1989.
Lakshmikantham, V. and Ladas, G.
[1]   *Differential Equations in Abstract Spaces*, Academic Press, New York 1972.
Lakshmikantham, V. and Leela, S.
[1]   *Differential and Integral Inequalities*, Vol. I and II, Academic Press, New York 1969.
[2]   *Nonlinear Differential Equations in Abstract Spaces*, Pergamon Press, New York 1981.
Lakshmikantham, V., Leela, S. and Martynyuk, A.A.
[1]   *Stability Analysis of Nonlinear Systems*, Marcel Dekker, New York 1989.

Lakshmikantham, V., Matrosov, V.M. and Sivasundaram, S.
[1]   *Vector Lyapunov Functions and Stability Analysis of Nonlinear Systems*, Kluwer
      Academic Publishers, Netherlands 1991.
Lakshmikantham, V., Oguztoreli, M.N. and Vatsala, A.S.
[1]   Monotone iterative technique for partial differential equations, *J. Math. Analy.
      Appl.* **102** (1984), 393-398.
Lakshmikantham, V. and Pandit, S.G.
[1]   The method of upper, lower solutions and hyperbolic partial differential equations,
      J. Math. Analy. Appl. **105** (1985), 466-477.
Lakshmikantham, V. and Rao, M.R.M.
[1]   *Theory of Integrodifferential Equations*, Gordon and Breach Science Publishers,
      UK 1995.
Lakshmikantham, V., Sivasundaram, S. and Kaymakcalan, B.
[1]   *Dynamic Systems on Measure Chains*, Kluwer Academic Publishers, Dordrecht
      1996.
Lakshmikantham, V., Shahzad, N. and Sivasundaram, S.
[1]   Nonlinear variation of parameters formula for dynamic systems on measure chains,
      Dynamics of Continuous, Discrete and Impulsive Systems **1** (1995), 255-265.
Lakshmikantham, V. and Trigiante, D.
[1]   *Theory of Difference Equations and Numerical Analysis*, Academic Press, New
      York 1987.
Lakshmikantham, V., Wen, L. and Zhang, B.G.
[1]   *Theory of Functional Differential Equations with Unbounded Delay*, Kluwer
      Academic Publishers, London 1994.
Lakshmikantham, V. and Zhang, B.G.
[1]   Monotone iterative technique for delay differential equations, *Applicable Analy.* **22**
      (1986), 227-233.
Leela, S., McRae, F.A. and Sivasundaram, S.
[1]   Controllability of impulsive differential equations, *J. Math. Analy. Appl.* **177**
      (1993), 24-30.
Leela, S. and Zouyousefain, M.
[1]   Stability results for difference equations of Volterra type, *Appl. Math and
      Computation* **36** (1990), 51-61.
Levin, J.J. and Nohel, J.A.
[1]   On a system of integro-differential equations occurring in reactor dynamics II,
      Arch. Rat. Mech. Anal. **11** (1962), 210-243.
Li, L.M.
[1]   Stability of linear neutral delay-differential systems, *Bull. Austral. Math. Soc.* **38**
      (1988), 339-344.
Lord, M.E.
[1]   Stability and asymptotic equivalence of perturbation of nonlinear systems of
      differential equations, *Rend. Sem. Mat. Univ. Padova* **67** (1982), 1-11.
Lord, M.E. and Mitchell, A.R.
[1]   A new approach to the method of nonlinear variation of parameters, *J. Appl. Math.
      Comp.* **4** (1978), 95-105.
Luca, N. and Talpalaru, P.
[1]   Stability and asymptotic behavior of a class of discrete systems, *Ann. Mat. Pure
      Appl.* **112** (1977), 351-382.

Malkin, I.G.
[1]   Stability in the case of constantly acting disturbances, *Prikl. Mat. Meh.* **8** (1944), 241-245. (In Russian)
[2]   Some problems of the theory of non-oscillations, Moscow (1956). (In Russian)

Mamedov, Ya.D.
[1]   One-sided estimations of the solutions of differential equations with delay arguments in Banach spaces, *Trudy Sem. Teor. Differential Uravneniis Otkbn. Argumentom Univ. Druzby Norodov Patrisa Lumumby* **6** (1968), 135-147.

Markus, L.
[1]   Controllability of nonlinear processes, *SIAM J. on Control* **3** (1965).

Marlin, J.A. and Struble, R.A.
[1]   Asymptotic equivalence of nonlinear systems, *J. Diff. Eqns.* **6** (1969), 578-596.

Martin Jr., R.H.
[1]   *Nonlinear Operators and Differential Equations in Banach Spaces*, John Wiley and Sons, New York 1976.

Miller, R.K.
[1]   On the linearization of Volterra integral equations, *J. Math. Analy. Appl.* **23** (1968), 198-206.
[2]   *Nonlinear Volterra Integral Equations*, Benjamin, Menlo Park, CA 1971.

Mil'man, V.D. and Myshkis, A.D.
[1]   Random impulses in linear dynamical systems, approximate methods of solutions of differential equations, *Publ. House Acad. Sci. Ukr. SSR*, Kiev (1963), 64-81.

Moore, R.E.
[1]   *Methods and Applications of Interval Analysis*, SIAM Studies in Appl. Math., Philadelphia 1979.

Murthy, K.N., Howell, G.W. and Sivasundaram, S.
[1]   Two (multi-) point nonlinear Lyapunov systems. Existence and uniqueness, *J. Math. Analy. Appl.*, (to appear).
[2]   Two point boundary value problem associated with a system of first order nonlinear impulse differential equations, *Appl. Analy.* **54** (1993), 303.

Murthy, K.N., Prasad, K.R. and Srinivas, M.A.S.
[1]   On the method of upper and lower solutions of matrix Riccati differential equations, *J. of Math. Analy. and Appl.* **147** (1990), 12-21.

Murthy, K.N. and Shaw, M.
[1]   Sensitivity analysis of nonlinear systems of differential equations, (to appear).

Murthy, K.N. and Sivasundaram, S.
[1]   Existence and uniqueness of solutions to three point boundary value problem associated with nonlinear first order systems of differential equations, *J. Math. and Appl.* **173** (1993), 158-164.
[2]   Existence, uniqueness and conditioning three point boundary value problems associated with a system of first order nonlinear differential equations, *J. Math. Phys. Sci.* **26** (1992), 267-280.

Nohel, J.A.
[1]   Asymptotic equivalence of Volterra equations, *Ann. Math. Pure Appl.* **96** (1973), 339-347.

Oguztoreli, N.N.
[1]   *Time-Lag Control Systems*, Academic Press, New York 1969.

Onuchic, N.

[1]   Relationships among the solutions of two systems of ordinary differential equations, *Michigan Math. J.* **10** (1963), 129-139.

Pachpatte, B.G.

[1]   A nonlinear variation of constant method for summary difference equations, *Tamkang J. Math.* **8** (1977), 203-212.

[2]   On the behavior of solutions of a certain class of nonlinear integro-differential equations, *Analele Stiinfice ale Universitatit "A.I. I. Cuza"* **24** (1978), 77-86.

[3]   Perturbations of Bianchi type partial integro-differential equation, *Bull. of the Inst. Math. Acad. Sinica* **10** (1982), 347-356.

Pazy, A.

[1]   *Semigroup of Linear Operators and Applications to Partial Differential Equations*, Applied Mathematical Sciences, Springer-Verlag **44** 1983.

Piazza, G. and Trigiante, D.

[1]   Propagazione degli errari nella integrazione numerica di equazioni differenziali ordinarie, *Pubbl. IAC III, Roma* **120** (1977).

Rajlakshmi, S. and Sivasundaram, S.

[1]   Vector Lyapunov functions and techniques in perturbation theory, *J. of Math. Analy. Appl.*, (to appear).

Rao, M.R.M. and Srinivas, P.

[1]   Asymptotic behavior of solutions of Volterra integro-differential equations, *Proc. Amer. Math. Soc.* **94** (1985), 55-60.

Rao, M.R.M. and Sivasundaram, S.

[1]   Stability of Volterra system with impulsive effect, *J. Appl. Math. Stoch. Analy.* **4** (1991), 83-93.

Rashbaev, Z.M.

[1]   On the stability of first approximation of solutions of a system of differential equations with retarded arguments, *Izv. Akad. Nauk. SSSR* **5** (1971), 63-66.

Ruan, S.

[1]   Successive over relaxation iteration for the stability of large scale systems, *J. Math. Analy. Appl.* **146** (1990), 389-396.

Russell, D.L.

[1]   Numerical solutions of singular initial value problem, *SIAM J. Numerical Analy.* **7** (1970), 399-417.

Shah, S.M. and Wiener, J.

[1]   Advanced differential equations with piecewise constant arguments deviations, *Internat. J. Math. and Math. Sci.* **6**:4 (1983), 671-703.

Shanholt, G.A.

[1]   A nonlinear variation of constant formula for functional differential equations, *Math. Systems Theory* **6** (1973), 343-352.

Shendge, G.R.

[1]   A new approach to the stability theory of functional differential equations, *J. Math. Analy. Appl.* **95** (1983), 319-334.

Sheng, Q. and Agarwal, R.P.
[1]   Periodic solutions of higher order hyperbolic partial differential equations, *Pan-American Math. J.* **2** (1992), 1-22.
[2]   On nonlinear variation of parameter method for summary difference equations, *Dynamic Systems and Appl.* **2** (1993), 227-242.
[3]   Nonlinear variation of parameter methods for summary difference equation in several independent variables, *Appl. Math. Comp.*, (to appear).
[4]   Monotone methods for higher-order partial differential equations, *Computers & Math. with Appl.*, (to appear).
[5]   Existence and uniqueness of solutions of nonlinear n-point boundary value problems, (preprint).

Simeonov, P.S. and Bainov, D.D.
[1]   Differentiability of solutions of systems with impulse effect with respect to initial data and parameter, *Bull. Inst. Math. Acad. Sinica* **15** (1987), 251-269.

Spencer, J.
[1]   Relations between boundary value functions for a nonlinear differential equation and its variational equation, *Canad. Math. Bull.* **18** (1975), 269-276.

Strauss, A.
[1]   On the stability of a perturbed nonlinear system, *Proc. Amer. Math. Soc.* (1966), 803-807.

Strauss, A. and Yorke, J.A.
[1]   Perturbing uniform asymptotically stable nonlinear systems, *J. Diff. Eqns.* **6** (1969), 452-483.

Tsokos, C.P. and Padgett, W.J.
[1]   *Random Integral Equations with Applications to Life Sciences and Engineering*, Academic Press, New York 1974.

Uvah, J. and Vatsala, A.S.
[1]   Monotone iterative technique for nonlinear IVP of integro-differential systems with singular coefficients, *Appl. Analy.* **51** (1993), 129-138.

Vasundhara Devi, J.
[1]   A variation of the Lyapunov second method to impulsive differential equations, *J. Math. Analy. Appl.* **77** (1993), 190-200.

Wanner, G. and Reitberger, H.
[1]   On the perturbation formulas of Grobner and Alekseev, *Bull. Inst. Pol. Iasi* **XIX** (1973), 15-25.

Webb, G.F.
[1]   Asynchronous exponential growth in differential equations with homogeneous nonlinearities, *Differential Equations in Banach Spaces*, Lecture Notes in Pure and Applied Mathematics Series, Marcel Dekker, New York Vol. **148** (1993), 225-233.
[2]   Growth bounds of solutions of abstract nonlinear differential equations, *J. Diff. and Int. Eqns.* **7** (1994), 1145-1152.

Wiener, J. and Aftabizadeh, A.R.
[1]   Differential equations alternately of retarded and advanced type, *J. Math. Analy. Appl.* **129** (1988), 243-255.

Yamamoto, Y.
[1]   Controllability of nonlinear system, *J. of Optim. Theory and Appl.* **22** (1977), 41-49.

Yamamota, Y. and Sugiura, I.
[1] On controllability of nonlinear control systems, *J. SICE in Japan* **9** (1973),
[2] Some sufficient conditions for the observability of nonlinear systems, *J. of Optim. Theory and Appl.* **13** (1974).

Zaidman, S.
[1] *Topics in Abstract Differential Equations*, Longman Scientific and Technical, Essex, UK 1994.

Zhang, Yi, Zhang, Y. and Wang, M.
[1] The exponential stability of Volterra integro-differential equations, *J. Xinjiang Univ.* **10** (1993), 18-27.

Zhihong, G., Xiangcai, W. and Yongqing, L.
[1] Variation of the parameters formula and the problem of BIBO for singular measure differential systems with impulsive effect, *Appl. Math. & Comp.* **60** (1994), 153-169.

Index